Hydrogels and Aerogels for Functional Textiles

Hydrogels and Aerogels for Functional Textiles

From Sustainable Syntheses to Applications

Edited by Sheraz Ahmad, Faheem Ahmad, Yasir Nawab, and Farooq Azam

WILEY-VCH

Editors

Sheraz Ahmad
National Textile University
Faisalabad
PK, 37610

Faheem Ahmad
National Textile University
Faisalabad
PK, 37610

Yasir Nawab
National Textile University
Faisalabad
PK, 37610

Farooq Azam
National Textile University
Faisalabad
PK, 37610

Cover Design: Wiley
Cover Image: Courtesy of the Editors

Typesetting: Straive, Chennai, India

Library of Congress Card No.: applied for

British Library Cataloguing-in-Publication Data A catalogue record for this book is available from the British Library.

Bibliographic information published by the Deutsche Nationalbibliothek
The Deutsche Nationalbibliothek lists this publication in the Deutsche Nationalbibliografie; detailed bibliographic data are available on the Internet at <http://dnb.d-nb.de>.

Print ISBN: 978-3-527-35496-2
ePDF ISBN: 978-3-527-85140-9
ePub ISBN: 978-3-527-85139-3
oBook ISBN: 978-3-527-85141-6

Printed and bound by
CPI Group (UK) Ltd, Croydon, CR0 4YY

C9783527354962_100326

Contents

Foreword *xiii*
Preface *xv*

1 **Introduction** *1*
Farooq Azam, Yasir Nawab, and Sheraz Ahmad
1.1 Hydrogels and Aerogels *3*
1.2 Unique Properties Relevant to Textiles *3*
1.3 Why Hydrogels and Aerogels in Textiles? *4*
1.4 Scope and Structure of the Book *5*
1.4.1 Scope *5*
1.4.2 Structure *6*
1.5 Target Audience and Applications *7*
1.6 Challenges and Opportunities *7*
References *8*

2 **Science of Hydrogels and Aerogels** *11*
Zeynep Ulker and Can Erkey
2.1 Introduction *11*
2.2 Hydrogels: Relationship Between Function and Structure *11*
2.2.1 Science of Preparing Hydrogels *13*
2.2.1.1 Physically Cross-Linked Hydrogels *13*
2.2.2 Chemically Cross-Linked Hydrogels *15*
2.3 Aerogels: Relationship Between Structure and Functions *16*
2.3.1 Science of Preparing Aerogels *16*
2.4 Textile Innovations *20*
2.5 Conclusions *23*
References *24*

3 **Types of Hydrogels and Aerogels in Textiles** *35*
Talha Siddique, Faheem Ahmad, and Jibran Khaliq
3.1 Introduction *35*
3.2 Natural-Fiber Hydrogels and Aerogels *37*

3.2.1 Cotton-Based Hydrogels and Aerogels *37*
3.2.2 Silk-Based Systems *39*
3.2.3 Wool-Based Materials *39*
3.3 Synthetic-Fiber Hydrogels and Aerogels *40*
3.3.1 Polyester-Based Hydrogels and Aerogels *40*
3.3.2 Nylon-Based Hydrogels and Aerogels *41*
3.3.3 Acrylic-Based Hydrogels and Aerogels *41*
3.4 Blended Hydrogels and Aerogels in Textiles *42*
3.4.1 Natural–Synthetic Blends *43*
3.4.1.1 Cellulose–PVA *43*
3.4.1.2 Silk–PLA *44*
3.4.1.3 Wool–Polyurethane (PU) *44*
3.4.2 Composite Textile Substrates *45*
3.4.2.1 Laminated/Coated Fabrics *45*
3.4.2.2 Functional Coatings *45*
3.5 Conclusion *46*
References *46*

4 Synthesis Techniques for Hydrogels and Aerogels for Textiles *55*
Sana Sadaf and Faheem Ahmad
4.1 Introduction *55*
4.2 Synthesis Techniques for Hydrogels and Aerogels from Natural Fiber *57*
4.2.1 Synthesis from Cotton *57*
4.2.2 Synthesis of Hydrogels and Aerogels from Silk *62*
4.2.3 Synthesis from Wool *66*
4.3 Synthesis Techniques for Hydrogels and Aerogels from Synthetic Fibers *68*
4.3.1 Synthesis from Polyester *68*
4.3.2 Acrylic Fiber-Based Hydrogels *71*
4.4 Blended-Fiber Hydrogels and Aerogels *73*
4.5 Conclusion *74*
References *76*

5 Functional Properties of Hydrogels and Aerogels in Textiles *81*
Mohamed Assal, Ayoub Nadi, Abdeslam el Bouari, and Omar Cherkaoui
5.1 Introduction *81*
5.2 Moisture Management *82*
5.3 Thermal Insulation Properties of Aerogel- and Hydrogel-Based Textiles *84*
5.4 Antimicrobial and Antibacterial Properties *86*
5.5 Flame Retardancy *87*

5.6 Biodegradability *89*
5.7 Conclusion *91*
References *91*

6 Applications in Sustainable Fashion *95*
Bushra Mushtaq, Nauman Ali Choudhry, and Abher Rasheed
6.1 Introduction *95*
6.1.1 The Need for Sustainable Fashion Innovation *95*
6.1.2 Role of Hydrogels and Aerogels in Sustainable Textiles *96*
6.1.3 Chapter Scope and Objectives *96*
6.2 Eco-friendly Clothing *97*
6.3 Biodegradable Fashion Accessories *98*
6.3.1 The Sustainability Challenges in Fashion Accessories *98*
6.3.2 Hydrogel-Based Leather Alternatives *98*
6.3.3 Aerogel-Based Functional Accessories *100*
6.3.3.1 Jewelry *100*
6.3.3.2 Belts and Straps *100*
6.3.3.3 Eyewear *100*
6.3.3.4 3D-Printed Aerogel Accessories *100*
6.3.4 Smart and Responsive Biodegradable Accessories *101*
6.4 Sustainable Sportswear *101*
6.4.1 The Environmental Impact of Conventional Sportswear *101*
6.4.2 Performance Enhancement by Hydrogel *101*
6.4.2.1 Moisture Management Systems *101*
6.4.2.2 Temperature Control *101*
6.4.2.3 Biomechanical Support *102*
6.4.3 Aerogel in Athletic Apparel *102*
6.4.3.1 Ultralight Insulation *102*
6.4.3.2 Increased Breathability *102*
6.4.3.3 Impact Protection *102*
6.4.4 Sustainable Manufacturing Breakthrough *102*
6.4.4.1 Waterless Dyeing *102*
6.4.4.2 Closed-Loop Recycling *103*
6.4.4.3 3D Knitting *103*
6.5 Hydrogel and Aerogel-Infused Footwear *103*
6.5.1 Environmental Impact of Conventional Footwear *103*
6.5.2 Application of Hydrogel in Footwear *103*
6.5.2.1 Advanced Cushioning System *103*
6.5.2.2 Moisture Management and Hygiene *104*
6.5.3 Application of Aerogel in Sustainable Footwear *104*
6.6 Sustainable Manufacturing Breakthrough *105*
6.7 Future Perspectives and Challenges *105*
6.8 Conclusion *106*
References *106*

7 Plant Fiber-Based Hydrogels and Aerogels for Biomedical Applications *111*
Atike Köken and Ayşe Çelik Bedeloğlu
7.1 Introduction *111*
7.1.1 Plant Fiber-Based Hydrogels and Aerogels: Synthesis Methods, Properties, and Applications *112*
7.1.2 Properties and Application Areas of Plant-Based Hydrogels and Aerogels *112*
7.2 Wound Dressings *113*
7.2.1 Hydrogel–Textile Composites for Wound Dressings and Applications *114*
7.2.2 Aerogels for Wound Dressings and Applications *117*
7.3 Drug Delivery Systems *120*
7.3.1 Hydrogels for Wound Dressings and Applications *121*
7.3.2 Aerogels for Wound Dressings and Applications *121*
7.4 Smart Bandages *122*
7.4.1 Materials Used in Smart Bandages and Mechanisms *123*
7.4.2 Applications *124*
7.4.3 Future Perspective *125*
7.5 Wearable Electronics and Health Monitors *125*
7.5.1 Materials Used and Mechanism *126*
7.5.2 Applications *126*
7.6 Sensor Applications *127*
7.7 Wearable Electronics and Health Monitors *128*
7.8 Future Perspective *129*
References *129*

8 Animal Fiber-Based Hydrogels and Aerogels for Biomedical Applications *135*
Bushra Mushtaq, Muhammad Amber Fareed, and Muhammad Sohail Zafar
8.1 Introduction *135*
8.1.1 Definition of Hydrogel and Aerogel *135*
8.1.2 Animal Fiber-Derived Biopolymers and Their Advantages *136*
8.1.3 Historical Development *136*
8.1.4 Current Market Landscape and Clinical Impact *137*
8.1.5 Multidisciplinary Convergence *137*
8.2 Preparation and Fabrication Methods *138*
8.2.1 Extraction and Purification *138*
8.2.1.1 Silk Fibroin Extraction (*Bombyx mori* and Recombinant Spider Silk) *138*
8.2.1.2 Keratin Extraction (Wool/Human Hair) *138*
8.2.1.3 Collagen Extraction (Type I from Bovine Tendon) *139*
8.2.2 Advanced Hydrogel Fabrication *140*
8.2.2.1 Physical Cross-Linking *140*
8.2.2.2 Chemical Cross-Linking *140*

8.2.3 Advanced Aerogel Fabrications 140
8.2.3.1 Sol–Gel Transition Control 140
8.2.3.2 Drying Technology Comparison 141
8.2.3.3 Freeze-Drying (Lyophilization) 141
8.2.3.4 Supercritical Drying 141
8.2.3.5 Ambient Pressure Drying 142
8.3 Properties and Characterization 142
8.3.1 Physicochemical Properties 142
8.3.1.1 Swelling Behavior 142
8.3.1.2 Porosity and Pore Size 143
8.3.2 Mechanical Properties 143
8.3.3 Thermal Stability 143
8.3.4 Biodegradability and Enzymatic Degradation 144
8.3.5 Surface Morphology and Microstructure 144
8.3.6 Biocompatibility and Cytotoxicity 144
8.3.7 Water Vapor Transmission Rate 145
8.3.8 Cross-Linking and Network Structure 145
8.3.9 Summary Table of Key Properties and Methods 145
8.4 Biomedical Applications 146
8.4.1 Wound Healing and Skin Regeneration 146
8.4.2 Tissue Engineering Scaffolds 147
8.4.3 Drug Delivery Systems 147
8.4.4 Injectable and In Situ-Forming Gels 148
8.4.5 Biosensors and Diagnostic Devices 148
8.4.6 Hemostatic and Antibacterial Applications 148
8.4.7 Ophthalmic and Dental Applications 148
8.4.8 Summary Table of Biomedical Applications 149
8.5 Conclusion 149
8.6 Future Trends 149
References 150

9 Advancements in Hydrogel and Aerogel Textile Technology *157*
Syeda Rubab Batool, Rahima Hussain, and Muhammad Anwaar Nazeer
9.1 Introduction 157
9.2 Fundamentals of Hydrogel and Aerogel Materials 158
9.2.1 Hydrogels 158
9.2.2 Aerogels 160
9.2.3 Comparison and Complementarity in Textile Applications 162
9.3 Fabrication and Integration Methods 163
9.3.1 Hydrogel–Textile Integration Techniques 163
9.3.2 Aerogel Textile Integration Techniques 166
9.4 Conductive Hydrogel and Aerogel 167
9.4.1 Material Design and Composites 167
9.4.2 Electrical and Electrochemical Properties 170

9.4.3 Applications for E-Textiles *171*
9.4.3.1 Hydrogel in E-Textiles *171*
9.4.3.2 Aerogel in E-Textiles *172*
9.5 Responsive and Adaptive Textile Systems *173*
9.5.1 Stimuli-Responsive Hydrogels and Aerogels *173*
9.5.1.1 Physical Stimuli *173*
9.5.1.2 Chemical Stimuli *175*
9.5.2 Adaptive Clothing Systems *176*
9.6 Sustainability and Biocompatibility *177*
9.7 Challenges *179*
9.8 Conclusions *180*
References *180*

10 Applications of Aerogels in Protective Clothing *187*
Abdul Salam, Farooq Azam, Yasir Nawab, and Jibran Khaliq
10.1 Introduction *187*
10.2 Mechanism of Thermal Insulation *188*
10.3 Types of Aerogels *189*
10.3.1 Inorganic *189*
10.3.2 Organic Aerogels *191*
10.4 Impact Resistance and Chemical Protection of Aerogels *192*
10.4.1 Use of Aerogels in Ballistic-Resistant Textiles *193*
10.4.2 Laboratory Test Methods *194*
10.4.3 Chemical Protection *195*
10.5 Aerogels as Lightweight Protective Gear *197*
10.6 Hybrid Applications and Smart Integration of Aerogels in Protective Textiles *198*
10.7 Manufacturing Techniques and Textile Integration *200*
10.7.1 Synthesis of Aerogels *201*
10.7.2 Aerogel–Textile Integration Techniques *201*
10.7.3 Structural Challenges in Aerogel–Textile Composites *202*
10.8 Conclusion *203*
References *203*

11 Challenges and Solutions in the Use of Hydrogels and Aerogels for Textiles *209*
Farooq Azam and Shahid Rasul
11.1 Introduction *209*
11.2 Environmental Impact Assessment *210*
11.3 Recycling and Upcycling Strategies *211*
11.3.1 Recycling *212*
11.3.2 Upcycling *212*
11.4 Biocompatibility Concerns in the Use of Hydrogels and Aerogels for Textiles *214*
11.5 Consumer Acceptance and Education *215*

11.6 Economic Feasibility and Life Cycle Analysis *216*
11.7 Conclusion *218*
References *218*

12 Future Horizon *223*
Bushra Mushtaq and Sheraz Ahmad
12.1 Sustainable Hydrogel and Aerogel Textile Innovations *223*
12.2 Collaborations and Interdisciplinary Research *224*
12.3 Potential Market Trends *224*
12.4 Regulations and Safety Standards *225*
12.5 Conclusion *225*
References *226*

13 Conclusion *227*
Yasir Nawab and Sheraz Ahmad

Index *229*

Foreword

Textiles are among humanity's oldest technologies, yet they continue to evolve with remarkable dynamism in response to the modern challenges they face. Today, as industries seek solutions that combine performance, sustainability, and intelligence, the integration of novel materials such as hydrogels and aerogels represents a groundbreaking advancement. This book comes at a juncture when the textile industry is under significant pressure to limit its environmental footprint while addressing the requirements of adaptive, high-performance goods in health care and fashion, as well as protective and wearable electronics. The authors of this volume have successfully used an interdisciplinary approach to the subject, covering the fundamentals of science, as well as its practical applications and perspectives. This work is powerful because it is extensive and comprehensive. The book provides an invaluable road map for both researchers and practitioners, starting with chapters on the nature of materials and methods of producing them and continuing through to detailed case studies in biomedical textiles, sustainable fashion, and protective clothing. Notably, it is not afraid of the practicalities of cost, durability, and recyclability, but rather offers new ways out and new approaches. The book will serve as a reference for scholars, students, and professionals establishing the next generation of smart and sustainable textiles. It is a valuable and timely addition to the discipline, and I am confident it will foster more innovation and interdisciplinary collaboration.

Prof. Mohd Sapuan Sali
Universiti Putra Malaysia

Preface

The textile industry is entering a transformative era where sustainability, multifunctionality, and smart integration are no longer aspirations but urgent necessities. Hydrogels and aerogels are both the most promising classes of emerging materials to spearhead this revolution, featuring exceptional versatility, lightweight structures, and adaptive properties. This book, *Hydrogels and Aerogels for Advanced Textiles*, has been conceptualized to provide a comprehensive resource that bridges the scientific foundations of these materials with their growing applications in textiles, health care, protective clothing, and sustainable fashion.

All the chapters encompass a comprehensive range of knowledge, beginning with the science and synthesis of hydrogels and aerogels and then proceeding to their combination with natural and synthetic fibers, their functional properties, and their applications in various industries. Later sections discuss more specialized areas, such as biomedical textiles based on plant fibers and animal fibers, how these materials can be used in smart and protective clothing, and how they can be transformed into eco-fashion. No less significant, the issues hindering commercialization, including cost, recyclability, and biocompatibility, are presented in this volume, along with recommendations on how these challenges can be addressed in the future and research directions.

This book is designed for a diverse readership, including researchers, material scientists, textile engineers, graduate students, and industry professionals seeking to explore the intersections of advanced materials and textile engineering. Through the unification of the scholarly fields of various studies, including chemistry, engineering, materials science, biomedical research, and sustainability, this work is expected to serve as an inspiration to advance further innovations and cooperation.

We are thankful to our coauthors, collaborators, and institutions that helped us make this effort possible. We also wish to thank the larger scientific community, whose commitment to sustainable and workable textiles has remained an inspiration in the work presented herein.

We hope that this book serves not only as a reference but also as a catalyst for pioneering research and practical innovations that will shape the textiles of tomorrow.

Sheraz Ahmad, Faheem Ahmad
Farooq Azam and Yasir Nawab
National Textile University
Faisalabad, Pakistan
6th February 2026

1

Introduction

Farooq Azam[1], Yasir Nawab[1,2], and Sheraz Ahmad[1]

[1]*National Textile University, School of Engineering & Technology, Department of Textile Technology, Sheikhupura Road 37610, Faisalabad, Pakistan*
[2]*University of Kamalia, QAED Academy, Rajana Rd, Kamalia, Pakistan*

In ancient civilizations, linen, cotton, and silk were some of the materials that were central in the lives of ancient civilizations, and it was not just a matter of daily living as it contributed to the cultural identity and global trade. The Middle Ages were the period when wool was dominant, and the Renaissance period when mechanized looms were introduced, which formed the base of industrialization. Major innovations were experienced in the 19th century, when people came up with synthetic dyes and sewing machines became very common. The 20th century saw a revolution in synthetic fibers, the introduction of technical textiles, and the adoption of digital printing technologies. The 21st century has changed gears to functional textiles that is the integration of functionality and sustainability principles to traditional materials—stretching the limits of textile innovation as illustrated in Figure 1.1. This historical timeline shows how textiles have been woven into the history of humanity and have always been adjusting to the changing needs of society and the new technological opportunities. The long-term path of textile evolution depicts the active balance of human inventiveness and the advancement of technology that has constantly changed not only the making but also the functionality of cloths [1].

With the changing global needs and the advancement in technology, textiles can no longer be used in their conventional roles as mere sources of basic protection and beauty. They now have to offer multifunctionality, responsiveness, and, most importantly, sustainability. The rising consumer awareness and the demand in the industry especially in the sector like healthcare, sportswear, fashion, and protective clothing are driving the textile industry to a transformational change. Traditionally, textiles were used to protect the human body against the elements, including cold, wind, and rain. Gradually, aesthetic and comfort characteristics were added. But nowadays, applications require a lot more—textiles now need to be able to interact with their surroundings, react to outside stimuli, and are often designed to be smart and sensor-connected. These demands have led to the development of smart textiles and

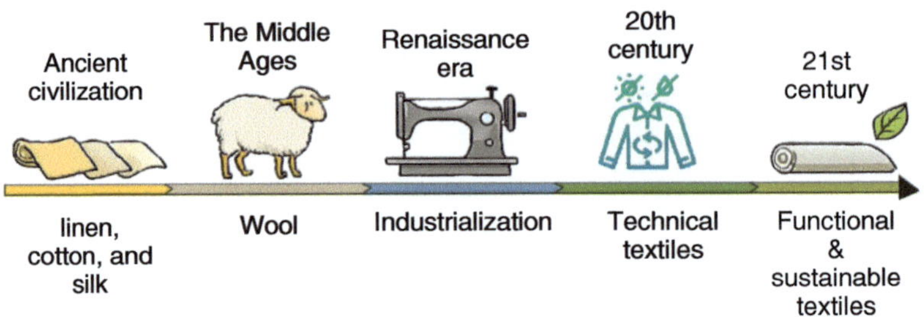

Figure 1.1 Historical timeline of textile evolution.

electronic textiles (e-textiles) that include components such as sensors, actuators, and control units to be used in real-time health monitoring, personal protection, communication, and even energy harvesting [2, 3].

Conventional textiles are characterized by lack of functionality, inefficient moisture regulation, inadequate thermal insulation, and high environmental burden because of chemical processing and the use of nonbiodegradable textiles. These materials are generally not responsive to external stimulus and provide little comfort in extreme weather. Unlike, the textiles in which hydrogels and aerogels are incorporated meet these limitations by adding novel properties of high moisture retention, temperature regulation, lightweight insulation, and responsiveness to stimuli. Hydrogels increase comfort and skin compatibility because of their ability to absorb water, and aerogels offer outstanding thermal insulation and breathability at minimal weight. Collectively, these smart materials are used to create high-performance, durable, and multifunctional textiles that can be used in contemporary applications in healthcare, sports, and extreme conditions [4, 5].

In this respect, hydrogels and aerogels have become revolutionary materials capable of revolutionizing the way textiles are used. The materials have exceptional properties that are highly valued, such as:

- Low density and high porosity, leading to lightweight, breathable, and insulating fabrics.
- Biocompatibility and biodegradability, which are in line with the aims of sustainable and environmentally friendly fashion.
- Thermal insulation, which provides benefits in protective clothing and cold-weather gear.
- Responsiveness to stimuli, including pH, temperature, and moisture responsiveness—allowing the textile to behave adaptively.
- Controlled release and moisture retention, which can be useful in wound dressing and drug delivery systems.
- Electrical conductivity, especially of carbon-based aerogels is opening new horizons in wearable electronics.

Additionally, natural fibers such as cotton, silk, and wool can be used to make hydrogels and aerogels, which further improve their sustainability rating and alignment with the circular economy. As the prospect of bio-fabrication, nanotechnology, three-dimensional (3D) printing, and engineering of fiber surfaces gains popularity, advanced materials are finding their way into textile matrices without sacrificing softness and wearability [6, 7].

The integration of these multifunctional materials is a paradigm change—it allows the development of not just smart and practical but also eco-friendly textiles. Given the push of regulatory agencies such as the European Union to implement measures that will make textiles sustainable, the production of hydrogel- and aerogel-absorbing textiles should be instrumental in reaching both performance and planetary objectives.

1.1 Hydrogels and Aerogels

Hydrogels are 3D networks of hydrophilic polymer chains that are either physically or chemically cross-linked. Their distinguishing feature is their ability to absorb and retain large amounts of water, leading to high swelling ratios. The cross-linking method strongly influences hydrogel properties: chemically cross-linked hydrogels form covalent bonds, offering higher mechanical stability, whereas physically cross-linked hydrogels rely on weaker intermolecular interactions and are typically less robust [8, 9]. The polymer type, water content, and cross-linking degree all play crucial roles in determining swelling, gel strength, degradation rate, and stability. Hydrogels can be synthesized from natural polymers such as alginate, chitosan, and cellulose—preferred for their biodegradability and biocompatibility—or from synthetic polymers, which generally provide higher mechanical strength but lack inherent biocompatibility [10, 11].

Aerogels are a special class of nanoporous materials, in which the liquid phase of a gel is replaced by gas without collapsing the structure. They are characterized by extremely high porosity, ultralow density, and very large surface areas. Early aerogels were silica-based, but more recent developments have introduced hybrid and biopolymer-based aerogels synthesized from polysaccharides and proteins [12, 13]. These materials can be tailored for specific functions such as biocompatibility, biodegradability, or thermal insulation by adjusting the synthesis parameters and structure.

1.2 Unique Properties Relevant to Textiles

The integration of hydrogels and aerogels into textiles enables the development of multifunctional fabrics with enhanced performance, comfort, and sustainability. The 3D hydrophilic polymer network of hydrogels provides exceptional water absorption and retention capacity, which is highly beneficial for moisture management. This property enhances comfort in sportswear, healthcare textiles, and daily wear by regulating humidity close to the skin. Hydrogels can also be engineered as "smart" textiles by incorporating stimuli-responsive functionalities that react to changes in pH, temperature, or humidity, enabling adaptive apparel with improved responsiveness and wearability. Aerogels contribute unique advantages through

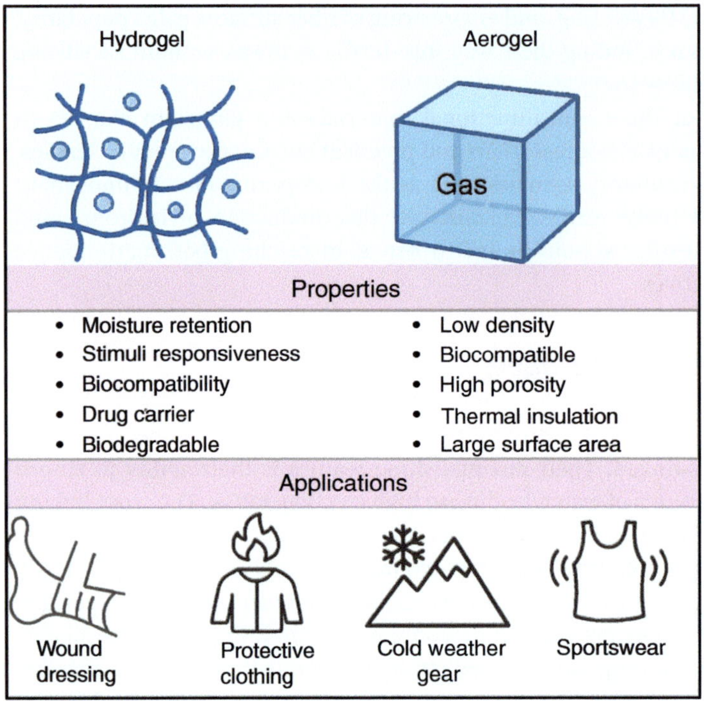

Figure 1.2 Key properties and textile applications of hydrogel and aerogel materials.

their nanostructured porosity and ultralightweight characteristics. Their very low thermal conductivity makes them excellent insulators, particularly valuable in protective clothing, extreme cold-weather garments, and high-performance technical textiles as shown in Figure 1.2. Beyond insulation, aerogels can impart flame-retardant properties and structural stability while maintaining breathability and flexibility of the base fabric [14, 15].

1.3 Why Hydrogels and Aerogels in Textiles?

The increasing demand for multifunctional and sustainable textiles has highlighted the need for materials that go beyond conventional fibers. Hydrogels and aerogels, due to their biocompatibility, biodegradability, and eco-friendly preparation routes, align well with the goals of sustainable textile development. Many hydrogel composites are synthesized from natural polymers such as alginate, chitosan, gelatin, and cellulose, which are abundant, nontoxic, and derived from renewable sources. Their incorporation into textiles not only reduces dependence on petroleum-derived synthetic fibers but also supports the global shift towards bio-based materials. Similarly, aerogels prepared from biopolymers such as chitosan or cellulose can be designed as lightweight, high-performance insulators while retaining biodegradability [16–18].

Waste minimization by means of multifunctionality is another point of sustainability. Textile composites based on hydrogel and aerogel can carry out multiple functions at the same time, which include moisture management, thermal insulation, flame retardancy, and antimicrobial protection, whereas the same functions require multiple chemical finishes to be applied that may emit toxic effluents. Furthermore, their porosity and high surface area may enable sophisticated roles, such as oil–water separation and chemical resistance, providing environmentally friendly solutions to protective and filtration textiles [19, 20]. Therefore, the new high-performance soft materials are promising in the textile of the next generation that is both sustainable and not compromised in their performances.

Hydrogels and aerogels will confer special properties that make textiles more comfortable, safe, and versatile. High-water absorption and retention capacity: hydrogels control moisture by absorbing sweat or exudates of a wound and slowly releasing it, enhancing the breathability and comfort of wearers of medical dressings, sportswear, and ordinary clothes. Their responsiveness to stimuli allows the fabrics to react to temperature, humidity, or pH and makes adaptive textiles that can be used in healthcare monitoring or performance wear. Moreover, the textile-reinforced hydrogel composites address the typical weakness of the mechanical strength, providing better flexibility, durability, and even conductivity when supplemented with functional fillers such as silver ions [17, 21].

These properties are supplemented by the high thermal insulating properties, ultralightweight structure, and nanoporous morphology of aerogels. They provide body heat in harsh environments and are becoming more popular in protective clothing, cold-weather clothing, and industrial insulation systems. Coatings and composites based on aerogels also provide a flame retardancy, UV protection, and sound absorption as well as water repellency and extend the utility of textiles considerably. As an example, the addition of aluminum hydroxide aerogel to fabrics decreases flammability without loss of flexibility, and silica aerogels added to nonwovens enhance sound absorption and thermal stability [22–24]. In combination, hydrogels and aerogels can transform textiles beyond passive materials into multifunctional systems that can be used to deliver comfort, protection, and environmental responsibility. Their incorporation is one of the important milestones in the way of connecting sustainability and high-performance textile engineering.

1.4 Scope and Structure of the Book

1.4.1 Scope

The book examines the science, synthesis, functional properties, and applications of hydrogels and aerogels to textiles and links basic concepts with industrial developments. It explores their compositions, preparation techniques and types, noting natural, synthetic, hybrid and responsive systems, and different aerogel types and compositions. It focuses on synthesis of natural and synthetic fibers, functional properties of moisture management, thermal insulation, flame retardance, antimicrobial property, and biodegradability, and their disruptive nature in sustainable fashion, biomedical textiles, smart wearables, and protective garments. The book also discusses the issues of the present-environmental impact, recycling, biocompatibility, and economic feasibility, as well as the future of sustainable innovation, market trends, interdisciplinary research, and regulation standards. As a reference and guide, it provides useful knowledge to researchers, textile engineers, material scientists, and other stakeholders in the industry that are interested in developing next-generation hydrogel- and aerogel-based textiles.

1.4.2 Structure

This book offers a complete insight into hydrogels and aerogels in textiles, their science and technology, synthesis processes, functional capabilities, and the variety of applications in fashion, healthcare, protection, and smart systems. Starting with the scientific background, the initial chapters describe the structural properties, methods of preparation, and property-tuning approaches of hydrogels and aerogels, defining how these versatile materials can be designed to perform specific textile functions by using various sources of fibers, natural, synthetic, or a mixture of both. Based on this background, the synthesis methods are explained on natural- and synthetic-based systems with a focus on their incorporation in textile substrates to develop high-performance fabrics with tunable moisture management, insulation, and durability. The book then moves to functional applications, demonstrating how hydrogels and aerogels enhance breathability, microbial resistance, flame suppression, and biodegradability, and how hybrid systems can combine the responsiveness of hydrogels with the insulation of aerogels for adaptive, sustainable textiles. In the context of sustainable fashion, these materials are presented as solutions to pressing environmental challenges by enabling biodegradable, recyclable, and energy-efficient apparel and accessories. Biomedical applications are explored through plant- and animal-derived hydrogel and aerogel systems, which support advanced wound healing, tissue engineering, drug delivery, and wearable biosensors, highlighting their role as next-generation biomaterials. Their integration into smart and electronic textiles is discussed with emphasis on wearable monitors, drug-delivery dressings, energy storage, and extreme-environment protection, demonstrating their ability to transform fabrics into intelligent, multifunctional systems. Specific focus is given to aerogels in protective clothing, where their superior thermal insulation, flame resistance, and lightweight robustness make them indispensable for military, industrial, and space applications. The book also addresses the major challenges hindering commercialization, such as high production costs, environmental concerns, and recycling difficulties, while proposing sustainable solutions through bio-based polymers, green chemistry, and circular economy models. Finally, it looks toward the future horizon, highlighting interdisciplinary collaboration, advances in additive manufacturing, and emerging market trends in eco-fashion, medical textiles, and protective gear. Collectively, the chapters illustrate how hydrogels and aerogels can act as enablers of next-generation textiles that are sustainable, intelligent, and adaptive, bridging scientific innovation with ecological and societal needs.

1.5 Target Audience and Applications

This book is intended for a diverse readership of researchers, graduate and undergraduate students, and industry professionals engaged in textiles, materials science, polymer chemistry, biomedical engineering, and sustainable design. For researchers and students, it provides a strong scientific foundation on the chemistry, structure, and synthesis of hydrogels and aerogels, along with insights into their integration into textile systems. For industry professionals, it highlights practical applications, case studies, and future market opportunities where these advanced materials can transform conventional practices. The applications discussed in this book span a wide spectrum: in fashion and sustainable apparel, hydrogels and aerogels enable biodegradable, recyclable, and performance-enhancing fabrics; in biomedical textiles, they support wound healing, tissue engineering, drug delivery, and wearable biosensors; in protective clothing, they provide superior thermal insulation, flame resistance, and impact protection for defense, industrial, and extreme environments. Additionally, the book explores their role in smart textiles and e-textiles, where their responsiveness and multifunctionality can support energy harvesting, sensing, and adaptive comfort. Collectively, the content aims to inspire interdisciplinary innovation, bridging academia and industry for the development of next-generation sustainable textiles.

1.6 Challenges and Opportunities

Despite their remarkable promise, hydrogel- and aerogel-based textiles face several challenges that limit their large-scale adoption. High production costs, particularly those associated with energy-intensive drying techniques and specialized crosslinking processes, remain major barriers. Environmental concerns also arise from the use of nonbiodegradable polymers, toxic solvents, and the difficulty of recycling hybrid textile–gel composites. Moreover, issues related to mechanical fragility, long-term stability, and biocompatibility require careful evaluation, especially for wearable and medical applications where safety is critical. However, these challenges also open the door to significant opportunities. Advances in green chemistry, bio-based polymers, additive manufacturing, and dynamic covalent chemistry are paving the way for cost-effective, scalable, and eco-friendly production methods. The growing global focus on circular economy models and sustainable product lifecycles further supports the development of recyclable and regenerative textile systems. With rising demand for smart fashion, biomedical solutions, and protective gear, the potential for hydrogel and aerogel textiles is vast. By fostering interdisciplinary collaboration and aligning innovation with environmental and regulatory frameworks, these materials are poised to redefine the textile industry, transforming it into a sector that is not only technologically advanced but also ecologically responsible.

References

1 Hossain, M.T., Shahid, M.A., Limon, M.G.M. et al. (2024). Techniques, applications, and challenges in textiles for a sustainable future. *Journal of Open Innovation: Technology, Market, and Complexity* 10 (1): 100230. https://linkinghub.elsevier.com/retrieve/pii/S2199853124000246.

2 Zhang, Y., Xia, X., Ma, K. et al. (2023). Functional textiles with smart properties: their fabrications and sustainable applications. *Advanced Functional Materials* 33 (33): https://onlinelibrary.wiley.com/doi/10.1002/adfm.202301607.

3 Aali Mohammadi, R., Shirazi, M., Moaref, R. et al. (2022). Protective smart textiles for sportswear. In: *Protective Textiles from Natural Resources*, 317–345. Elsevier https://linkinghub.elsevier.com/retrieve/pii/B9780323904773000250.

4 Azam, F., Ahmad, F., Ulker, Z. et al. (2022). The role and applications of aerogels in textiles. *Advances in Materials Science and Engineering* 2022: 1–22. https://www.hindawi.com/journals/amse/2022/2407769/.

5 Azam, F., Ahmad, F., Ahmad, S. et al. (2022). Preparation and characterization of alginate hydrogel fibers reinforced by cotton for biomedical applications. *Polymers (Basel)* 14 (21): 4707. https://www.mdpi.com/2073-4360/14/21/4707.

6 Linhares, T., Carneiro, V.H., Pessoa de Amorim, M.T., and Durães, L. (2023). A comparative thermoacoustic insulation study of silica aerogels reinforced with reclaimed textile fibres: cotton, polyester and wool. *Gels* 9 (7): 548. https://www.mdpi.com/2310-2861/9/7/548.

7 Chen, G., Yang, Z., Pan, H. et al. (2025). A review of hydrogel fiber: design, synthesis, applications, and futures. *Chemical Reviews* 125 (13): 5991–6056. https://pubs.acs.org/doi/10.1021/acs.chemrev.5c00159.

8 Zainal, S.H., Mohd, N.H., Suhaili, N. et al. (2021). Preparation of cellulose-based hydrogel: a review. *Journal of Materials Research and Technology* 10: 935–952. https://linkinghub.elsevier.com/retrieve/pii/S2238785420320895.

9 Zhang, Y.S. and Khademhosseini, A. (2017). Advances in engineering hydrogels. *Science (80-)* 356 (6337): https://www.science.org/doi/10.1126/science.aaf3627.

10 Adelnia, H., Ensandoost, R., Shebbrin Moonshi, S. et al. (2022). Freeze/thawed polyvinyl alcohol hydrogels: present, past and future. *European Polymer Journal* 164: 110974. https://linkinghub.elsevier.com/retrieve/pii/S0014305721007084.

11 Zhao, L., Zhou, Y., Zhang, J. et al. (2023). Natural polymer-based hydrogels: from polymer to biomedical applications. *Pharmaceutics* 15 (10): 2514. https://www.mdpi.com/1999-4923/15/10/2514.

12 Wang, Y., Su, Y., Wang, W. et al. (2019). The advances of polysaccharide-based aerogels: preparation and potential application. *Carbohydrate Polymers* 226: 115242. https://linkinghub.elsevier.com/retrieve/pii/S0144861719309099.

13 Ulker, Z. and Erkey, C. (2014). An emerging platform for drug delivery: aerogel based systems. *Journal of Controlled Release* 177: 51–63. https://linkinghub.elsevier.com/retrieve/pii/S0168365913009759.

14 Neumann, M., di Marco, G., Iudin, D. et al. (2023). Stimuli-responsive hydrogels: the dynamic smart biomaterials of tomorrow. *Macromolecules* 56 (21): 8377–8392. https://pubs.acs.org/doi/10.1021/acs.macromol.3c00967.

15 Alwin, S. and Sahaya Shajan, X. (2020). Aerogels: promising nanostructured materials for energy conversion and storage applications. *Materials for Renewable and Sustainable Energy* 9 (2): 7. http://link.springer.com/10.1007/s40243-020-00168-4.

16 Wu, M., Shao, Z., Zhao, N. et al. (2023). Biomimetic, knittable aerogel fiber for thermal insulation textile. *Science (80-)* 382 (6677): 1379–1383. https://www.science.org/doi/10.1126/science.adj8013.

17 Jiang, S., Deng, J., Jin, Y. et al. (2023). Breathable, antifreezing, mechanically skin-like hydrogel textile wound dressings with dual antibacterial mechanisms. *Bioactive Materials* 21: 313–323. https://linkinghub.elsevier.com/retrieve/pii/S2452199X22003577.

18 Ahmed, M.S., Islam, M., Hasan, M.K., and Nam, K.W. (2024). A comprehensive review of radiation-induced hydrogels: synthesis, properties, and multidimensional applications. *Gels* 10 (6): 381. https://www.mdpi.com/2310-2861/10/6/381.

19 Bhuiyan, M.A.R., Wang, L., Shaid, A. et al. (2020). Silica aerogel-integrated nonwoven protective fabrics for chemical and thermal protection and thermophysiological wear comfort. *Journal of Materials Science* 55 (6): 2405–2418. http://link.springer.com/10.1007/s10853-019-04203-2.

20 Liu, H., Shang, J., Wang, Y. et al. (2022). Ag/AgCl nanoparticles reinforced cellulose-based hydrogel coated cotton fabric with self-healing and photo-induced self-cleaning properties for durable oil/water separation. *Polymer (Guildford)* 255: 125146. https://linkinghub.elsevier.com/retrieve/pii/S0032386122006346.

21 Zaman, S.u., Mushtaq, B., Ahmad, F. et al. (2023). Development of conductive cotton non-woven alginate hydrogel composite for smart textiles. *Journal of Polymers and the Environment* 31 (9): 3998–4006. https://link.springer.com/10.1007/s10924-023-02847-5.

22 Talebi, Z., Soltani, P., Habibi, N., and Latifi, F. (2019). Silica aerogel/polyester blankets for efficient sound absorption in buildings. *Construction and Building Materials* 220: 76–89. https://linkinghub.elsevier.com/retrieve/pii/S095006181931445X.

23 Thennarasu, P., Senthilkumar, M., and Prabunathan, P. (2021). Functionalization of cotton fabric with titania silica aerogel using rice husk as precursor for enhanced UV shielding. *Materials Today: Proceedings* 42: 795–803. https://linkinghub.elsevier.com/retrieve/pii/S2214785320389616.

24 Venkataraman, M., Mishra, R., Kotresh, T.M. et al. (2016). Aerogels for thermal insulation in high-performance textiles. *Textile Progress* 48 (2): 55–118. http://www.tandfonline.com/doi/full/10.1080/00405167.2016.1179477.

2

Science of Hydrogels and Aerogels

Zeynep Ulker[1] and Can Erkey[2]

[1]*Gebze Technical University, Department of Mechanical Engineering, Cumhuriyet Mah., 2254. Sk., No: 2, Gebze, Kocaeli 41400 Turkey*
[2]*Koç University, Chemical and Biological Engineering Department,, Rumelifeneri Yolu 34450, Istanbul Turkey*

2.1 Introduction

Gels consist of a three-dimensional (3D) network comprising a solid structure and a gas or a fluid medium within their pores [1]. In general, gels with liquids (water) as the medium are called hydrogels, and those with gas as the medium are denoted as aerogels [1, 2]. Researchers are increasingly investigating both hydrogels and aerogels in various applications. While they have similar properties, they present different challenges. Hydrogels display comparable properties to natural tissues regarding mechanical characteristics such as viscoelasticity [3]. They typically have large macropores, which enable good diffusion properties. On the other hand, aerogels are mesoporous structures having ultra porosities (up to 99%), large pore volumes, and very high surface areas [4]. One should be familiar with the basics of hydrogel and aerogel science to prepare a material suitable for a specific application. One of the applications in which they are increasingly investigated is textile innovations, where they can impart different properties to fabrics that traditional textiles cannot offer.

2.2 Hydrogels: Relationship Between Function and Structure

Either physically or chemically cross-linked hydrophilic polymer chains form hydrogels, constructing a 3D network [5, 6]. They can retain large amounts of water, leading to very high swelling ratios. Various methods such as physical or chemical cross-linking are employed to prepare hydrogels as illustrated in Figure 2.1 [7]. Chemically cross-linked hydrogels display covalent bonds between polymer chains [8]. However, with physical cross-linking, only physical interactions are present among the polymer chains. The mechanical strength of conventional

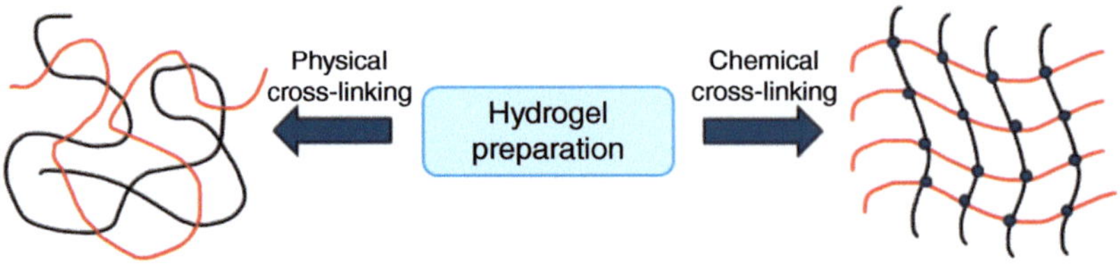

Figure 2.1 Diagram for the preparation of hydrogels by different cross-linking methods.

physically cross-linked hydrogels is usually low [9, 10]. Thus, researchers are trying to engineer hydrogels with enhanced physicochemical and mechanical properties.

The choice of the polymer used as the precursor, the type of cross-linking, and the cross-linking degree can affect the final properties of the synthesized hydrogels [11, 12]. It is crucial to understand how synthesis parameters affect the resulting mechanical properties to select appropriate preparation conditions and subsequent application areas [13]. Low cross-linking degrees are usually associated with larger pore sizes, leading to fragile materials with faster degradation rates [12]. Conversely, higher cross-linking degrees lead to stronger networks with less swelling capacity. Thus, key properties of the hydrogels, such as the gel strength and the swelling capacity, are inversely correlated [14]. Each factor, such as the polymer type, water content, and degree of cross-linking, should be considered separately to prepare hydrogels for a specific application.

Hydrogels are prepared from a variety of polymers, including natural or synthetic ones. Different characteristics of each polymer are reflected in the final properties of the hydrogel. Natural hydrogels are synthesized using biocompatible and biodegradable polymers usually preferred in biomedical applications [15]. Synthetic polymers, alternatively, are also used to synthesize hydrogels by the reaction of small organic monomers [16]. They are usually superior compared to natural hydrogels in terms of mechanical properties. However, they lack the biodegradability and the biocompatibility of the natural ones.

Natural polymer hydrogels are mostly synthesized from polysaccharides, polynucleotides, and polypeptides [17]. Those polymers are easier to access, abundant, less costly, nontoxic, and have other attractive biological properties. They are obtained from various natural sources and can be classified as neutral, cationic, and anionic, depending on their charge [12, 17]. Natural polymer-based hydrogels are preferred, particularly in biomedical applications, including various types of drug delivery systems, tissue engineering, and wound dressings [18, 19]. Alginate, chitosan, and cellulose can be given as examples of natural polymers used to prepare hydrogels [15, 20, 21]. Complex purification or extraction processes are needed to obtain most of the natural polymers used to synthesize hydrogels [15]. The difficulty of such processes and the differences in the sources from which polymers are extracted lead to variable final properties.

Hydrogels have become useful materials for various applications during the last century. They exhibit soft physical properties like natural tissues, making them ideal candidates, especially for biomedical applications [3, 22]. As an important hydrogel subclass, stimuli-responsive hydrogels are also attracting increased attention. When hydrogels are used as "smart" materials, they can react to small environmental changes. Environmental stimuli can be physical (light, sound, temperature, or electric or magnetic fields) or chemical (pH, ionic strength, or humidity). Triggered deformations can be swelling, shrinking, or shape morphing, as well as bending, twisting, folding, or moving [3].

2.2.1 Science of Preparing Hydrogels

Whether natural or synthetic polymers are used for the synthesis, hydrogels are prepared employing one of the two main methods, such as physical and chemical cross-linking [12]. The choice of the method leads to the formation of hydrogels having distinct characteristics due to their nature and the occurrence of cross-links.

2.2.1.1 Physically Cross-Linked Hydrogels

Physical cross-linking involves hydrogen bonding, van der Waals forces, hydrophobic interactions, or coordination bonds rather than covalent bonding [12]. Due to the weaker nature of those forces, physical cross-links can be reversible under certain conditions, in contrast to chemical cross-links, which are rather harder to break. This property leads to the preparation of stimuli-responsive hydrogels that can react to various external physical or chemical stimuli. However, these hydrogels usually have a shorter lifespan, leading to a faster degradation rate [12]. On the other hand, it was also seen that using dual or multiple noncovalent cross-links to enhance the mechanical performance of the physical hydrogels was more effective because of the complementary effects of different noncovalent bonds [23].

Alginate is used to prepare physically cross-linked hydrogels by ionic interactions. It is soluble in water and can easily form a gel when multivalent (especially divalent) metal cations such as Ca^{2+}, Cu^{2+}, and Zn^{2+} are present [24]. The change in the concentration of the cross-linking ion can lead to different structural and mechanical properties, including elastic modulus, surface pore size, and swelling/degradation behavior [25]. In a recent study, different cross-linking ions were used to prepare hydrogels [26]. Investigations were conducted to determine how several ions affected the resulting mechanical and swelling properties. It was seen that using trivalent ions such as Fe^{3+} resulted in more compact networks with a higher elastic modulus. However, the uncontrollable gelation kinetics due to the high water solubility of these ions leading to uneven hydrogel structures with poor mechanical properties required different synthesis techniques for alginate hydrogels [25]. One of the approaches is to slow down and control gelation using compounds having lower water solubility, such as calcium sulfate ($CaSO_4$) and calcium carbonate ($CaCO_3$), which allow the formation of cross-links over a longer period [27]. $CaCO_3$ in the alginate solution does not immediately cause gelation, as calcium carbonate is not water-soluble at neutral pH. However, adding glucono-δ-lactone helps dissociate calcium ions, which then leads to a slower gelation [28].

Another important physical interaction affecting the stability, structure, and properties of hydrogels is hydrogen bonding [29]. These interactions occur when a positively charged hydrogen atom and a negatively charged acceptor atom form an electrostatic bond. In a recent study, sodium alginate/carboxymethyl chitosan hydrogel beads were prepared by physical cross-linking through hydrogen bonding and electrostatic interactions with citric acid [30]. Synthesized hydrogel beads have shown lower swelling ratios compared to similar hydrogels that are prepared through only electrostatic interactions [31] or ionic cross-linking (Ca^{2+}) [32]. They were also pH sensitive.

Attractive forces between hydrophobic molecules, causing hydrophobic interactions, are also used to form physically cross-linked hydrogels [29, 33]. It was shown that the presence of nystatin provided more hydrodynamic/mechanical stability to the network through the formation of hydrophobic interactions and hydrogen bonding [33]. Among the investigated mechanical properties, physical hydrogels were shown to have the highest value for elongation at break. It also demonstrated higher swelling compared to chemical hydrogels with the highest cross-linking.

Another method of physical cross-linking is crystallization via the cooling of a polymer under certain conditions without the need for a cross-linker [34]. Hydrogels prepared with this method primarily consist of polymer chains physically entangled [35]. The length, temperature, and cycle number of the freeze–thaw process; the molecular weight; and the concentration of the polymer can significantly affect the final properties of the synthesized hydrogels [14, 36]. In the study of Mori et al., polyvinyl alcohol (PVA) hydrogels were synthesized using two successive freeze–thaw cycles [37]. It was suggested that the resulting PVA hydrogels had high water contents (around 90 wt%), good mechanical properties with 0.6 MPa tensile strength, and 130% elongation at break. Using a higher number of freeze–thaw cycles leads to higher crystalline regions, which yield stronger and more stable hydrogels [38]. Peppas and Scott have shown that the gels that were prepared with five freezing–thawing cycles were stronger and more rigid compared to those formed with fewer cycles [39]. Swelling studies were also conducted to conclude that the ingress of water first increased with time and gradually dropped, indicating molecular densification. Another strategy to enhance the mechanical properties of PVA hydrogels prepared by the freeze–thaw method was the addition of ethylene glycol, which significantly lowers the freezing point of the mixed solvent [35]. In this way, PVA molecule chains can self-assemble into many microcrystalline domains at low temperatures, enhancing the mechanical properties of the resulting hydrogels.

Self-assembly is another useful tool for forming physical hydrogels. The use of self-assembly yields hydrogels showing properties consistent with a physically cross-linked network [40]. The gelation is usually performed using chemical groups found on native biomolecules without the need for an external functional group. Suitable biopolymers can be mixed to form a self-assembly composite hydrogel through ionic, hydrogen, or hydrophobic attractions [41]. Chen et al. prepared polyvinyl alcohol tannic acid hydrogels (PVA-TA) via self-assembly method [42]. It has been shown that the hydrogen bonding between the constituents enabled the formation of self-assembly and furthermore caused the crystallization of the PVA chains. The change in the concentration of the two components resulted in hydrogels having diverse morphologies, from microspheres to oriented porous structures. Most of the prepared hydrogels had favorable mechanical properties. The tensile strength and the elastic modulus of the self-assembly hydrogels are gradually reduced when a higher concentration of PVA-TA is used.

2.2.2 Chemically Cross-Linked Hydrogels

Chemical hydrogels consist of a covalently bonded network, with a more stable structure and better mechanical properties [29]. They can withstand external environmental factors compared to physical hydrogels and thus show improved mechanical performance [43]. Swelling properties and biodegradability can also be modified via covalent cross-linking [17]. Methods such as cross-linking by small molecules, ionizing radiation, or free radical mechanisms can be used to prepare covalently linked hydrogels [17].

Covalent cross-linking can be achieved via mono- or bifunctional small molecules like formaldehyde, glutaraldehyde, and genipin [17]. Although chemically cross-linked hydrogels are attractive due to their superior mechanical properties, covalent cross-linking agents might be toxic, which then necessitates their thorough removal after the formation of the gels [27]. Thus, desired functional groups were also introduced to polymers to prepare hydrogels in situ. The Schiff base mechanism, Michael addition, or disulfide bonding can be used for that purpose. Ito et al. designed novel dextran-based hydrogels by modifying carboxymethyl dextran with a hydrozide group and modifying carboxymethylcellulose and dextran with an aldehyde group without the need for a cross-linking agent [44]. Resulting hydrogels had different swelling properties depending on the polymer that was used, but they were both found to be successful in preventing peritoneal adhesion. Alginate is a frequently used polymer to easily prepare physically cross-linked hydrogels, as described previously. However, those hydrogels lack the functional group needed for cell interaction and have limited mechanical properties. Bouhadir et al. have synthesized new polymers using the part of the alginate molecules responsible for gelation, thereby creating alginate derivatives [45]. The resulting hydrogels had variable mechanical properties and improved cell adhesion due to the linkage of a cell adhesion peptide to the synthesized polymer. These covalently cross-linked networks also led to a degradation rate based on the cross-linking density [46].

Another way of preparing covalently cross-linked hydrogels is through ionizing radiation [47, 48]. Among those cases, photo cross-linking sometimes needs possibly toxic photo initiators. However, in a recent study, the authors prepared strong hydrogels from alginate at different methacrylation degrees with diacrylated poly(ethylene glycol) (PEGDA) [49]. They have employed electron beam irradiation at varying doses without a photo initiator. It was concluded that hydrogels with a higher storage modulus were obtained by adding PEGDA into alginate with a high degree of methacrylation. It was also seen that more cross-linked networks led to a higher storage modulus, which then restricted the swelling ratio.

The free radical polymerization can also be used to prepare chemical hydrogels starting with the low molecular weight monomers and cross-linking agents. The reactions are initiated with various small molecules called initiators, including potassium persulfate (KPS) and ammonium persulfate (APS).

All in all, hydrogels possess different properties depending on their synthesis methods. Physical cross-linking enables the formation of more relaxed networks having greater swelling ratios. Furthermore, it can be used to form stimuli-responsive hydrogels, which can change their properties with respect to changing environmental conditions. Conversely, chemical cross-linking offers a more stable network with a higher mechanical strength. Thus, preparation methods along with the choice of polymers should be based on the intended application, which will necessitate the requirement of different kinds of physical and structural properties.

2.3 Aerogels: Relationship Between Structure and Functions

Kistler was the first to synthesize aerogels, a special class of nanoporous materials, who then defined them as materials retaining their pore structure intact upon exchanging their liquid phase with a gas [50]. They have distinctive properties such as high porosity, high specific surface area, and very low density, which make them appealing and suitable for different application areas [51–54]. Research on aerogels has increased dramatically since their first invention in 1930s, particularly in the past 20–30 years [55] and there are many aerogel based products in the market today, primarily for thermal insulation and barrier coatings. The first aerogels were commonly obtained from alkali silicates, metal alkoxides, or low molecular weight organic precursors. Since then, the use of biopolymers from polysaccharides or proteins to prepare novel hybrid or composite aerogels has gained more attention [56–58].

Their physical and chemical properties, such as their high surface area, very low density, and especially their pore structure, are crucial for selecting the right application area. While biocompatibility and biodegradability are more desirable for biomedical applications, low thermal conductivity is more desirable in thermal insulation applications. Those properties are dependent upon various factors that can be tailored through the synthesis procedure.

2.3.1 Science of Preparing Aerogels

The sol–gel process is generally employed to synthesize gels from any organic and/or inorganic precursor that can form a stable 3D network in the selected solvent [59]. The procedure involves several consequent stages, such as gel formation, solvent exchange, and finally drying [60, 61]. Sol–gel processing can be divided into two sub-methods to produce various aerogels [62]. The first one can be used to synthesize the earliest researched and most used inorganic aerogels [63, 64] and some organic aerogels [65, 66]. Among inorganic aerogels, silica aerogels were the most extensively researched ones based on their attractive properties, such as high porosity, large surface area, and low thermal conductivity [67]. Those materials are mostly prepared by the sol–gel process, where metal alkoxides such as tetraethylorthosilicate (TEOS) or tetramethylorthosilicate (TMOS) are used as chemical precursors [54]. Precursors are primarily hydrolyzed in an appropriate solvent, the rate of which can be sped up with the addition of proper catalysts. Following hydrolysis processes, molecules go through condensation reactions to create a network, which leads to the formation of a gel. Aging and solvent exchange processes are conducted to enhance the mechanical strength of the wet gels and to replace the pore liquid with a suitable solvent (usually alcohol) for the subsequent drying step. Drying these wet gels, usually called alcogels, without destroying their structural integrity, is the final stage of the synthesis. To achieve this, supercritical drying, freeze-drying, or ambient pressure drying is widely used.

Similarly, the dissolution of the precursor in a solution, followed by the regeneration of the polymer network with physical or chemical cross-linking, is also used to prepare aerogels by the sol–gel process [62]. Biopolymer aerogels, including cellulose, alginate, starch, carrageenan, and pectin, are typically synthesized with this technique [68–71].

It is necessary to understand the mechanism behind each step of the synthesis and corresponding synthesis parameters to tailor and enhance the properties of the resulting aerogels for a specific application [62]. The 3D network of the aerogel and thereby related physiochemical properties are affected by the choice of the precursor, synthesis route, reaction parameters, and different drying methods.

Bedilo and Klabunde studied the effect of different precursor concentrations in zirconia aerogels [72]. Smaller pores with lower pore volumes and higher densities were obtained with increasing precursor concentrations, which favored olation reactions. Furthermore, a higher specific surface area (450 m^2/g) was obtained with the highest precursor concentration. Along similar lines, Groult and Budtova synthesized different pectin aerogels by modifying synthesis parameters [69]. They have seen that using higher polymer concentrations led to the polymer network being more resistant to shrinkage throughout the synthesis process. Moreover, the change in the non-solvent type from ethanol to acetone also resulted in slightly larger pores with less agglomerated pore walls. Pectin aerogels with lower densities and higher specific surface areas were prepared with acetone.

Along with modifying the concentration of the precursor or the type of solvent, the choice of the type and the amount of catalyst also affect the properties of the resulting aerogels. Resorcinol-formaldehyde aerogels can be prepared by the polycondensation of resorcinol and formaldehyde with acid or base catalysts. It was seen that the use of lower pH values resulted in the synthesis of less branched structures. Furthermore, a more interconnected network with smaller particles was obtained with higher pH values, leading to higher surface areas [73–75].

After gelation, the resulting wet gels are subjected to an aging step during which the structure and the properties of the gel change. During this step, wet gels are kept inside their pore liquid mainly to render those fragile wet gels stronger to avoid shrinkage during subsequent steps of the synthesis. Upon gelation, not all the precursors added initially might have been completely used [76]. The remaining unreacted monomers might continue reacting during aging, even if the gelation has already occurred. Haereid et al. aged wet silica gels in ethanol/water solutions with varying concentrations at different temperatures (20–70 °C) and times (2 and 24 hours) [77]. The linear shrinkage observed after supercritical drying was decreased from 29% to 2% using 20 vol.% water/ethanol for 24 hours at 60 °C. In another study, Iswar et al. studied the effect of different aging times (from 2 to 24 hours) and temperatures (55, 65, and 75 °C) on the physiochemical properties of silica aerogels when aged in the pore liquid [78]. They have also used two different drying techniques, such as ambient pressure drying and supercritical fluid drying. It was concluded that the density and the linear shrinkage of the aerogels dried by ambient pressure were reduced. On the other hand, pore size and pore volumes increased with increasing aging time and temperature. However, there was no significant change for aerogels that were supercritically dried.

The subsequent step, the solvent exchange, involves replacing the pore liquid with an appropriate solvent for drying. During this process, any unreacted monomers, oligomers, and the solvent that are employed in the synthesis are removed from the network while pore-filling liquid is exchanged with the selected solvent [79]. This process is crucial to obtain aerogels with intact structures at the end of the drying, as was also explained in Kistler's initial work [50]. This method has been performed in the same way since Kistler's time, either by immersing the synthesized wet gels in an excess of solvent (one step) or by carrying out a multistep process using solvent/water mixtures with increasing solvent concentration [80–82]. As a one-step solvent exchange process usually results in higher shrinkage values, multistep solvent exchange is the commonly used method [80, 83]. Apart from the duration, the temperature at which the solvent exchange is also shown to be important by various researchers. Cardea et al. used a low-temperature solvent exchange for chitosan aerogels and observed a lower shrinkage value, which helped preserve the 3D network [84]. Performing the solvent exchange process at more stable gel conditions (gelation temperature) was thought to dictate this process. On the other hand, the network of silica gels was shown to crumble when high temperatures were used for solvent exchange [85]. Subrahmanyam et al. tested the effect of 15 different solvents using a one-step solvent exchange process [83]. Volume shrinkage values less than 90% were obtained only with the use of six of the selected solvents, such as methanol, DMSO, glycerol, propylene glycol, ethylene glycol, and DMF, again showing the necessity of a multistep solvent exchange process. As a high shrinkage value leads to higher densities, which in turn affect other aerogel properties such as pore volume, surface area, and thermal conductivity, the number of solvent exchange steps should also be carefully controlled during the synthesis of all types of aerogels [62].

The last stage of the process, which significantly affects the porous structure and thereby physicochemical properties, is the drying step. During this step, the solvent filling the pore is exchanged with air while retaining the gel structure intact. As aerogels are nanostructured materials consisting of nanosized pores, capillary stresses arising from the gas–liquid boundaries can be very significant, which, in turn, makes the drying more difficult. The porous structure concerning size and shape can be destroyed because of these high capillary stresses within the pores [86].

Wet gels are dried using three different methods, such as ambient pressure drying, freeze-drying, and supercritical drying [87]. The parameters of the drying, along with the drying method itself, significantly affect the final aerogel properties. Among these three methods, ambient pressure drying seems to be a proper choice that would reduce the cost and energy consumption associated with other drying techniques while offering a simpler and safer implementation [88, 89]. Ambient pressure drying involves the removal of the pore solvent at ambient conditions ranging from room temperatures up to 200 °C [87]. The evaporation of the solvent from the nanosized pores creates significant capillary stress, which then damages the 3D structure of the gels by extensive shrinkage or cracking. To prevent the negative effects of this capillary tension, the surface of the gels is modified through solvent exchange with the use of low surface tension solvents. Einarsrud and Haereid were able to prepare low-density monolithic silica gels by using ambient pressure drying [90]. The gels were strengthened during aging, where they were immersed in silane solutions, which caused additional silica precipitation to the already formed network [90–92]. Increased strength then prevents collapse of the pores during drying at ambient conditions. Another method is the use of a co-precursor or silylating agents such as methyltrimethoxysilane (MTMS) or trichloromethyl silane (TCMS), along with the silica precursor to modify the surface chemistry and cause the springback effect, which can be described as the re-expansion of the silylated silica gels during drying [89, 93, 94].

Freeze-drying, alternatively, removes the solvent from the pores by sublimation. The wet gels are initially frozen below the solvent's freezing point. Then, the pressure is lowered below the sublimation pressure [87, 95]. When the drying is complete, the system returns to its initial ambient condition, resulting in dried gels commonly referred to as cryogels [96, 97]. Freeze-drying can also damage the porous structure, causing crystal growth and stress development within the pores. Moreover, the solvent should be selected carefully [87, 98]. If water is used, it can expand during freezing, damaging the network. On the other hand, the use of alcohol might also render the process difficult, as it can have extremely low freezing points. Hybrid alginate cryogels were prepared by Barros et al. using gelatin, gellan gum, carboxymethyl cellulose, and lignin by gelation with pressurized carbon dioxide [96]. The authors have used two freezing regimes in which they either used rapid freezing in liquid nitrogen or slow freezing at 193 K before submitting the wet gels to the freezing process. Slowly frozen samples were more fragile and had irregular shapes. As the solvent was water, its slow freezing resulted in large ice crystals, which then damaged the structure while growing [99]. On the other hand, rapidly frozen samples had higher porosity and a more homogeneous pore structure. In addition to the freezing rate, the freezing temperature also has an impact on the porous structure of aerogels [100]. The pore volume and the pore size of aerogels decreased when the freezing temperature increased from −60 to −20 °C. It was concluded that higher temperatures result in slower nucleation rates, leading to the formation of larger crystals, which then causes the disappearance of the large mesopores.

The last method involves supercritical fluids to extract the liquid from the pores of the gels. When the pressure and the temperature of a fluid are above its corresponding critical pressure and temperature, the fluid reaches its supercritical state. Supercritical carbon dioxide ($scCO_2$) is the most commonly used fluid for this technique as it has comparably mild critical conditions, such as 74 bar and 31 °C. This drying method results in the production of aerogel having higher pore volumes and higher surface areas compared to other drying methods [87].

Ciftci et al. prepared cellulose nanofiber aerogels with freeze-drying and supercritical drying [101]. Aerogels obtained using supercritical drying had a more homogeneous porous structure consisting of thinner fibers that did not aggregate. Similarly, the BET (Brunauer, Emmett and Teller) surface area of freeze-dried samples ranged from 16 to 20 m^2/g compared to supercritically dried ones having 72 to 115 m^2/g surface areas. Zhao et al. found similar results when they compared supercritical drying and freeze-drying while preparing zirconia aerogels [102]. Aerogels dried with $scCO_2$ were transparent monoliths. Moreover, they had a mesoporous structure with a 640 m^2/g surface area, whereas the freeze-dried ones were opaque, white powders having a microporous structure with a smaller surface area (around 400 m^2/g). In another study, polyurea aerogels were synthesized using the same drying methods [103]. Although denser structures with aggregation of fibers were observed for freeze-dried samples compared to more homogeneous fibrillar structures of supercritically dried samples, it was concluded that physicochemical characteristics of the resulting aerogels were not significantly affected by the drying method.

Rodriguez-Dorado et al. used and compared three drying methods to prepare alginate aerogels [97]. Cryogels and xerogels had lower textural properties compared to aerogels, which were supercritically dried. Sc-dried ones had smoother surfaces compared to the partially collapsed structures of freeze-dried ones. In addition, the use of supercritical drying led to the synthesis of aerogels with higher surface area than freeze-drying. Oven-dried gels had lost their porosity and thus were not subjected to measurements.

2.4 Textile Innovations

Since ancient times, textiles have been used to make clothes. However, their significance, as well as their application in several areas such as filtration, protection, and medical applications grew because of their functional qualities. Among various applications in which textiles are used, the demand for technical textiles with unique features—like fire resistance, heat resistance, and antibacterial properties—is increasing significantly. The fiber type, fabrication method, and structure have a major impact on the textile fabrics' properties [104]. Furthermore, fabric porosity affects the comfort of the finished product and improves thermal insulation and filtering abilities. Although woven, knitted, and nonwoven fabrics can be used to obtain porous textiles, only macroporous structures were achieved [104]. As nanoporous structures would be more desirable in such applications, there are different methods to provide textiles with the required porosities. Among those methods, electrospinning is an advanced fiber production technique used to produce continuous fibers with diameters ranging from microns to nanometers [105, 106]. Another idea is to make a textile composite using a material that is highly porous but will not damage the final properties of the material. In light of this idea, various hydrogel and aerogel textile composites were developed for different textile applications [104, 107–109]. Due to their unique structures and properties, hydrogels and aerogels are increasingly appealing in textile innovations. Although most of the applications concerning hydrogel-based textiles are thought to be in the biomedical field, especially in wound dressings [110–112] and the ones comprising aerogel-based textiles in thermal insulation [113–115], there are many different fields in which those materials with high capabilities are investigated.

Hydrogel-based composites are commonly investigated for wound dressing applications. Wang et al. coated a thermoresponsive hydrogel, the sol–gel transition behavior of which is temperature-controlled, on a nonwoven fabric [116]. The hybrid system showed high potential as a controlled drug delivery system and was able to manage moisture and exudate levels. Through the removal of sweat, blood, or other bodily fluids, the function of the samples was to keep the topical skin area clean, breathable, and comfortable. Ahmad et al. prepared a hydrogel textile composite incorporating an alginate hydrogel in the hydroentangled cotton nonwoven by a simple sol–gel method [117]. Synthesized samples had high flexibility along with good mechanical strength. Increasing the alginate concentration reduced the air permeability due to a higher cross-linked network with smaller pores. However, it also increased the wound absorption capacity (up to 650%) when the drying temperature was increased. In another study, gelatin hydrogel fibers were knitted using knitting needles, followed by the cross-linking of fibers [111]. The resulting glyhydrogel textile dressing showed excellent breathability, antibacterial, and stretchability properties.

The poor mechanical strength of hydrogels can be given as one of the obstacles preventing their widespread use. Zhang et al. prepared a poly(vinyl alcohol) hydrogel composite reinforced with cotton fibers, which had improved mechanical properties compared to individual components such as pure cotton or neat PVA hydrogel [118]. The resulting samples displayed high tensile strength (5.7 MPa), high compressive strength (12.6 MPa), and high fracture energy (23,000 J/m^2) in addition to high water content and biocompatibility. The authors have also tested the inclusion of another hydrophilic compound, sodium alginate, and concluded that sodium alginate improved swelling properties but reduced the stiffness and toughness of the composite. In another work, authors have again used cotton nonwoven fabric to strengthen hydrogels to prepare conductive fabrics [119]. The hybrid material was synthesized by immersing the fabric in an alginate solution, followed by immersion in a silver nitrate solution. The hydrophilic network of the hydrogel facilitated the flow of electrons, while the addition of silver ions gave the necessary conductivity.

In the work of Liu et al., a cellulose-based hydrogel combined with Ag/AgCl nanoparticles was prepared [120]. The composite hydrogel displayed good mechanical properties along with anti-swelling, self-healing, and self-cleaning properties.

It was further tested for oil/water separation capabilities after coated on cotton fabric through the sol–gel process. The samples had good oil/water separation capabilities for different mixtures with a separation efficiency up to 99.5%. Even after 50 cycles, the composite still had a separation efficiency greater than 99.0%. With a similar purpose, Azam et al. prepared strong aerogel fibers reinforced with cotton fabric [52]. The resulting samples exhibited good oleophobicity underwater and excellent separation efficiency up to 99.4%.

The high water content of hydrogels serves as the phase transformation agent to encourage evaporation [108, 121]. When they are exposed to fires, heat energy dissipates, resulting in the evaporation of this water, leading to the investigation of hydrogels as fire-resistant materials [108]. However, the poor mechanical properties, inflammability, and low melting points necessitate the preparation of composite hydrogels for this application. Nie et al. prepared a hydrogel/fabric composite flame-retardant material with viscose nonwoven fabric as rigid reinforcement [108]. The hydrogel served as the coating and enhanced the thermal insulation ability of the fabric. It was concluded that the resulting samples had the potential to be used in firefighting and personal protection areas with a simple preparation method.

Along with hydrogels, aerogels can also be incorporated into textiles to make fire-resistant materials. Aluminum hydroxide aerogel was prepared using the sol–gel method on polysulfonamide fabrics to enhance properties like thermal insulation and flame resistance [122]. The best result was obtained with an aerogel having a comparably higher density (0.645 g/cm^3) with 293 m^2/g surface area, having the smallest pore volume. The vertical burning damage length was reduced from 4.90 to 1.30 cm with the addition of this aluminum hydroxide aerogel. Aerogel fibers with porous interiors were also fabricated by extruding a poly(amic acid) hydrogel, followed by freeze-drying and imidization by heating in argon or nitrogen [123]. Then, these fibers were used to weave fire-retardant and thermally insulating fabrics with a high modulus and elasticity.

Aerogel textile composites are also good candidates for sound absorption due to their high surface areas combined with high porosities [107]. The sound absorption characteristics of silica aerogel polyester blankets were examined in the study of Talebi et al. [124]. Using ambient pressure drying subsequent to sol–gel method, silica aerogel was synthesized on the nonwoven fabrics to prepare composite blankets. It was shown that composite blankets had higher sound absorption coefficients than neat nonwoven samples at all frequency ranges. The low bulk density of the aerogel, along with smaller pore sizes and greater tortuosity values of blankets when combined with aerogels were thought to be the reason for this improvement. In another study, the ballistic efficiency of silica aerogels cross-linked with isocyanate was investigated in aramid fabrics [125]. It was concluded that all samples containing aerogels decreased the number of fabrics perforated by 72%, showing the potential of this type of aerogel for energy-absorbing applications. If only aramid fabric were used, seven to eight layers of fabric were perforated, compared to two perforated layers for aerogel-containing samples. Another interesting application in which aerogel textile composites are used is UV shielding [126]. In the study of Thennerasu et al., a hybrid silica-titania aerogel was synthesized with minimum titania addition and then further surface modified with a silane [126]. Then, it was coated onto cotton fabric using the pad-dry cure method and tested for UV shielding capabilities. The hybrid coating provided an enhanced UV shielding ability compared to the neat cotton fabric, as indicated by a UPF value of 16.6. Furthermore, samples exhibited improved thermal stabilities with the help of heat-resistant nanohybrids.

Aerogels can also be prepared as hydrophobic depending on the precursor or the surface modification technique. This attribute can be used to test them as water-repellent agents. A knitted polyester fabric intended for outerwear was impregnated with a hydrophobic silica aerogel by padding with a binder [127]. The hybrid fabric was more resistant to water spray and water droplets and had an increased water contact angle up to 170°. Furthermore, the appearance or the tensile strength was not affected by the addition of silica aerogel. Another composite fabric was prepared by combining a water-based silica aerogel synthesized through ambient pressure drying with a nanofibrous polyester material [128]. The resulting water contact angle was measured as 130.1° compared to 103.40° without the addition of the aerogel. The repellency ability of aerogels was also used to prepare textile aerogel composites with chemical resistance [129]. Considering this, the adsorption capacity of a silica aerogel incorporated viscose/polyester nonwoven fabric was investigated for seven solvents. The hybrid exhibited increased chemical resistance with a greater amount of chemical absorption, indicating that the material might be an alternative to activated carbon used at present due to the aerogel's higher porosity.

Among all applications involving aerogel-based textiles, thermal insulation stands out [114, 130]. After the production of a flexible aerogel blanket in the early 1990s, it started to be used in thermal insulation applications [131]. Moreover, Aspen Aerogels has produced various silica aerogel-based composites such as Spaceloft®, Cryogel®, and Pyrogel® for industrial use. The thermal conductivity of a PET/silica aerogel blanket was determined to be around 0.011–0.015 W/mK with a working temperature range of −200 to 650 °C [114]. Although silica aerogel is the most commonly used one in that application, there is also some research concerning organic and/or bio-aerogels. Wu et al. prepared chitosan aerogel fibers and then encapsulated them with a stretchable layer, resulting in a similar core-shell structure to that of a polar bear hair [115]. It was seen that the fiber showed an increased strain up to 1000% (compared to 2% with traditional aerogel fibers) and kept its thermal insulation property after 10,000 stretching cycles.

2.5 Conclusions

Usual methods to obtain porous textile structures are based on weaving, knitting, and nonwoven manufacturing. Nonwoven textiles have higher porosities among all the fabric types produced using conventional methods. Electrospinning is an alternative technique for the manufacture of nanoporous fibers. However, the slow production rate, coupled with fibers having low strength, limited the use of this method. Thus, producing textile-based composites incorporating nanoporous materials has gained more interest. Among porous structures, aerogels and hydrogels were shown to be compatible with the production processes of textiles. Moreover, with their unique properties, aerogels and hydrogels offer various possibilities for forming textile-based composites. Aerogels, with their very low densities and high porosities, are very good candidates for thermal insulation applications. Furthermore, their lightweight makes them ideal for textile innovations used in space applications where weight matters. On the other hand, hydrogels can hold a significant amount of water inside their 3D structure and release it when necessary, which is very useful in skincare textiles, especially in wound management. The fact that they can be stimuli-sensitive helps to prepare smart textiles with various functionalities such as controlled drug release, smart wetting, or protection against extreme differences in environmental conditions.

References

1 Tofanica, B.-M., Belosinschi, D., and Volf, I. (2022). Gels, aerogels and hydrogels: a challenge for the cellulose-based product industries. *Gels* 8 (8): 497. https://doi.org/10.3390/gels8080497.

2 Yamauchi, A. (2001). GelsIntroduction. In: *Gels Handbook*, 4–12. Elsevier https://doi.org/10.1016/B978-012394690-4/50075-X.

3 Neumann, M., di Marco, G., Iudin, D. et al. (2023). Stimuli-responsive hydrogels: the dynamic smart biomaterials of tomorrow. *Macromolecules* 56 (21): 8377–8392. https://doi.org/10.1021/acs.macromol.3c00967.

4 Karaaslan, M.A., Kadla, J.F., and Ko, F.K. (2016). Lignin-based aerogels. In: *Lignin in Polymer Composites*, 67–93. Elsevier https://doi.org/10.1016/B978-0-323-35565-0.00005-9.

5 Ho, T.-C., Chang, C.C., Chan, H.P. et al. (2022). Hydrogels: properties and applications in biomedicine. *Molecules* 27 (9): 2902. https://doi.org/10.3390/molecules27092902.

6 Zainal, S.H., Mohd, N.H., Suhaili, N. et al. (2021). Preparation of cellulose-based hydrogel: a review. *Journal of Materials Research and Technology* 10: 935–952. https://doi.org/10.1016/j.jmrt.2020.12.012.

7 Zafar, S., Hanif, M., Azeem, M. et al. (2022). Role of crosslinkers for synthesizing biocompatible, biodegradable and mechanically strong hydrogels with desired release profile. *Polymer Bulletin* 79 (11): 9199–9219. https://doi.org/10.1007/s00289-021-03956-8.

8 Akhtar, M.F., Hanif, M., and Ranjha, N.M. (2016). Methods of synthesis of hydrogels … a review. *Saudi Pharmaceutical Journal* 24 (5): 554–559. https://doi.org/10.1016/j.jsps.2015.03.022.

9 Lin, J., Zheng, S.Y., Xiao, R. et al. (2020). Constitutive behaviors of tough physical hydrogels with dynamic metal-coordinated bonds. *Journal of the Mechanics and Physics of Solids* 139: 103935. https://doi.org/10.1016/j.jmps.2020.103935.

10 Zhang, Y.S. and Khademhosseini, A. (2017). Advances in engineering hydrogels. *Science* 356 (6337): https://doi.org/10.1126/science.aaf3627.

11 Okay, O. and Lozinsky, V.I. (2014). Synthesis and structure–property relationships of cryogels. In: 103–157. https://doi.org/10.1007/978-3-319-05846-7_3.

12 Priya, A.S., Premanand, R., Ragupathi, I. et al. (2024). Comprehensive review of hydrogel synthesis, characterization, and emerging applications. *Journal Of Composites Science* 8 (11): 457. https://doi.org/10.3390/jcs8110457.

13 Wang, Z., Zhong, D., Xiao, R., and Qu, S. (2024). Correlation between synthesis parameters and hyperelasticity of hydrogels: experimental investigation and theoretical modeling. *Journal of the Mechanics and Physics of Solids* 190: 105733. https://doi.org/10.1016/j.jmps.2024.105733.

14 Adelnia, H., Ensandoost, R., Shebbrin Moonshi, S. et al. (2022). Freeze/thawed polyvinyl alcohol hydrogels: present, past and future. *European Polymer Journal* 164: 110974. https://doi.org/10.1016/j.eurpolymj.2021.110974.

15 Zhao, L., Zhou, Y., Zhang, J. et al. (2023). Natural polymer-based hydrogels: from polymer to biomedical applications. *Pharmaceutics* 15 (10): 2514. https://doi.org/10.3390/pharmaceutics15102514.

16 Bao, Z., Xian, C., Yuan, Q. et al. (2019). Natural polymer-based hydrogels with enhanced mechanical performances: preparation, structure, and property. *Advanced Healthcare Materials* 8 (17): https://doi.org/10.1002/adhm.201900670.

17 Bashir, S., Hina, M., Iqbal, J. et al. (2020). Fundamental concepts of hydrogels: synthesis, properties, and their applications. *Polymers* 12 (11): 2702. https://doi.org/10.3390/polym12112702.

18 Hu, Y., Zhang, Z., Li, Y. et al. (2018). Dual-crosslinked amorphous polysaccharide hydrogels based on chitosan/alginate for wound healing applications. *Macromolecular Rapid Communications* 39 (20): https://doi.org/10.1002/marc.201800069.

19 Piyakulawat, P., Praphairaksit, N., Chantarasiri, N., and Muangsin, N. (2007). Preparation and evaluation of chitosan/carrageenan beads for controlled release of sodium diclofenac. *AAPS PharmSciTech* 8 (4): 120. https://doi.org/10.1208/pt0804097.

20 Garcia-Garcia, A., Muñana-González, S., Lanceros-Mendez, S. et al. (2024). Biodegradable natural hydrogels for tissue engineering, controlled release, and soil remediation. *Polymers* 16 (18): 2599. https://doi.org/10.3390/polym16182599.

21 Nanda, D., Behera, D., Pattnaik, S.S., and Behera, A.K. (2025). Advances in natural polymer-based hydrogels: synthesis, applications, and future directions in biomedical and environmental fields. *Discover Polymers* 2 (1): 6. https://doi.org/10.1007/s44347-025-00017-5.

22 Koetting, M.C., Peters, J.T., Steichen, S.D., and Peppas, N.A. (2015). Stimulus-responsive hydrogels: theory, modern advances, and applications. *Materials Science and Engineering: R: Reports* 93: 1–49. https://doi.org/10.1016/j.mser.2015.04.001.

23 Chang, X., Geng, Y., Cao, H. et al. (2018). Dual-crosslink physical hydrogels with high toughness based on synergistic hydrogen bonding and hydrophobic interactions. *Macromolecular Rapid Communications* 39 (14): https://doi.org/10.1002/marc.201700806.

24 Choi, J.Y., Joo, Y.J., Kang, R.J. et al. (2024). Effect of spray-type alginate hydrogel dressing on burn wounds. *Gels* 10 (2): 152. https://doi.org/10.3390/gels10020152.

25 Liu, G., Ye, S., Li, Y. et al. (2025). Copper ions-photo dual-crosslinked alginate hydrogel for angiogenesis and osteogenesis. *Journal of Biomedical Materials Research, Part A* 113 (1): https://doi.org/10.1002/jbm.a.37790.

26 Malektaj, H., Drozdov, A.D., and deClaville Christiansen, J. (2023). Mechanical properties of alginate hydrogels cross-linked with multivalent cations. *Polymers* 15 (14): 3012. https://doi.org/10.3390/polym15143012.

27 Lee, K.Y. and Mooney, D.J. (2012). Alginate: properties and biomedical applications. *Progress in Polymer Science* 37 (1): 106–126. https://doi.org/10.1016/j.progpolymsci.2011.06.003.

28 Crow, B.B. and Nelson, K.D. (2006). Release of bovine serum albumin from a hydrogel-cored biodegradable polymer fiber. *Biopolymers* 81 (6): 419–427. https://doi.org/10.1002/bip.20442.

29 Hong, F., Qiu, P., Wang, Y. et al. (2024). Chitosan-based hydrogels: from preparation to applications, a review. *Food Chemistry* 21: 101095. https://doi.org/10.1016/j.fochx.2023.101095.

30 Jing, H., Huang, X., du, X. et al. (2022). Facile synthesis of pH-responsive sodium alginate/carboxymethyl chitosan hydrogel beads promoted by hydrogen bond. *Carbohydrate Polymers* 278: 118993. https://doi.org/10.1016/j.carbpol.2021.118993.

31 Lv, X., Zhang, W., Liu, Y. et al. (2018). Hygroscopicity modulation of hydrogels based on carboxymethyl chitosan/alginate polyelectrolyte complexes and its application as pH-sensitive delivery system. *Carbohydrate Polymers* 198: 86–93. https://doi.org/10.1016/j.carbpol.2018.06.058.

32 Lin, Y.-H., Liang, H.F., Chung, C.K. et al. (2005). Physically crosslinked alginate/N,O-carboxymethyl chitosan hydrogels with calcium for oral delivery of protein drugs. *Biomaterials* 26 (14): 2105–2113. https://doi.org/10.1016/j.biomaterials.2004.06.011.

33 Enache, A.-C., Cojocaru, C., Samoila, P. et al. (2022). Evaluation of physically and/or chemically modified chitosan hydrogels for proficient release of insoluble nystatin in simulated fluids. *Gels* 8 (8): 495. https://doi.org/10.3390/gels8080495.

34 Madduma-Bandarage, U.S.K. and Madihally, S.V. (2021). Synthetic hydrogels: synthesis, novel trends, and applications. *Journal of Applied Polymer Science* 138 (19): https://doi.org/10.1002/app.50376.

35 Wu, F., Gao, J., Xiang, Y., and Yang, J. (2023a). Enhanced mechanical properties of PVA hydrogel by low-temperature segment self-assembly vs. freeze–thaw cycles. *Polymers* 15 (18): 3782. https://doi.org/10.3390/polym15183782.

36 Hassan, C.M. and Peppas, N.A. (2000). Structure and applications of poly(vinyl alcohol) hydrogels produced by conventional crosslinking or by freezing/

thawing methods. In: *Biopolymers · PVA Hydrogels, Anionic Polymerisation Nanocomposites*, 37–65. Berlin, Heidelberg: Springer Berlin Heidelberg https://doi.org/10.1007/3-540-46414-X_2.

37 Mori, Y., Tokura, H., and Yoshikawa, M. (1997). Properties of hydrogels synthesized by freezing and thawing aqueous polyvinyl alcohol solutions and their applications. *Journal of Materials Science* 32 (2): 491–496. https://doi.org/10.1023/A:1018586307534.

38 Holloway, J.L., Lowman, A.M., and Palmese, G.R. (2013). The role of crystallization and phase separation in the formation of physically cross-linked PVA hydrogels. *Soft Matter* 9 (3): 826–833. https://doi.org/10.1039/C2SM26763B.

39 Peppas, N.A. and Scott, J.E. (1992). Controlled release from poly (vinyl alcohol) gels prepared by freezing-thawing processes. *Journal of Controlled Release* 18 (2): 95–100. https://doi.org/10.1016/0168-3659(92)90178-T.

40 Webber, M.J. and Pashuck, E.T. (2021). (Macro)molecular self-assembly for hydrogel drug delivery. *Advanced Drug Delivery Reviews* 172: 275–295. https://doi.org/10.1016/j.addr.2021.01.006.

41 Rahmatpour, A. and Alizadeh, A.H. (2024). Biofilm hydrogel derived from physical crosslinking (self-assembly) of xanthan gum and chitosan for removing Cd^{2+}, Ni^{2+}, and Cu^{2+} from aqueous solution. *International Journal of Biological Macromolecules* 266: 131394. https://doi.org/10.1016/j.ijbiomac.2024.131394.

42 Chen, Y.-N., Jiao, C., Zhao, Y. et al. (2018). Self-assembled polyvinyl alcohol–tannic acid hydrogels with diverse microstructures and good mechanical properties. *ACS Omega* 3 (9): 11788–11795. https://doi.org/10.1021/acsomega.8b02041.

43 Eivazzadeh-Keihan, R., Noruzi, E.B., Mehrban, S.F. et al. (2022). Review: the latest advances in biomedical applications of chitosan hydrogel as a powerful natural structure with eye-catching biological properties. *Journal of Materials Science* 57 (6): 3855–3891. https://doi.org/10.1007/s10853-021-06757-6.

44 Ito, T., Yeo, Y., Highley, C.B. et al. (2007). Dextran-based in situ cross-linked injectable hydrogels to prevent peritoneal adhesions. *Biomaterials* 28 (23): 3418–3426. https://doi.org/10.1016/j.biomaterials.2007.04.017.

45 Bouhadir, K.H., Hausman, D.S., and Mooney, D.J. (1999). Synthesis of cross-linked poly(aldehyde guluronate) hydrogels. *Polymer* 40 (12): 3575–3584. https://doi.org/10.1016/S0032-3861(98)00550-3.

46 Lee, K.Y., Bouhadir, K.H., and Mooney, D.J. (2000). Degradation behavior of covalently cross-linked poly(aldehyde guluronate) hydrogels. *Macromolecules* 33 (1): 97–101. https://doi.org/10.1021/ma991286z.

47 Ahmed, M.S., Islam, M., Hasan, M.K., and Nam, K.W. (2024). A comprehensive review of radiation-induced hydrogels: synthesis, properties, and multidimensional applications. *Gels* 10 (6): 381. https://doi.org/10.3390/gels10060381.

48 Yang, J., Rao, L., Wang, Y. et al. (2022). Recent advances in smart hydrogels prepared by ionizing radiation technology for biomedical applications. *Polymers* 14 (20): 4377. https://doi.org/10.3390/polym14204377.

49 Mignon, A., Zimmer, J., Gutierrez Cisneros, C. et al. (2023). Electron-beam-initiated crosslinking of methacrylated alginate and diacrylated poly(ethylene glycol) hydrogels. *Polymers* 15 (24): 4685. https://doi.org/10.3390/polym15244685.

50 Kistler, S.S. (1932). Coherent expanded-aerogels. *The Journal of Physical Chemistry* 36 (1): 52–64. https://doi.org/10.1021/j150331a003.

51 Alwin, S. and Sahaya Shajan, X. (2020). Aerogels: promising nanostructured materials for energy conversion and storage applications. *Materials for Renewable and Sustainable Energy* 9 (2): 7. https://doi.org/10.1007/s40243-020-00168-4.

52 Azam, F., Ahmad, F., Ahmad, S. et al. (2024). Alginate-cotton blended aerogel fibers: synthesis, characterization, and oil/water separation. *International journal of Environmental Science and Technology* 21 (5): 5065–5078. https://doi.org/10.1007/s13762-023-05329-2.

53 Caps, R. and Fricke, J. (2004). Aerogels for thermal insulation. In: *Sol-Gel Technologies for Glass Producers and Users*, 349–353. Boston, MA: Springer US https://doi.org/10.1007/978-0-387-88953-5_46.

54 Ulker, Z. and Erkey, C. (2014). An emerging platform for drug delivery: aerogel based systems. *Journal of Controlled Release* 177: 51–63. https://doi.org/10.1016/j.jconrel.2013.12.033.

55 García-González, C.A., Budtova, T., Durães, L. et al. (2019). An opinion paper on aerogels for biomedical and environmental applications. *Molecules* 24 (9): 1815. https://doi.org/10.3390/molecules24091815.

56 Betz, M., García-González, C.A., Subrahmanyam, R.P. et al. (2012). Preparation of novel whey protein-based aerogels as drug carriers for life science applications. *The Journal of Supercritical Fluids* 72: 111–119. https://doi.org/10.1016/j.supflu.2012.08.019.

57 Wang, Y., Su, Y., Wang, W. et al. (2019). The advances of polysaccharide-based aerogels: preparation and potential application. *Carbohydrate Polymers* 226: 115242. https://doi.org/10.1016/j.carbpol.2019.115242.

58 Zhao, S., Malfait, W.J., Guerrero-Alburquerque, N. et al. (2018). Biopolymer aerogels and foams: chemistry, properties, and applications. *Angewandte Chemie International Edition* 57 (26): 7580–7608. https://doi.org/10.1002/anie.201709014.

59 Smirnova, I. and Gurikov, P. (2017). Aerogels in chemical engineering: strategies toward tailor-made aerogels. *Annual Review of Chemical and Biomolecular Engineering* 8 (1): 307–334. https://doi.org/10.1146/annurev-chembioeng-060816-101458.

60 Abdul Khalil, H.P.S., Adnan, A.S., Yahya, E.B. et al. (2020). A review on plant cellulose nanofibre-based aerogels for biomedical applications. *Polymers* 12 (8): 1759. https://doi.org/10.3390/polym12081759.

61 Smirnova, I. and Gurikov, P. (2018). Aerogel production: current status, research directions, and future opportunities. *The Journal of Supercritical Fluids* 134: 228–233. https://doi.org/10.1016/j.supflu.2017.12.037.

62 Payanda Konuk, O., Alsuhile, A.A.A.M., Yousefzadeh, H. et al. (2023). The effect of synthesis conditions and process parameters on aerogel properties. *Frontiers in Chemistry* 11: https://doi.org/10.3389/fchem.2023.1294520.

63 Meti, P., Wang, Q., Mahadik, D.B. et al. (2023). Evolutionary progress of silica aerogels and their classification based on composition: an overview. *Nanomaterials* 13 (9): 1498. https://doi.org/10.3390/nano13091498.

64 Pierre, A.C. and Rigacci, A. (2011). SiO_2 aerogels. In: *Aerogels Handbook*, 21–45. New York, NY: Springer New York https://doi.org/10.1007/978-1-4419-7589-8_2.

65 Mulik, S. and Sotiriou-Leventis, C. (2011). Resorcinol–formaldehyde aerogels. In: *Aerogels Handbook*, 215–234. New York, NY: Springer New York https://doi.org/10.1007/978-1-4419-7589-8_11.

66 Pekala, R.W. (1989). Organic aerogels from the polycondensation of resorcinol with formaldehyde. *Journal of Materials Science* 24 (9): 3221–3227. https://doi.org/10.1007/BF01139044.

67 Gurav, J.L., Jung, I.K., Park, H.H. et al. (2010). Silica aerogel: synthesis and applications. *Journal of Nanomaterials* 2010 (1): https://doi.org/10.1155/2010/409310.

68 Ganesan, K. and Ratke, L. (2014). Facile preparation of monolithic κ-carrageenan aerogels. *Soft Matter* 10 (18): 3218.https://doi.org/10.1039/c3sm52862f–3224.

69 Groult, S. and Budtova, T. (2018). Tuning structure and properties of pectin aerogels. *European Polymer Journal* 108: 250–261. https://doi.org/10.1016/j.eurpolymj.2018.08.048.

70 Long, L.-Y., Weng, Y.-X., and Wang, Y.-Z. (2018). Cellulose aerogels: synthesis, applications, and prospects. *Polymers* 10 (6): 623. https://doi.org/10.3390/polym10060623.

71 Zhu, F. (2019). Starch based aerogels: production, properties and applications. *Trends in Food Science & Technology* 89: 1–10. https://doi.org/10.1016/j.tifs.2019.05.001.

72 Bedilo, A.F. and Klabunde, K.J. (1997). Synthesis of high surface area zirconia aerogels using high temperature supercritical drying. *Nanostructured Materials* 8 (2): 119–135. https://doi.org/10.1016/S0965-9773(97)00011-1.

73 Feng, Y.N., Miao, L., Tanemura, M. et al. (2008). Effects of further adding of catalysts on nanostructures of carbon aerogels. *Materials Science and Engineering: B* 148 (1–3): 273–276. https://doi.org/10.1016/j.mseb.2007.09.059.

74 Horikawa, T., Hayashi, J., and Muroyama, K. (2004). Controllability of pore characteristics of resorcinol–formaldehyde carbon aerogel. *Carbon* 42 (8–9): 1625–1633. https://doi.org/10.1016/j.carbon.2004.02.016.

75 Pekala, R.W. and Schaefer, D.W. (1993). Structure of organic aerogels. 1. Morphology and scaling. *Macromolecules* 26 (20): 5487–5493. https://doi.org/10.1021/ma00072a029.

76 Ahmad, S., Ahmad, S., and Sheikh, J.N. (2023). Silica centered aerogels as advanced functional material and their applications: a review. *Journal of Non-Crystalline Solids* 611: 122322. https://doi.org/10.1016/j.jnoncrysol.2023.122322.

77 Haereid, S., Nilsen, E., Ranum, V., and Einarsrud, M.A. (1997). Thermal and temporal aging of two step acid-base catalyzed silica gels in water/ethanol solutions. *Journal of Sol-Gel Science and Technology* 8 (1/3): 153–157. https://doi.org/10.1023/A:1026400617168.

78 Iswar, S., Malfait, W.J., Balog, S. et al. (2017). Effect of aging on silica aerogel properties. *Microporous and Mesoporous Materials* 241: 293–302. https://doi.org/10.1016/j.micromeso.2016.11.037.

79 Schwan, M., Nefzger, S., Zoghi, B. et al. (2021). Improvement of solvent exchange for supercritical dried aerogels. *Frontiers in Materials* 8: https://doi.org/10.3389/fmats.2021.662487.

80 García-González, C.A., Alnaief, M., and Smirnova, I. (2011). Polysaccharide-based aerogels—promising biodegradable carriers for drug delivery systems. *Carbohydrate Polymers* 86 (4): 1425–1438. https://doi.org/10.1016/j.carbpol.2011.06.066.

81 Gurikov, P., S. P., R., Griffin, J.S. et al. (2019). *110th Anniversary:* solvent exchange in the processing of biopolymer aerogels: current status and open questions. *Industrial & Engineering Chemistry Research* 58 (40): 18590–18600. https://doi.org/10.1021/acs.iecr.9b02967.

82 Robitzer, M., David, L., Rochas, C. et al. (2008). Nanostructure of calcium alginate aerogels obtained from multistep solvent exchange route. *Langmuir* 24 (21): 12547–12552. https://doi.org/10.1021/la802103t.

83 Subrahmanyam, R., Gurikov, P., Dieringer, P. et al. (2015). On the road to biopolymer aerogels—dealing with the solvent. *Gels* 1 (2): 291–313. https://doi.org/10.3390/gels1020291.

84 Cardea, S., Pisanti, P., and Reverchon, E. (2010). Generation of chitosan nanoporous structures for tissue engineering applications using a supercritical fluid assisted process. *The Journal of Supercritical Fluids* 54 (3): 290–295. https://doi.org/10.1016/j.supflu.2010.05.014.

85 Bangi, U.K.H., Dhere, S.L., and Venkateswara Rao, A. (2010). Influence of various processing parameters on water-glass-based atmospheric pressure dried aerogels for liquid marble purpose. *Journal of Materials Science* 45 (11): 2944–2951. https://doi.org/10.1007/s10853-010-4287-9.

86 Bisson, A., Rigacci, A., Lecomte, D. et al. (2003). Drying of silica gels to obtain aerogels: phenomenology and basic techniques. *Drying Technology* 21 (4): 593–628. https://doi.org/10.1081/DRT-120019055.

87 Şahin, İ., Özbakır, Y., İnönü, Z. et al. (2017). Kinetics of supercritical drying of gels. *Gels* 4 (1): 3. https://doi.org/10.3390/gels4010003.

88 Ebrahimi, A., Dahrazma, B., and Adelifard, M. (2020). Facile and novel ambient pressure drying approach to synthesis and physical characterization of cellulose-based aerogels. *Journal of Porous Materials* 27 (4): 1219–1232. https://doi.org/10.1007/s10934-020-00901-4.

89 Maleki, H., Durães, L., García-González, C.A. et al. (2016). Synthesis and biomedical applications of aerogels: possibilities and challenges. *Advances in Colloid and Interface Science* 236: 1–27. https://doi.org/10.1016/j.cis.2016.05.011.

90 Einarsrud, M.-A. and Haereid, S. (1994). Preparation of transparent, monolithic silica xerogels with low density. *Journal of Sol-Gel Science and Technology* 2 (1–3): 903–906. https://doi.org/10.1007/BF00486373.

91 Haereid, S., Einarsrud, M.-A., and Scherer, G.W. (1994). Mechanical strengthening of TMOS-based alcogels by aging in silane solutions. *Journal of Sol-Gel Science and Technology* 3 (3): 199–204. https://doi.org/10.1007/BF00486558.

92 Hæreid, S., Dahle, M., Lima, S., and Einarsrud, M.A. (1995). Preparation and properties of monolithic silica xerogels from TEOS-based alcogels aged in silane solutions. *Journal of Non-Crystalline Solids* 186: 96–103. https://doi.org/10.1016/0022-3093(95)00039-9.

93 Venkateswara Rao, A., Nilsen, E., and Einarsrud, M.-A. (2001). Effect of precursors, methylation agents and solvents on the physicochemical properties of silica aerogels prepared by atmospheric pressure drying method. *Journal of Non-Crystalline Solids* 296 (3): 165–171. https://doi.org/10.1016/S0022-3093(01)00907-3.

94 Zemke, F., Scoppola, E., Simon, U. et al. (2024). Springback effect of ambient-pressure-dried silica aerogels: nanoscopic effects of silylation revealed by *in situ* synchrotron X-ray scattering. *Nanoscale Advances* 6 (1): 111–125. https://doi.org/10.1039/D3NA00584D.

95 Simón-Herrero, C., Caminero-Huertas, S., Romero, A. et al. (2016). Effects of freeze-drying conditions on aerogel properties. *Journal of Materials Science* 51 (19): 8977–8985. https://doi.org/10.1007/s10853-016-0148-5.

96 Barros, A., Quraishi, S., Martins, M. et al. (2016). Hybrid alginate-based cryogels for life science applications. *Chemie Ingenieur Technik* 88 (11): 1770–1778. https://doi.org/10.1002/cite.201600096.

97 Rodríguez-Dorado, R., López-Iglesias, C., García-González, C.A. et al. (2019). Design of aerogels, cryogels and xerogels of alginate: effect of molecular weight, gelation conditions and drying method on particles' micromeritics. *Molecules* 24 (6): 1049. https://doi.org/10.3390/molecules24061049.

98 Walker, R.C., Potochniak, A.E., Hyer, A.P., and Ferri, J.K. (2021). Zirconia aerogels for thermal management: review of synthesis, processing, and properties information architecture. *Advances in Colloid and Interface Science* 295: 102464. https://doi.org/10.1016/j.cis.2021.102464.

99 Kang, H.-W., Tabata, Y., and Ikada, Y. (1999). Fabrication of porous gelatin scaffolds for tissue engineering. *Biomaterials* 20 (14): 1339–1344. https://doi.org/10.1016/S0142-9612(99)00036-8.

100 Pan, Q., Liu, Q., Shi, Y. et al. (2025). A facile strategy for freeze-drying preparation of silica aerogel from sodium silicate. *Ceramics International* 51 (4): 5342–5350. https://doi.org/10.1016/j.ceramint.2024.11.508.

101 Ciftci, D., Ubeyitogullari, A., Huerta, R.R. et al. (2017). Lupin hull cellulose nanofiber aerogel preparation by supercritical CO_2 and freeze drying. *The Journal of Supercritical Fluids* 127: 137–145. https://doi.org/10.1016/j.supflu.2017.04.002.

102 Zhao, Z., Chen, D., and Jiao, X. (2007). Zirconia aerogels with high surface area derived from sols prepared by electrolyzing zirconium oxychloride solution: comparison of aerogels prepared by freeze-drying and supercritical CO_2 (l) extraction. *The Journal of Physical Chemistry C* 111 (50): 18738–18743. https://doi.org/10.1021/jp075150b.

103 Członka, S., Bertino, M.F., Kośny, J., and Shukla, N. (2018). Freeze-drying method as a new approach to the synthesis of polyurea aerogels from isocyanate and water. *Journal of Sol-Gel Science and Technology* 87 (3): 685–695. https://doi.org/10.1007/s10971-018-4769-9.

104 Azam, F., Ahmad, F., Ulker, Z. et al. (2022). The role and applications of aerogels in textiles. *Advances in Materials Science and Engineering* 2022: 1–22. https://doi.org/10.1155/2022/2407769.

105 Islam, M.S., Ang, B.C., Andriyana, A., and Afifi, A.M. (2019). A review on fabrication of nanofibers via electrospinning and their applications. *SN Applied Sciences* 1 (10): 1248. https://doi.org/10.1007/s42452-019-1288-4.

106 Wang, P., Lv, H., Cao, X. et al. (2023). Recent progress of the preparation and application of electrospun porous nanofibers. *Polymers* 15 (4): 921. https://doi.org/10.3390/polym15040921.

107 McNeil, S.J. and Gupta, H. (2022). Emerging applications of aerogels in textiles. *Polymer Testing* 106: 107426. https://doi.org/10.1016/j.polymertesting.2021.107426.

108 Nie, Y., Mugaanire, I.T., Guo, Y. et al. (2021). A hybrid hydrogel/textile composite as flame-resistant dress. *Progress in Natural Science: Materials International* 31 (1): 33–40. https://doi.org/10.1016/j.pnsc.2020.11.009.

109 Sayed, U. and Deshmukh, I. (2021). Hydrogels for textile applications - review. *International Journal of Advanced Science and Engineering* 7 (4): https://doi.org/10.29294/IJASE.7.4.2021.1968-1974.

110 Azam, F., Ahmad, F., Ahmad, S. et al. (2023). Synthesis and characterization of natural fibers reinforced alginate hydrogel fibers loaded with diclofenac sodium for wound dressings. *International Journal of Biological Macromolecules* 241: 124623. https://doi.org/10.1016/j.ijbiomac.2023.124623.

111 Jiang, S., Deng, J., Jin, Y. et al. (2023). Breathable, antifreezing, mechanically skin-like hydrogel textile wound dressings with dual antibacterial mechanisms. *Bioactive Materials* 21: 313–323. https://doi.org/10.1016/j.bioactmat.2022.08.014.

112 Türkoğlu, G.C., Sarıışık, A.M., and Karavana, S.Y. (2021). Development of textile-based sodium alginate and chitosan hydrogel dressings. *International Journal of Polymeric Materials and Polymeric Biomaterials* 70 (13): 916–925. https://doi.org/10.1080/00914037.2020.1765364.

113 Li, M., Chen, X., Li, X. et al. (2022). Ultralight aerogel textiles based on aramid nanofibers composites with excellent thermal insulation and electromagnetic shielding properties. *Composites Communications* 35: 101346. https://doi.org/10.1016/j.coco.2022.101346.

114 Venkataraman, M., Mishra, R., Kotresh, T.M. et al. (2016). Aerogels for thermal insulation in high-performance textiles. *Textile Progress* 48 (2): 55–118. https://doi.org/10.1080/00405167.2016.1179477.

115 Wu, M., Shao, Z., Zhao, N. et al. (2023b). Biomimetic, knittable aerogel fiber for thermal insulation textile. *Science* 382 (6677): 1379–1383. https://doi.org/10.1126/science.adj8013.

116 Wang, X., Hu, H., Yang, Z. et al. (2014). Smart hydrogel-functionalized textile system with moisture management property for skin application. *Smart Materials and Structures* 23 (12): 125027. https://doi.org/10.1088/0964-1726/23/12/125027.

117 Ahmad, F., Nosheen, A., Azam, F. et al. (2024). Hydroentangled waste cotton nonwoven based alginate hydrogel wound dressing for high wound exudates. *Heliyon* 10 (18): e37952. https://doi.org/10.1016/j.heliyon.2024.e37952.

118 Zhang, R., Wu, Y., Lin, P. et al. (2020). Extremely tough hydrogels with cotton fibers reinforced. *Advanced Engineering Materials* 22 (11): https://doi.org/10.1002/adem.202000508.

119 Zaman, S.u., Mushtaq, B., Ahmad, F. et al. (2023). Development of conductive cotton nonwoven alginate hydrogel composite for smart textiles. *Journal of Polymers and the Environment* 31 (9): 3998–4006. https://doi.org/10.1007/s10924-023-02847-5.

120 Liu, H., Shang, J., Wang, Y. et al. (2022). Ag/AgCl nanoparticles reinforced cellulose-based hydrogel coated cotton fabric with self-healing and photo-induced self-cleaning properties for durable oil/water separation. *Polymer* 255: 125146. https://doi.org/10.1016/j.polymer.2022.125146.

121 Illeperuma, W.R.K., Rothemund, P., Suo, Z., and Vlassak, J.J. (2016). Fire-resistant hydrogel-fabric laminates: a simple concept that may save lives. *ACS Applied Materials & Interfaces* 8 (3): 2071–2077. https://doi.org/10.1021/acsami.5b10538.

122 Jiang, Y., Zhang, L., Xu, H. et al. (2017). Preparation and characterization of thermal protective aluminum hydroxide aerogel/PSA fabric composites. *Journal of Sol-Gel Science and Technology* 82 (2): 370–379. https://doi.org/10.1007/s10971-017-4323-1.

123 Wang, Y., Cui, Y., Shao, Z. et al. (2020). Multifunctional polyimide aerogel textile inspired by polar bear hair for thermoregulation in extreme environments. *Chemical Engineering Journal* 390: 124623. https://doi.org/10.1016/j.cej.2020.124623.

124 Talebi, Z., Soltani, P., Habibi, N., and Latifi, F. (2019). Silica aerogel/polyester blankets for efficient sound absorption in buildings. *Construction and Building Materials* 220: 76–89. https://doi.org/10.1016/j.conbuildmat.2019.06.031.

125 Ayten, A.İ., Taşdelen, M.A., and Ekici, B. (2020). An experimental investigation on ballistic efficiency of silica-based crosslinked aerogels in aramid fabric. *Ceramics International* 46 (17): 26724–26730. https://doi.org/10.1016/j.ceramint.2020.07.147.

126 Thennarasu, P., Senthilkumar, M., and Prabunathan, P. (2021). Functionalization of cotton fabric with titania silica aerogel using rice husk as precursor for enhanced UV shielding. *Materials Today: Proceedings* 42: 795–803. https://doi.org/10.1016/j.matpr.2020.11.348.

127 Altay, P., Atakan, R., and Özcan, G. (2021). Silica aerogel application to polyester fabric for outdoor clothing. *Fibers and Polymers* 22 (4): 1025–1032. https://doi.org/10.1007/s12221-021-0420-4.

128 Mazrouei-Sebdani, Z., Khoddami, A., Hadadzadeh, H. et al. (2016). The effect of the nano-structured aerogel powder on the structural parameters, water repellency, and water vapor/air permeability of a fibrous polyester material. *Materials Chemistry and Physics* 177: 99–111. https://doi.org/10.1016/j.matchemphys.2016.04.002.

129 Bhuiyan, M.A.R., Wang, L., Shaid, A. et al. (2020). Silica aerogel-integrated nonwoven protective fabrics for chemical and thermal protection and thermophysiological wear comfort. *Journal of Materials Science* 55 (6): 2405–2418. https://doi.org/10.1007/s10853-019-04203-2.

130 Peng, L., Su, B., Yu, A., and Jiang, X. (2019). Review of clothing for thermal management with advanced materials. *Cellulose* 26 (11): 6415–6448. https://doi.org/10.1007/s10570-019-02534-6.

131 Fesmire, J.E., Ancipink, J.B., Swanger, A.M. et al. (2017). Thermal conductivity of aerogel blanket insulation under cryogenic-vacuum conditions in different gas environments. *IOP Conference Series: Materials Science and Engineering* 278: 012198. https://doi.org/10.1088/1757-899X/278/1/012198.

3

Types of Hydrogels and Aerogels in Textiles

Talha Siddique[1], Faheem Ahmad[2], and Jibran Khaliq[3]

[1] *University of Agriculture, Institute of Animal and Dairy Sciences, University Main Road, 38000 Faisalabad, Pakistan*
[2] *National Textile University, School of Engineering and Technology, Sheikhupura Road, 38000, Faisalabad, Pakistan*
[3] *Northumbria University, Faculty of Engineering and Environment, Department of Mechanical and Construction Engineering, Newcastle upon Tyne, NE1 8ST, UK*

3.1 Introduction

A type of polymer material known as hydrogel is characterized by its three-dimensional (3D) network structure. It is composed of a significant number of hydrophilic groups, including carboxyl and hydroxyl groups, and it can hold a substantial amount of water inside its structure [1, 2]. Hydrogels are composed of a 3D network. The hydrophilic groups that hydrogels contain, such as $-NH_2$, $-COOH$, $-OH$, $-CONH_2$, $-CONH$, and $-SO_3H$, allow them to not only absorb a significant amount of water but also cause them to inflate in water. Hydrogels are frequently included in fibers, fabrics, or coatings in the textile industry for the purpose of providing functions such as moisture management, cooling effects, wound healing, or medication administration, respectively [3–5].

Meanwhile, unlike hydrogels, aerogels possess exceptionally high porosity, allowing the pores to occupy almost 90% of the material's volume. This renders them exceptionally lightweight. The pores are exceptionally diminutive (2–50 nm), resulting in significantly large specific surface areas [6, 7]. The term "aerogel" refers to a gel that contains gas components rather than liquid ones. It has extremely low density and poor thermal conductivity due to its 3D nanoporous structure. These materials are extremely similar to hydrogels, which are also created by assembling composites, inorganic materials, and organic polymers in a 3D network [8]. What distinguishes these two constructions is the drying process. To create aerogels, supercritical drying is used. By doing this, the adhesion forces that would otherwise exist between liquids and solids as well as the surface tension between liquids and gases are avoided. The gel structures become more compact as a result [9, 10].

Hydrogels and aerogels are increasingly becoming prevalent in textiles due to their distinctive structural and functional properties. Hydrogels are recognized for their moisture-retaining properties and biocompatibility, whereas aerogels are characterized by their extremely low density and effective thermal conductivity. The combination of lightweight properties and performance has increased the demand for their application in textile systems, particularly for advanced functions such as thermal insulation, water and fire resistance, soundproofing, and environmental remediation [11–13].

The classification of hydrogels is influenced by the polymers used, their source, the cross-linking method, their responsiveness to stimuli, and their ionic charge. The polymers used in the hydrogels can be synthetic, natural, or a hybrid of both [14]. Aerogels comprise both natural components, such as cellulose, and synthetic components, including silica or polymers, each serving a specific function. It is fascinating to contemplate the utilization of hydrogel or aerogel elements to enhance fabrics in composite or hybrid systems. These systems can do several tasks concurrently [15].

This chapter aims to examine the utilization of hydrogels and aerogels in textiles by classifying them based on their fiber origin (natural, synthetic, or blended) (as depicted in Figures 3.1 and 3.2) and exploring how their composition influences critical performance attributes such as biodegradability, thermal insulation, and moisture management.

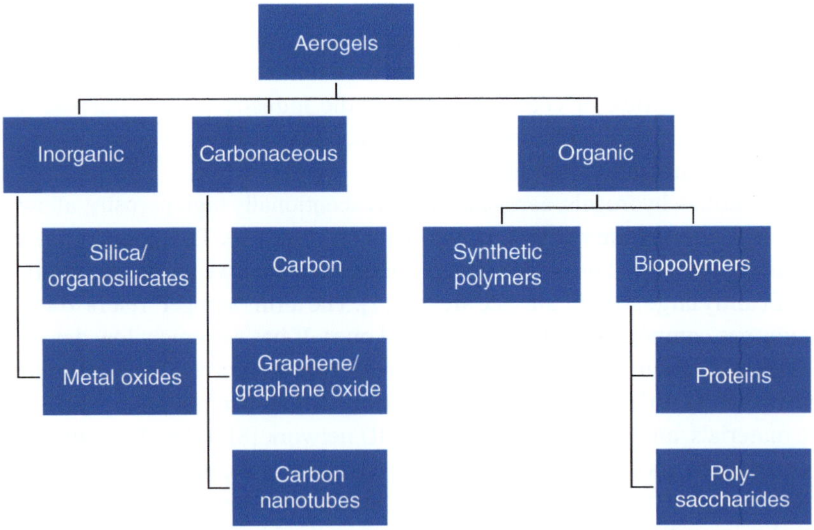

Figure 3.1 The main types of aerogels. *Source:* [15] / with permission of Elsevier.

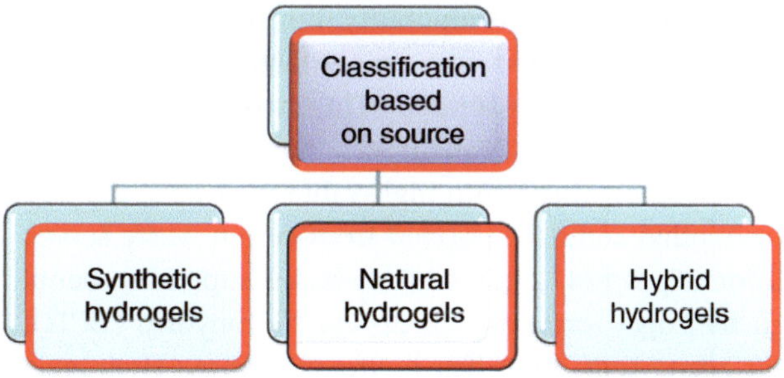

Figure 3.2 Types of hydrogels. *Source:* Adapted from [14].

3.2 Natural-Fiber Hydrogels and Aerogels

Hydrogels made from natural polymers are known as natural polymer hydrogels. These natural polymers are made up of polysaccharides, polynucleotides, and polypeptides. Natural polymers come from a range of natural sources and are neutral, cationic, and anionic in nature [16]. These polymers are readily available, cheap, plentiful, nontoxic, and biodegradable. They also have other desirable biological characteristics. There has always been a lot of interest in the connection between structure and function, especially when it comes to naturally occurring biologically active substances. The development of materials for use in biomedical technology has advanced significantly over the last few decades as a result of improvements in structural and functional ingredients. Because natural polymers are made up of covalently bound monomeric units, their structures are bigger and more distinct. Natural polymer hydrogels have a wide range of biological uses, including wound healing, tissue engineering, and controlled and targeted medication release [17, 18].

Recently, various types of absorbents based on carbon and silica have been developed, including graphene and carbon nanotubes (CNTs). Nonetheless, they pose challenges for the climate, the precursors incur high costs, and the preparation process is quite complex to comprehend. It is environmentally friendly, cost-effective, biodegradable, and readily available. These are several reasons that have garnered significant interest from specialists globally in cellulose-based aerogel [19, 20]. Cotton, rice straw, sugarcane bagasse, newspaper, and wood represent a selection of natural materials used in the production of cellulose for cellulose-based aerogels [21].

3.2.1 Cotton-Based Hydrogels and Aerogels

Cellulose, a natural polymer, is present in the cell walls of plants. The primary function is to provide rigidity to plant cells. It comprises as much as 96.5% cellulose, establishing it as the predominant source [22]. Cotton cellulose, an organic polymer, is used in the production of garments, socks, towels, tents, and coffee paper due to its accessibility, environmental sustainability, biocompatibility, and cost-effectiveness [23]. Cotton waste generated from manufacturing can reach up to 10 tons monthly and is deemed unrecoverable. This waste is converted into low-value products such as oil absorbers, mushroom substrates, and rugs. Cotton waste recovery transforms cotton waste into high-value materials, thereby decreasing its volume. Carboxymethyl cellulose derived from cotton waste serves as a superabsorbent material for wipes [24, 25]. Cellulose-based superabsorbents have been developed through hydrogel formation using epichlorohydrin (ECH), a base-catalyzed cross-linking agent that etherifies hydroxyl groups. Hydrogels reinforced with cellulose fiber (CF) exhibit improved mechanical stability due to the natural polymer's high crystallinity and biodegradability [26]. Cross-linking of polymer chains results in the formation of a network structure. Gelation results from chemical cross-linking and hydrogen bonding, either intramolecular or intermolecular, involving the hydroxyl groups of cellulose and the solvent. The electrostatic interaction between molecules facilitates water absorption [27]. Physical cross-linking leads to self-assembled chain entanglements in the cellulose backbone, resulting in irreversible gelation upon heating [28].

Moreover, due to its significant cellulose content, cotton serves as a valuable component for the enhancement of structural materials [29] and can also be transformed into an innovative aerogel material [30]. The process of fabricating the CF aerogels is illustrated in Figure 3.3.

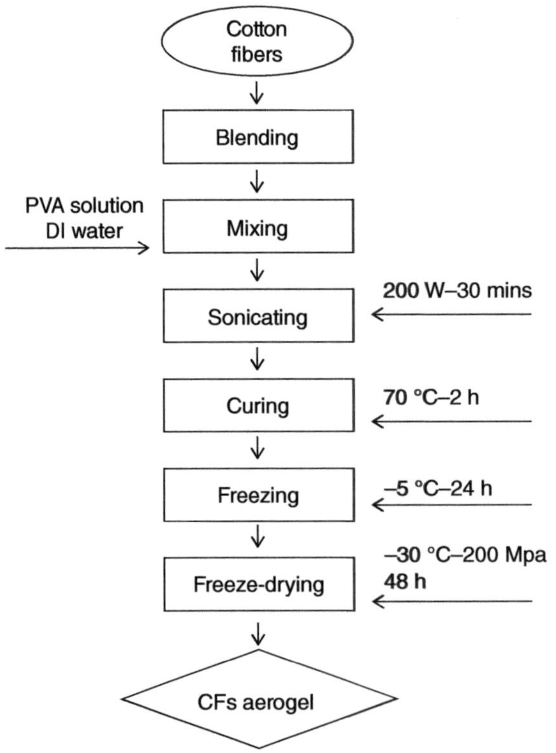

Figure 3.3 Fabrication steps for CF aerogels. *Source:* [31] / Nga Hoang Nguyen Do.

3.2.2 Silk-Based Systems

Silk fibroin hydrogels have garnered significant interest from researchers and inventors due to their exceptional properties and diverse applications. Silk fibroin, a protein derived from silkworm and spider silk, is used in the production of these hydrogels. They occupy a distinctive role at the convergence of medical, materials, and biological sciences [32, 33]. Their properties in biomaterials are notable, encompassing biocompatibility, strength, and water absorption [34]. Biocompatibility is advantageous for medicinal delivery systems, tissue engineering, and regenerative medicine. These hydrogels not only promote cell adhesion and proliferation but also enable the controlled release of medications, thereby broadening the possibilities for targeted therapies [35].

Silk fibroin hydrogels have numerous potential applications, with biomedicine being one of them. Hydrogels have applications across various fields, including wound dressings, bioadhesives, and wearable electronics [36, 37].

Recently Maleki et al. [38] reported a novel trend in the field: the synthesis of silk fibroin as aerogels and their application in mechanically preserving the delicate structure of silica and silsesquioxane aerogels [38, 39]. Aerogels composed of ultralight silk fibroin, referred to as AeroSF, have been developed with a specific gravity of 0.02 g/cm^2. Due to its very low thermal conductivity of 0.028 W/mK, it presents potential applications as a thermal insulator. Polymethylsilsesquioxane (PMSQ), a silica-based hybrid polymer known for its flexibility and low thermal conductivity, was incorporated to enhance the mechanical and insulation performance of the aerogel. A novel filter has been developed, integrating superhydrophobic and oleophilic PMSQ-silk fibroin properties. This filter effectively separates oil from water/oil mixtures in a continuous manner. The effective utilization of additive manufacturing technologies, such as 3D printing, has led to the emergence of new technical sectors, including multipore biomaterials and customized anisotropic thermal insulators, through the production of silica-silk fibroin aerogels with intentionally modified structures and morphologies [40].

3.2.3 Wool-Based Materials

Keratin is a macromolecule investigated for hydrogel development owing to its abundant availability, ability to promote cell adhesion, biocompatibility in vivo, and rapid wound healing properties [41, 42]. Keratins are intermediate filament proteins present in various biological structures, including hair, nails, skin, fur, wool, horns, hooves, and avian tissues such as beaks and feathers. Keratin forms disulfide bonds with associated proteins owing to its elevated cysteine residue content. Keratins in avian and mammalian species exhibit differences in molecular weight, secondary structures, and amino acid composition [43]. Feather keratins exhibit a β-sheet structure, which contributes to their mechanical strength and stability. Feather keratins exhibit consistent -sheet conformation and a molecular weight of 10 kDa, whereas mammalian keratins are diverse proteins characterized by an α-helical conformation and a molecular weight ranging from 10 to 75 kDa [44]. Feather keratins do not possess integrin binding sites that facilitate cellular adhesion, in contrast to mammalian keratins [45]. These modifications influence the production of hydrogel scaffolds as well as their physical and biological properties.

Loh et al. [46] conducted a study that involves the creation of wool aerogels through the combination of wool waste fibers and a cross-linking binder known as polyvinyl alcohol (PVA). The aerogels produced exhibit low densities ranging from 0.004 to 0.023 g/cm^3, alongside a high porosity of 97.73–99.63%. The aerogels demonstrate a remarkable degree of hydrophobicity, exhibiting contact angles of 138° after being coated with methyltriethoxysilane (MTEOS).

3.3 Synthetic-Fiber Hydrogels and Aerogels

Synthetic-fiber hydrogels and aerogels are a class of materials made from synthetic polymers [47, 48]. Synthetic fibers such as acrylic, nylon, and polyester are incorporated into or coated with polymer networks to produce hydrogels and aerogels. Natural polymers are extensively used in the biomedical sector due to their biocompatibility, biodegradability, and recognizable components by the body. However, they lack the necessary mechanical qualities that should correspond to the intended application site. The incorporation of natural polymers into the human body may provoke immunological and inflammatory reactions [49]. Synthetic polymer hydrogels and aerogels lack inherent bioactivity; nonetheless, they can be engineered to possess favorable mechanical properties and other advantageous characteristics [50, 51].

3.3.1 Polyester-Based Hydrogels and Aerogels

Recent focus has intensified on altering the surfaces of textile-based materials across various domains, including filtration, medical textiles, smart textiles, flame-retardant fabrics, and textile catalysts [52, 53]. Polyester, or polyethylene terephthalate, is the predominant fiber in this context. Due to its physical and chemical attributes, such as affordability, stability, softness, lightness, elasticity, and wrinkle resistance, PET is an excellent material for mass production. The permanent functionalization of PET necessitates covalent bonding with the surface owing to its hydrophobic properties and lack of functional groups [54–56]. One method to enhance the use of polyester fibers was to use adhesive polyelectrolytes rich in functional groups, such as polyvinylamine (PVAm). These coatings possess antimicrobial qualities, serve as a foundation for subsequent surface modifications, and have demonstrated efficacy in extracting metals from contaminated groundwater, among other applications [57–60].

The combination of aerogels with standard polyester base materials results in a functional textile known as an aerogel polyester chip. Aerogels excel at thermal insulation, moisture absorption, and air permeability due to their very porous and lightweight structure, comprising more than 90% air. Combining polyester with aerogel produces lightweight and thermal-efficient textile material [61].

The production of aerogel polyester chips involves two processes. Initially, aerogel particles are combined with polyester resin to create a composite material. The subsequent step is to decompose the composite into smaller fragments. Subsequently, these chips can be transformed into polyester fibers that incorporate aerogels. Textiles composed of these fibers exhibit superior insulation, moisture absorption, and breathability compared with those fabricated from traditional polyester [62].

3.3.2 Nylon-Based Hydrogels and Aerogels

Synthetic polyamide nylon possesses a backbone abundant in amide (-CO-NH-) groups. This enhances its compatibility with hydrophilic polymer networks in terms of chemical and hydrogen bonding. Nylon may be efficiently coated with hydrogel and used in composites. Hydrogel matrices, when used as clothing substrates, provide superior moisture absorption and enhanced comfort [63]. Nylon, conversely, possesses many amide functionalities that facilitate its integration with other materials. Hydrogels capable of absorbing and retaining water can be included into nylon compression and athletic apparel, facilitating unrestricted and comfortable movement [64]. Aerogel-based composites also use nylon alongside hydrogels. Research indicates that nylon-6 composites supplemented with silica aerogel exhibit reduced thermal conductivity compared with pure nylon despite postheat processing methods such as extrusion and compression molding. Consequently, significant opportunities exist for future research on heat management in technical apparel [65].

The amide functions of nylon fibers facilitate adhesion to hydrophilic hydrogel matrices at the interface from a molecular perspective. These traits serve as significant fulcrums. Composites constructed from these fibers and aerogel phases, such as silica, demonstrate enhanced performance in moisture retention and thermal insulation [63, 65].

3.3.3 Acrylic-Based Hydrogels and Aerogels

The presence of anionic carboxyl groups in acrylic acid makes the hydrogels responsive to pH changes. Hydrogels containing these groups exhibit greater swelling at basic pH levels and reduced swelling at acidic pH levels compared with hydrogels with nonionic or neutral pendant groups. At basic pH, carboxylate ions form as carboxylic groups release a proton. The swelling rise is attributed to the electrostatic repulsion of these ions. Due to their exceptional adhesive qualities, hydrogels derived from acrylic acid have been extensively manufactured and used across various domains, including the biomedical sector. Ultraviolet (UV) light was used to merge poly(acrylic acid) and glycidyl methacrylate dextran, resulting in the formation of hydrogels as a copolymer. Upon the addition of dextranase to the solution at pH 7.4, the hydrogels produced exhibited swelling that was responsive to pH [66]. According to Lee et al. [67], the hydrogel composed of glycerol was synthesized through the polymerization of acrylic acid, facilitated by benzoyl peroxide and Novozym 435. This hydrogel exhibited greater inflation

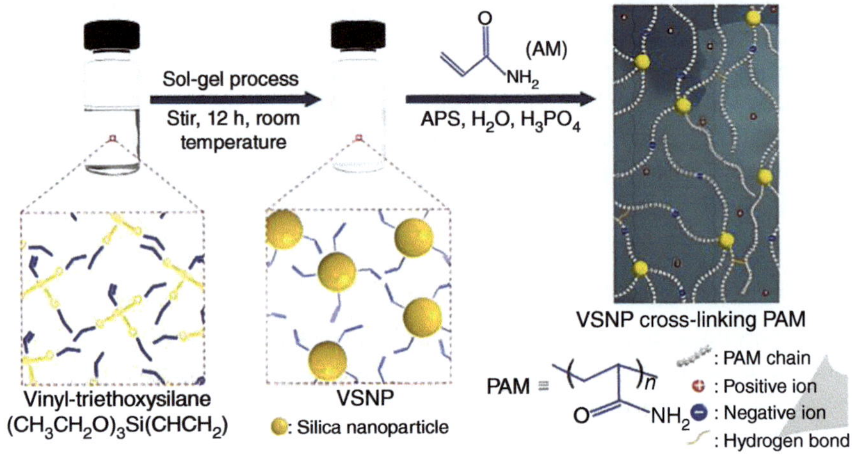

Figure 3.4 Synthesis of vinyl hybrid silica nanoparticles (VSNPs) from vinyl-triethoxysilane nanoparticles, followed by the production of vinyl-hybrid silica nanoparticle–polyacrylamide (VSNP-PAM) from phosphoric acid (proton source), acrylamide (AM, primary monomer), and ammonium persulfate (APS, initiator). *Source:* [70] / John Wiley & Sons.

than any other [67], regardless of the pH level. In addition, the cross-linking agent *N,N′*-methylenebisacrylamide enabled the synthesis of poly(acrylic acid) hydrogel through free radical copolymerization [68]. Nho et al. [69] developed poly(acrylic acid) hydrogels and examined their physical properties, including swelling ratio, mucoadhesion, and gel contents, by subjecting an acrylic acid solution to an electron beam with a maximum energy of 75 kGy (Figure 3.4).

Hina et al. [71] synthesized novel cross-linked hydrogels of self-healing poly(acrylamide) clay via the free radical method [71]. Poly(acrylamide) hydrogels exhibit limited strength and demonstrate poor stability in aqueous environments. To enhance their properties, they have been copolymerized with diverse monomers. Free radical polymerization was used to synthesize a hydrogel from poly(acrylamide-co-acrylic acid). A rheological analysis of the hydrogels indicated that copolymer hydrogels exhibited superior mechanical strength [72]. A separate study used foaming, synchronous polymerization, and poly(acrylamide-co-acrylic acid)/poly(acrylamide) to produce highly porous interpenetrating network hydrogels [73].

3.4 Blended Hydrogels and Aerogels in Textiles

Ensuring compatibility between natural and biodegradable synthetic polymer blends is essential before their use. The primary rationale for amalgamating two or more polymers is to create a novel, enhanced material [74]. This is accomplished by using a polymer with superior properties to alter or enhance the characteristics of the inferior polymer. An inexhaustible array of natural polymers exists, particularly those derived from lignocellulosic sources. They are the optimal candidates for integration as they can be tailored to accommodate advanced technology and are safe for living organisms. The utilization of natural polymer technologies facilitates the adoption of environmentally sustainable materials. Nevertheless, we must not overlook that synthetic polymers can function similarly to natural polymers. Before regeneration, polymer mixing typically involves melting or dissolving the polymers to achieve either uniform or nonuniform dispersion [75]. Numerous categories of regenerated materials exist, including hydrogel, aerogel, cryogel, xerogel, among others. These are produced using a combination of natural and inorganic components. It is commonly believed that hydrogels can be regenerated, whereas aerogels cannot always be regenerated. Hydrogels and aerogels serve multiple functions, notably in the fields of medicine and agriculture. They serve as appropriate alternatives to synthetic materials that are nonbiodegradable as they may be molded and are nontoxic to living organisms. Their characteristics are comparable with or superior to those of synthetic items. They are classified as smart materials due to their ability to react to external stimuli. The multitude of products derived from them underscores the significance of integrating natural and biodegradable synthetic polymers to enhance their durability. Polymer mixes provide composites applicable in various domains. This is due to the strong adhesion properties of mixed polymers. This section concentrates on the polymer amalgamation of biodegradable hydrogels and aerogels derived from lignocellulose (including cellulose, hemicellulose, and lignin) and synthetic substances (such as polyurethane, polylactic acid, polyvinyl alcohol, and polyethylene glycol), along with various other polymeric materials used in different industries [76, 77].

3.4.1 Natural–Synthetic Blends

3.4.1.1 Cellulose–PVA

Hydrogels derived from PVA are highly suitable for biomedical applications due to their excellent biocompatibility and superior mechanical properties [78–80]. Various methods exist for the fabrication of these hydrogels, including cryogelation (repeated freezing and thawing in aqueous solutions), chemical cross-linking of polyvinyl alcohol (PVA), noncryogenic physical gelation of PVA, among others. The fabrication of PVA-based hydrogels using these approaches significantly enhances their material properties, including mechanical strength, biocompatibility, and chemical stability [79, 81]. In this context, it is beneficial to synthesize PVA-based hydrogels with adjustable molecular weight and appropriate supramolecular polymer connections for biomedical applications. PVA is distinctive among other materials for hydrogel production due to its water solubility, biodegradability, responsiveness to external stimuli, and ability to form stable networks via diverse cross-linking techniques [82, 83].

Recent studies have extensively examined the incorporation of cellulose into PVA hydrogels. The incorporation of microsized cellulose and bacterial cellulose into PVA-based hydrogels enhances their storage modulus, tensile strength, and compressive strength [84]. The abundance of hydroxyl groups on the cellulose surface allows for various cross-linking methods to achieve the desired superficial characteristics. Numerous studies have focused on cellulose nanoparticles due to their distinctive properties, including biocompatibility, high surface area, significant water retention capacity, and crystalline structure. Cellulose nanostructures, such as cellulose nanocrystals (CNCs), enhance hydrogels by modifying their properties. Acid treatment of cellulose microfibrils typically results in the formation of rod-shaped nanoparticles known as CNCs. Their crystallinity exceeds that of other cellulose-derived materials. CNCs exhibit low density, high axial stiffness, and significant tensile strength [85, 86].

3.4.1.2 Silk–PLA

Polylactic acid (PLA), a biodegradable thermoplastic derived from renewable resources such as corn starch and sugarcane, is commonly used in composite fiber and hydrogel applications. This natural biopolymer exhibits superior mechanical strength, biodegradability, ease of processing, and biocompatibility [87, 88]. Regenerated silk fibroin has been used to create various scaffolds that address diverse issues [89]. Electrospinning, hydrogels, and 3D printing represent various methods for creating these scaffolds. Bianza Moise Bakadia et al. [90] synthesized hydrogels through the freezing and thawing of a mixture comprising azithromycin (AZM), silk fibroin (SS), and polyvinyl alcohol (PVA), using Genping (GNP) as a cross-linking agent. The hydrogel demonstrated adhesion to skin, compatibility with keratinocytes and fibroblasts, and a gradual release of β and AZM during freeze-drying. The AZM-SS/PVA hydrogel demonstrates superior wound healing capabilities compared with the commercially available Tegaderm membrane dressing. It reduces the severity of systemic burns. Hao Wang et al. [91] developed a hydrogel catheter using silk fibroin, gelatin, and tyramine through a diffusion-driven cross-linking technique. The catheter can be customized to accommodate the patient's anatomy and requirements. Polylactic acid (PLA) is a biodegradable, hydrophobic polyester with numerous safe and environmentally acceptable applications [92]. Its compressive strength, closely analogous to that of authentic bone, can facilitate the regeneration of bone tissue. The degradation products of PLA during bone tissue regeneration demonstrate low cytotoxicity in local tissues, remarkable biocompatibility, and the results of significant research. PLA is used to manufacture a variety of products due to its ease of thermal processing, biocompatibility, and biodegradability [93].

3.4.1.3 Wool–Polyurethane (PU)

Keratinous wool composites consist of the fibers combined with polyurethane (PU) films or foams. They possess durability, insulation properties, and fire resistance. The incorporation of PU enhances the workability and flexibility of these composites, rendering them ideal for protective gear and cushioned liners. Recent research has demonstrated that incorporating renewable materials and using layered structures in waste wool and PU foams can reduce thermal conductivity and enhance durability [41–43].

3.4.2 Composite Textile Substrates

3.4.2.1 Laminated/Coated Fabrics

Researchers use commercially available aerogel fiber composites or combinations of these composites with additional materials to create experimental samples that can substitute traditional thermal insulation materials. Zhang et al. [94] investigated the impact of six different material composition schemes on the thermal protection effectiveness of firefighting suits, using aerogel fiber composites and commercially available microencapsulated phase change materials (PCMs) as thermal liners. Aspen Systems produced several of the aerogel fiber composites available for purchase in stores today. All evaluations of this application approach have demonstrated that aerogel and its synergistic components are effective in providing heat protection. The detachment and dispersion of aerogel particles from the fabric have complicated the reapplication of this technology. The descent of aerogel particles compromises the integrity of the aerogel materials and poses a health risk to anybody exposed to the airborne dust.

Aerogel fiber composites are often encased or laminated before use to prevent the dispersion of aerogel particles into the air. In lieu of the commonly used M400 Thinsulate coverall undergarment for cold-water diving suits, Nuckols et al. [95] used a prototype aerogel liner fabricated from Aspen Aerogel AR5401 panels, which was encased in Pertex nylon Oxford.

3.4.2.2 Functional Coatings

The incorporation of polymers into protective garment fabrics to enhance functionality is increasingly common [96]. Zábranský et al. [97] said that the coated layers have been enhanced with active nanoparticles that offer superior thermal protection. This has resulted in the creation of nanocomposite solutions for the forthcoming generation of thermal protective apparel. SiO_2 aerogel has garnered significant attention as a potential innovative material for thermal insulation due to its network structure, porosity, and low density [98]. Nonetheless, SiO_2 aerogel exhibits minimal capacity to impede heat transfer in the context of infrared radiation. Tang et al. [99] included chemicals into SiO_2 aerogel to inhibit infrared radiation. Loading SiO_2 aerogel onto fabrics and determining its application has consistently posed challenges. SiO_2 aerogel possesses exceptional thermal protection capabilities due to its efficacy. The improper combination of SiO_2 aerogel and fibers results in SiO_2 aerogel-coated fabrics that exhibit limited durability in practical applications [100]. The atypical drying rate of SiO_2 aerogel may result in the formation of fissures on the surfaces of coated fabrics. The additive possesses significant thermal conductivity; nevertheless, adhesives are typically used to address these issues. It remains challenging to identify the appropriate adhesive that will enhance the adhesion of the coated layer to the fibers while preserving the excellent thermal insulation properties of SiO_2 aerogel.

Lightweight, environmentally friendly blended hydrogel/aerogel textiles made of natural-synthetic polymer hybrids or laminated/coated fabric composites provide insulation, responsiveness, and comfort. In practical textile environments, dependable performance can be attained by interface engineering and network architecture. This is especially useful for managing substance behavior, expansion, and duration.

3.5 Conclusion

Recently, hydrogels and aerogels have emerged in the textile industry as versatile functional materials. These materials exhibit superior performance compared with conventional fabrics in several aspects. This chapter discusses the various types of fibers and their origins. The discussion focused on the ways in which natural, synthetic, and hybrid systems alter the properties and applications of materials. Hydrogels and aerogels are composed of natural fibers such as wool, silk, or cotton. They benefit the environment, regulate moisture, and provide user comfort. These materials are advantageous for the production of clothing and health-care products that are environmentally sustainable. Composite synthesis, photopolymerization, and grafting can modify synthetic fiber systems such as acrylic, nylon, and polyester to enhance properties, including flame resistance, antibacterial activity, and heat sensitivity. This remains accurate despite the common occurrence of hydrophobicity and other issues in such systems. Integrating various techniques may yield multifunctional fabrics that are environmentally sustainable. These technologies use natural or synthetic polymers, as well as laminated and coated substrates integrated with aerogels.

References

1 Zhao, X., Chen, X., Yuk, H. et al. (2021). Soft materials by design: unconventional polymer networks give extreme properties. *Chemical Reviews* 121 (8): 4309–4372. https://doi.org/10.1021/acs.chemrev.0c01088.

2 Ho, T.-C., Chang, C.-C., Chan, H.-P. et al. (2022). Hydrogels: properties and applications in biomedicine. *Molecules* 27 (9): 2902. https://doi.org/10.3390/molecules27092902.

3 Velmurugan, N., Kumar, G.G., Han, S.S. et al. (2009). Synthesis and characterization of potential fungicidal silver nano-sized particles and chitosan membrane containing silver particles. *Iranian Polymer Journal* 18: 383–392.

4 Jiao, T.-F., Zhou, J.-A., Zhou, J.-X. et al. (2011). Synthesis and characterization of chitosan-based Schiff base compounds with aromatic substituent groups. *Iranian Polymer Journal* 20: 123–136.

5 Gbenebor, O.P., Adeosun, S.O., Lawal, G.I. et al. (2017). Acetylation, crystalline and morphological properties of structural polysaccharide from shrimp exoskeleton. *Engineering Science and Technology* 20: 1155–1165. https://doi.org/10.1016/j.jestch.2017.05.002.

6 Pierre, A.-C. and Pajonk, G.-M. (2002). Chemistry of aerogels and their applications. *Chemical Reviews* 102: 4243–4266. https://doi.org/10.1021/cr0101306.

7 Venkataraman, M., Mishra, R., Kotresh, T.M. et al. (2016). Aerogels for thermal insulation in high-performance textiles. *Textile Progress* 48 (2): 55–118. https://doi.org/10.1080/00405167.2016.1179477.

8 Kistler, S.S. (1931, https://doi.org/10.1038/127741a0). Coherent expanded aerogels and jellies. *Nature* 127 (3211): 741.

9 García-González, C.A., Sosnik, A., Kalmár, J. et al. (2021). Aerogels in drug delivery: from design to application. *Journal of Controlled Release* 332: 40–63. https://doi.org/10.1016/j.jconrel.2021.02.012.

10 Hasan, M.A., Sangashetty, R., Esther, A.C.M. et al. (2017). Prospect of thermal insulation by silica aerogel: a brief review. *Journal of The Institution of Engineers (India): Series D* 98 (2): 297–304. https://doi.org/10.1007/s40033-017-0136-1.

11 Kabir, S.M.F., Sikdar, P.P., Haque, B. et al. (2018). Cellulose-based hydrogel materials: chemistry, properties and their prospective applications. *Progress in Biomaterials* 7 (3): 153–174. https://doi.org/10.1007/s40204-018-0095-0.

12 Akter, S., Azim, A.Y.M.A., and Al Faruque, M.A. (2014). Medical textiles: significance and future prospect in Bangladesh. *European Scientific Journal* 10.

13 Standard, O.C. (2017). Increased bending rigidity of wool fabric imparted by hybrid organic-inorganic sol–gel coatings. *Textile Research Journal* 87 (5): 607–616. https://doi.org/10.1177/0040517516635990.

14 Bashir, S., Hina, M., Iqbal, J. et al. (2020). Fundamental concepts of hydrogels: synthesis, properties, and their applications. *Polymers* 12: 2702. https://doi.org/10.3390/polym12112702.

15 McNeil, S.J. and Gupta, H. (2022). Emerging applications of aerogels in textiles. *Polymer Testing* 106: 107426. https://doi.org/10.1016/j.polymertesting.2021.107426.

16 Hu, Y., Zhang, Z., Li, Y. et al. (2018). Dual-crosslinked amorphous polysaccharide hydrogels based on chitosan/alginate for wound healing applications. *Macromolecular Rapid Communications* 39: e1800069. https://doi.org/10.1002/marc.201800069.

17 Kwiecień, I. and Kwiecień, M. (2018). Application of polysaccharide-based hydrogels as probiotic delivery systems. *Gels* 4: 47. https://doi.org/10.3390/gels4020047.

18 Gholamali, I. (2019). Stimuli-responsive polysaccharide hydrogels for biomedical applications: a review. *Regenerative Engineering and Translational Medicine* 1–24. https://doi.org/10.1007/s40883-019-00120-1.

19 Fan, J., Ifuku, S., Wang, M. et al. (2018). Robust nanofibrillated cellulose hydro/aerogels from benign solution/solvent exchange treatment. *ACS Sustainable Chemistry & Engineering* 6 (5): 6624–6634. https://doi.org/10.1021/acssuschemeng.8b00418.

20 Zou, L., Phule, A.D., Sun, Y. et al. (2020). Superhydrophobic and superoleophilic polyethylene aerogel coated natural rubber latex foam for oil-water separation application. *Polymer Testing* 85: 106451. https://doi.org/10.1016/j.polymertesting.2020.106451.

21 Shaid, A., Wang, L., and Padhye, R. (2016). The flammability of aerogel and PCM coated cotton fabric for firefighter's garment. *Proceedings of the 9th Textile Bioengineering and Informatics Symposium—6th Asian Protective Clothing Conference*, Melbourne, Australia (12–16 July 2016), pp. 61–68.

22 Liu, Y. (2018). *Chemical Composition and Characterization of Cotton Fibers.* Berlin, Germany: Springer.

23 Ahmed, E.M. (2015). Hydrogel: preparation, characterization, and applications: a review. *Journal of Advanced Research* 6 (2): 105–121. https://doi.org/10.1016/j.jare.2013.07.006.

24 Lacoste, C., Lopez-Cuesta, J.-M., and Bergeret, A. (2019). Development of a biobased superabsorbent polymer from recycled cellulose for diapers applications. *European Polymer Journal* 116: 38–44. https://doi.org/10.1016/j.eurpolymj.2019.03.013.

25 Bidgoli, H., Zamani, A., Jeihanipour, A., and Taherzadeh, M.J. (2014). Preparation of carboxymethyl cellulose superabsorbents from waste textiles. *Fibers and Polymers* 15 (3): 431–436. https://doi.org/10.1007/s12221-014-0431-5.

26 Chang, C., Duan, B., Cai, J., and Zhang, L. (2010). Superabsorbent hydrogels based on cellulose for smart swelling and controllable delivery. *European Polymer Journal* 46 (1): 92–100. https://doi.org/10.1016/j.eurpolymj.2009.04.033.

27 Kono, H. (2014). Characterization and properties of carboxymethyl cellulose hydrogels crosslinked by polyethylene glycol. *Carbohydrate Polymers* 106: 84–93. https://doi.org/10.1016/j.carbpol.2014.02.020.

28 Zhou, J., Chang, C., Zhang, R., and Zhang, L. (2007). Hydrogels prepared from unsubstituted cellulose in NaOH/urea aqueous solution. *Macromolecular Bioscience* 7 (6): 804–809. https://doi.org/10.1002/mabi.200700007.

29 Alomayri, T., Shaikh, F.U.A., and Low, I.M. (2013). Thermal and mechanical properties of cotton fabric-reinforced geopolymer composites. *Journal of Materials Science* 48: 6746–6752.

30 Yan, J., Choi, H.Y., Hong, Y.K., and Jeong, Y.G. (2018). Thermal insulation performance of cotton and PET-based hybrid fabrics impregnated with silica aerogel via a facile dip-dry process. *Fibers and Polymers* 19: 854–860.

31 Mai, T.N., Luu, P.T., Le, P.K. et al. (2019). Fabrication of cotton aerogels and its application in water treatment. *Proceedings of the 12th Regional Conference on Chemical Engineering (RCChE 2019): Transformation towards Smart, Resilient & Sustainable Communities*, Ho Chi Minh City, Vietnam (15–16 October 2019), 85 pp, ISBN: 978-604-67-1373-9.

32 Zhang, Z., Ortiz, O., Goyal, R., and Kohn, J. (2014). Biodegradable polymers. In: *Handbook of Polymer Applications in Medicine and Medical Devices*, Plastics Design Library (ed. K. Modjarrad and S. Ebnesajjad), 303–335. Oxford, UK: William Andrew Publishing https://doi.org/10.1016/B978-0-323-22805-3.00013-X.

33 Reizabal, A., Costa, C.M., Pérez-Álvarez, L. et al. (2023). Silk fibroin as sustainable advanced material: material properties and characteristics, processing, and applications. *Advanced Functional Materials* 33 (3): 2210764. https://doi.org/10.1002/adfm.202210764.

34 Zheng, H. and Zuo, B. (2021). Functional silk fibroin hydrogels: preparation, properties and applications. *Journal of Materials Chemistry B* 9 (5): 1238–1258. https://doi.org/10.1039/D0TB02099K.

35 Seib, F.P. (2018). Reverse-engineered silk hydrogels for cell and drug delivery. *Therapeutic Delivery* 9 (6): 469–487. https://doi.org/10.4155/tde-2018-0016.

36 Andrady, A.L. and Mark, J.E. (2007). Biodegradability of polymers. In: *Physical Properties of Polymers Handbook* (ed. J.E. Mark), 951–964. New York, NY: Springer https://doi.org/10.1007/978-0-387-69002-5_56.

37 Singh, V., Tripathi, D.K., Sharma, V.K. et al. (2023). Silk fibroin hydrogel: a novel biopolymer for sustained release of vancomycin drug for diabetic wound healing. *Journal of Molecular Structure* 1286: 135548. https://doi.org/10.1016/j.molstruc.2023.135548.

38 Maleki, H., Montes, S., Hayati-Roodbari, N. et al. (2018). Compressible, thermally insulating, and fire retardant aerogels through self-assembling silk fibroin biopolymers inside a silica structure—an approach towards 3D printing of aerogels. *ACS Applied Materials & Interfaces* 10 (26): 22718–22730. https://doi.org/10.1021/acsami.8b06778.

39 Maleki, H., Whitmore, L., and Hüsing, N. (2018). Novel multifunctional polymethylsilsesquioxane–silk fibroin aerogel hybrids for environmental and thermal insulation applications. *Journal of Materials Chemistry A* 6 (26): 12598–12612. https://doi.org/10.1039/C8TA02790B.

40 Maleki, H. and Huesing, N. (2021). Silica–silk fibroin hybrid (bio)aerogels: two-step versus one-step hybridization. *Journal of Sol-Gel Science and Technology* 98: 430–438. https://doi.org/10.1007/s10971-019-04933-4.

41 Verma, V., Verma, P., Ray, P., and Ray, A.R. (2008). Preparation of scaffolds from human hair proteins for tissue-engineering applications. *Biomedical Materials* 3 (2): 025007.

42 Hill, P., Brantley, H., and Van Dyke, M. (2010). Some properties of keratin biomaterials: kerateines. *Biomaterials* 31 (4): 585–593.

43 Bragulla, H.H. and Homberger, D.G. (2009). Structure and functions of keratin proteins in simple, stratified, keratinized and cornified epithelia. *Journal of Anatomy* 214 (4): 516–559.

44 Fraser, R.B.D., Mac Rae, T.P., and Rogers, G.E. (1972). *Keratins, Their Composition, Structure and Biosynthesis.* Springfield, IL: C.C. Thomas Publishing.

45 Tachibana, A., Furuta, Y., Takeshima, H. et al. (2002). Fabrication of wool keratin sponge scaffolds for long-term cell cultivation. *Journal of Biotechnology* 93 (2): 165–170.

46 Loh, J.W., Goh, X.Y., Nguyen, P.T.T. et al. (2022). Advanced aerogels from wool waste fibers for oil spill cleaning applications. *Journal of Polymers and the Environment* 30: 681–694. https://doi.org/10.1007/s10924-021-02234-y.

47 Paraskevopoulou, P., Chriti, D., Raptopoulos, G., and Anyfantis, G.C. (2019). Synthetic polymer aerogels in particulate form. *Materials* 12 (9): 1543. https://doi.org/10.3390/ma12091543.

48 Corkhill, P.H., Hamilton, C.J., and Tighe, B.J. (1989). Synthetic hydrogels VI. Hydrogel composites as wound dressings and implant materials. *Biomaterials* 10 (1): 3–10. https://doi.org/10.1016/0142-9612(89)90002-1.

49 Silva, A.K.A., Richard, C., Bessodes, M. et al. (2009). Growth factor delivery approaches in hydrogels. *Biomacromolecules* 10 (1): 9–18. https://doi.org/10.1021/bm8009974.

50 Lin, C.C. and Metters, A.T. (2006). Hydrogels in controlled release formulations: network design and mathematical modeling. *Advanced Drug Delivery Reviews* 58 (12–13): 1379–1408. https://doi.org/10.1016/j.addr.2006.09.004.

51 Peri, J.B. (1966). Infrared study of OH and NH2 groups on the surface of a dry silica aerogel. *The Journal of Physical Chemistry* 70 (9): 2937–2945. https://doi.org/10.1021/j100881a502.

52 Mittal, K.L. and Bahners, T. (2017). *Textile Finishing: Recent Developments and Future Trends*. Beverly, MA: Scrivener Publishing.

53 Fahmy, H.M., Eid, R.A.A., Nada, D.E., and El-Aziz, S.M.A. (2015). Functionalization of linen fabric using poly (N-vinyl-2-pyrrolidone). *Egyptian Journal of Chemistry* 58: 447–458.

54 Ballottin, D., Fulaz, S., Cabrini, F. et al. (2017). Antimicrobial textiles: biogenic silver nanoparticles against *Candida* and *Xanthomonas*. *Materials Science & Engineering C, Materials for Biological Applications* 75: 582–589. https://doi.org/10.1016/j.msec.2017.02.101.

55 Paul, R. (2015). Functional finishes for textiles: an overview. In: *Woodhead Publishing Series in Textiles* (ed. R. Paul), 1–14. Cambridge, UK: Woodhead Publishing.

56 Radetić, M. (2013). Functionalization of textile materials with silver nanoparticles. *Journal of Materials Science* 48: 95–107. https://doi.org/10.1007/s10853-012-6677-7.

57 Butola, B.S. (2008). Advances in functional finishes for polyester and polyamide-based textiles. In: *Woodhead Publishing Series in Textiles* (ed. M.L. Gulrajani), 325–353. Cambridge, UK: Woodhead Publishing.

58 Deopura, B.L. and Padaki, N.V. (2015). Synthetic textile fibers: polyamide, polyester and aramid fibers. In: *Woodhead Publishing Series in Textiles* (ed. B.L. Deopura, R. Alagirusamy, M. Joshi, and B. Gupta), 97–114. Cambridge, UK: Woodhead Publishing.

59 Perepelkin, K.E. (2001). Physicochemical nature and structural dependence of the unique properties of polyester fibers. *Fiber Chem.* 33: 340–352. https://doi.org/10.1023/A:1013034600106.

60 Timma, L.M., Lewald, L., Gier, F. et al. (2019). Nonfouling textiles with tunable antimicrobial activity based on a zwitterionic polyamine finish. *RSC Advances* 9: 9783–9791. https://doi.org/10.1039/C9RA00074H.

61 Linhares, T., Carneiro, V.H., Pessoa de Amorim, M.T., and Durães, L. (2023). A comparative thermoacoustic insulation study of silica aerogels reinforced with reclaimed textile fibres: cotton, polyester and wool. *Gels* 9 (7): 548. https://doi.org/10.3390/gels9070548.

62 Barrios, E., Fox, D., Li Sip, Y.Y. et al. (2019). Nanomaterials in advanced, high-performance aerogel composites: a review. *Polymers* 11 (4): 726. https://doi.org/10.3390/polym11040726.

63 Li, S., Wang, Y., Xia, J. et al. A flexible nylon-spandex fabric-supported PVA/PANI hydrogel sensor with high strength and conductivity. *ACS Applied Polymer Materials* 6 (15): 5127–5137. https://doi.org/10.1021/acsapm.4c00868.

64 Gürgen, S. and Yıldız, T. (2020). Stab resistance of smart polymer coated textiles reinforced with particle additives. *Composite Structures* 235: 111812. https://doi.org/10.1016/j.compstruct.2019.111812.

65 Krishnaswamy, S., Tinsley, L., Marchante, V. et al. (2018). Effect of extrusion and compression moulding on the thermal properties of nylon-6/silica aerogel composites. *Journal of Thermoplastic Composite Materials* 31 (7): 992–1009. https://doi.org/10.1177/0892705717728516.

66 Kim, I.-S. and Oh, I.-J. (2005). Drug release from the enzyme-degradable and pH-sensitive hydrogel composed of glycidyl methacrylate dextran and poly(acrylic acid). *Archives of Pharmacal Research* 28: 983–987. https://doi.org/10.1007/BF02973887.

67 Lee, A.L.S., Ordeñez, G., and Chakraborty, S. (2011). Novel glycerol cross-linked poly(acrylic acid) hydrogel for encapsulation and release of benzocaine. *Philippine Science Letters* 4: 81–87.

68 Amin, M.C.I.M., Ahmad, N., Halib, N., and Ahmad, I. (2012). Synthesis and characterization of thermo- and pH-responsive bacterial cellulose/acrylic acid hydrogels for drug delivery. *Carbohydrate Polymers* 88: 465–473. https://doi.org/10.1016/j.carbpol.2011.12.022.

69 Nho, Y.-C., Park, J.-S., and Lim, Y.-M. (2014). Preparation of poly(acrylic acid) hydrogel by radiation crosslinking and its application for mucoadhesives. *Polymers* 6: 890–898. https://doi.org/10.3390/polym6030890.

70 Huang, Y., Zhong, M., Shi, F. et al. (2017). An intrinsically stretchable and compressible supercapacitor containing a polyacrylamide hydrogel electrolyte. *Angewandte Chemie International Edition* 56: 9141–9145. https://doi.org/10.1002/anie.201705212.

71 Hina, M., Bashir, S., Kamran, K. et al. (2020). Synthesis and characterization of self-healable poly(acrylamide) hydrogel electrolytes and their application in fabrication of aqueous supercapacitors. *Polymer* 210: 123020. https://doi.org/10.1016/j.polymer.2020.123020.

72 Nesrinne, S. and Djamel, A. (2017, https://doi.org/10.1016/j.arabjc.2013.11.027). Synthesis, characterization and rheological behavior of pH sensitive poly(acrylamide-co-acrylic acid) hydrogels. *Arabian Journal of Chemistry* 10: 539–547.

73 Ao, H., Huang, M., Wu, J. et al. (2009). Synthesis and properties of poly(acrylamide-co-acrylic acid)/polyacrylamide superporous IPN hydrogels. *Polymers for Advanced Technologies* 20: 1044–1049. https://doi.org/10.1002/pat.1361.

74 Mirtaghavi, A., Luo, J., and Muthuraj, R. (2020). Recent advances in porous 3D cellulose aerogels for tissue engineering applications: a review. *Journal of Composites Science* 4: 152. https://doi.org/10.3390/jcs4040152.

75 Agarwal, S. (2020). Biodegradable polymers: present opportunities and challenges in providing a microplastic-free environment. *Macromolecular Chemistry and Physics* 221: 2000017. https://doi.org/10.1002/macp.202000017.

76 Amran, U., Zakaria, S., Chia, C. et al. (2019, https://doi.org/10.1007/s10570-019-02271-w). Polyols and rigid polyurethane foams derived from liquefied lignocellulosic and cellulosic biomass. *Cellulose* 26: 3231–3246.

77 Sun, X.-F., Zeng, Q., Wang, H., and Hao, Y. (2019, https://doi.org/10.1007/s10570-018-2180-x). Preparation and swelling behavior of pH/temperature responsive semi-IPN hydrogel based on carboxymethyl xylan and poly(N-isopropyl acrylamide). *Cellulose* 26: 1909–1922.

78 Alves, M.-H., Jensen, B.E.B., Smith, A.A.A., and Zelikin, A.N. (2011). Poly(vinyl alcohol) physical hydrogels: new vista on a long serving biomaterial. *Macromolecular Bioscience* 11: 1293–1313. https://doi.org/10.1002/mabi.201100145.

79 Hassan, C.M. and Peppas, N.A. (2000). Structure and applications of poly(vinyl alcohol) hydrogels produced by conventional crosslinking or by freezing/thawing methods. In: *Biopolymers · PVA Hydrogels, Anionic Polymerisation Nanocomposites*, 37–65. Berlin, Germany: Springer.

80 Lozinsky, V.I., Galaev, I.Y., Plieva, F.M. et al. (2003). Polymeric cryogels as promising materials of biotechnological interest. *Trends in Biotechnology* 21: 445–451. https://doi.org/10.1016/j.tibtech.2003.08.002.

81 Rumon, M.M.H., Rahman, M.S., Akib, A.A. et al. (2025). Progress in hydrogel toughening: addressing structural and crosslinking challenges for biomedical applications. *Discover Materials* 5: 5. https://doi.org/10.1007/s43939-025-00178-x.

82 Liang, X., Zhong, H.J., Ding, H. et al. (2024). Polyvinyl alcohol (PVA)-based hydrogels: recent progress in fabrication, properties, and multifunctional applications. *Polymers* 16: 2755. https://doi.org/10.3390/polym16192755.

83 Liang, X., Ding, H., Wang, Q. et al. (2021). Nature-inspired semi-IPN hydrogels with tunable mechanical properties and multi-responsiveness. *New Journal of Chemistry* 45: 861–871. https://doi.org/10.1039/D0NJ04675B.

84 Peng, H., Wang, S., Xu, H., and Hao, X. (2017). Preparation, properties and formation mechanism of cellulose/polyvinyl alcohol bio-composite hydrogel membranes. *New Journal of Chemistry* 41 (14): 6564–6573. https://doi.org/10.1039/C7NJ00845G.

85 Qiao, K., Zheng, Y., Guo, S. et al. (2015). Hydrophilic nanofiber of bacterial cellulose guided the changes in the micro-structure and mechanical properties of nf-BC/PVA composites hydrogels. *Composites Science and Technology* 118: 47–54. https://doi.org/10.1016/j.compscitech.2015.08.004.

86 Postek, M.T., Moon, R.J., Rudie, A.W., and Bilodeau, M.A. (2013). *Production and Applications of Cellulose*. Peachtree Corners, GA: Tappi Press.

87 Long, D.-P., Cheng, X., Hao, Z., and Sun, J. (2021). Genetic hybridization of highly active exogenous functional proteins into silk-based materials using "light-clothing" strategy. *Matter* 4 (6): 2039–2058.

88 Sahoo, J.K., Hasturk, O., Falcucci, T., and Kaplan, D.L. (2023). Silk chemistry and biomedical material designs. *Nature Reviews Chemistry* 7 (5): 302–318.

89 Pallotta, I., Kluge, J.A., Moreau, J. et al. (2014). Characteristics of platelet gels combined with silk. *Biomaterials* 35 (11): 3678–3687.

90 Bakadia, B.M., Lamboni, L., Ahmed, A.A.Q. et al. (2023). Antibacterial silk sericin/poly(vinyl alcohol) hydrogel with antifungal property for potential infected large burn wound healing: systemic evaluation. *Smart Materials in Medicine* 4: 37–58.

91 Wang, H., Wan, H., Wang, Q. et al. (2023). Engineered multifunctional silk fibroin/gelatin hydrogel conduit loaded with miR-29a@ZIF-8 nanoparticles for peripheral nerve regeneration. *Smart Materials in Medicine* 4: 480–492.

92 Tabata, H., Kato, Y., Yoshida, K. et al. (2012). Fabrication of polylactic acid (PLA) microcapsules and release of the internal hydrophilic substance under ultrasound irradiation. *IOP Conference Series: Materials Science and Engineering* 42 (1): 012012.

93 Fairag, R., Rosenzweig, D.H., Ramirez-Garcialuna, J.L. et al. (2019). Three-dimensional printed polylactic acid scaffolds promote bone-like matrix deposition in vitro. *ACS Applied Materials & Interfaces* 11 (17): 15306–15315.

94 Bardy, E., Mollendorf, J., and Pendergast, D. (2006). Thermal resistance and compressive strain of underwater aerogel–syntactic foam hybrid insulation at atmospheric and elevated hydrostatic pressure. *Journal of Physics D: Applied Physics* 39 (9): 1908–1918. https://doi.org/10.1088/0022-3727/39/9/028.

95 Nuckols, M.L., Chao, J.C., and Swiergosz, M.J. (2005). Manned evaluation of a prototype composite cold water diving garment using liquids and super insulation aerogel materials. *Proceedings of the 25th International Conference on Offshore Mechanics and Arctic Engineering*, 2005, pp. 27–34, doi: https://doi.org/10.1115/OMAE2006-92026.

96 Tian, W.L., Qi, L.H., Su, C.Q. et al. (2016). Effect of fiber transverse isotropy on effective thermal conductivity of metal matrix composites reinforced by randomly distributed fibers. *Composite Structures* 152: 637–644.

97 Zábranský, L., Buršíková, V., Souček, P. et al. (2017). Thermal stability of hard nanocomposite Mo-B-C coatings. *Vacuum* 138: 199–204. https://doi.org/10.1016/j.vacuum.2016.12.016.

98 Liu, Z.H., Ding, Y.D., Wang, F., and Deng, Z.P. (2016). Thermal insulation material based on SiO_2 aerogel. *Construction and Building Materials* 122: 548–555.

99 Tang, G.H., Zhao, Y., and Guo, J.F. (2016). Multi-layer graded doping in silica aerogel insulation with temperature gradient. *International Journal of Heat and Mass Transfer* 99: 192–200. https://doi.org/10.1016/j.ijheatmasstransfer.2016.03.093.

100 Rostami, J., Khandel, O., Sedighardekani, R. et al. (2021). Enhanced workability, durability, and thermal properties of cement-based composites with aerogel and paraffin coated recycled aggregates. *Journal of Cleaner Production* 297: 126518. https://doi.org/10.1016/j.jclepro.2021.126518.

4

Synthesis Techniques for Hydrogels and Aerogels for Textiles

Sana Sadaf and Faheem Ahmad

National Textile University, School of Engineering and Technology, Department of Textile Technology, Sheikhupura Road 37610, Faisalabad, Pakistan

4.1 Introduction

Hydrogels and aerogels have versatile structural and functional properties that make them an attractive option for use in the textile industry. Both have three-dimensional cross-linked structure with different compositions, porosities, and mechanical properties. The three-dimensional structure of hydrogels attributes for maintaining their structural integrity during the absorption and retention of huge volumes of liquid [1]. Along with high water absorption capacity, hydrogels have some other excellent features such as biocompatibility and stimuli-responsive nature due to which they have potential applications in health-care textiles (drug release, wound dressings, etc.) and smart textiles [2]. On the contrary, aerogels are highly porous and lightweight materials with exceptional thermal insulation properties [3]. Silica aerogels are most commonly used aerogels, but for the application of aerogels in textile industry, attempts have been conducted by researchers to synthesize aerogels from bio-based materials (cellulose, starch, chitosan, etc.), which can have potential applications in thermal protective clothing, breathable insulating fabrics, and wearable sensors [4, 5] (Figure 4.1).

Different material fabrication techniques, for example, electrospinning and coating can be used to integrate hydrogels and aerogels into textiles [7]. Due to the multifunctional and tunable properties of these integrated hydrogels and aerogels, there are potential possibilities for next-generation smart textiles. Hydrogels and aerogels can be synthesized from natural fibers, synthetic fibers, and blended materials. This chapter describes the synthesis methods of hydrogels and aerogels for textile applications from different raw materials (natural, synthetic, and blended) (Figure 4.2).

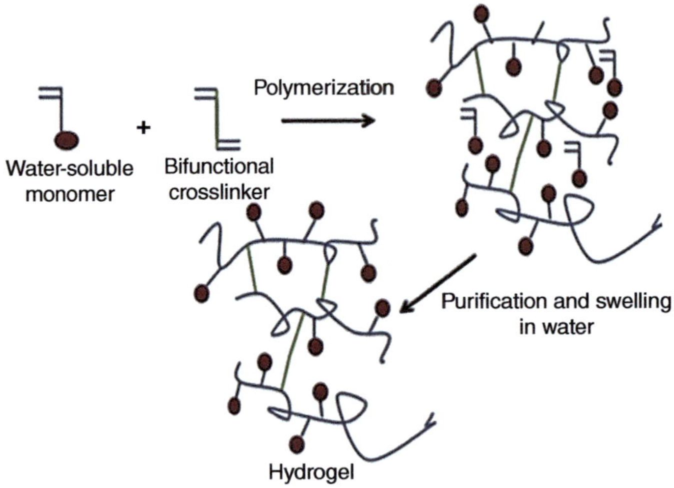

Figure 4.1 General scheme for hydrogel synthesis. *Source:* Petros et al., 2020 / John Wiley & Sons / CC BY 4.0.

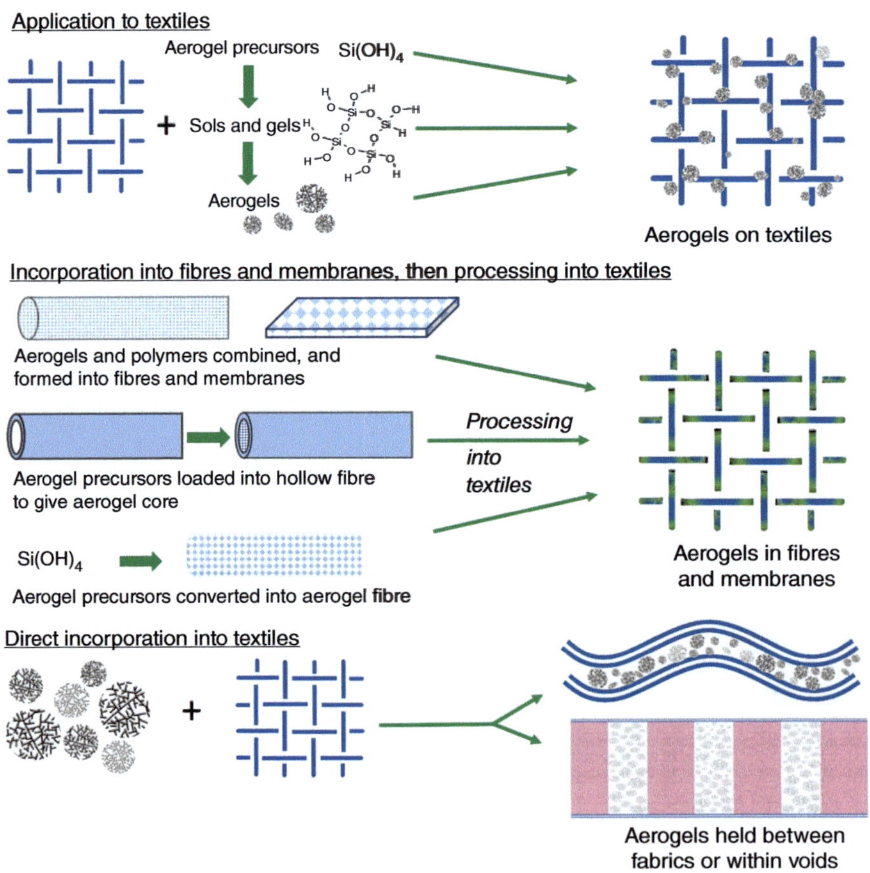

Figure 4.2 Application of aerogels in textile through different modes. *Source:* [8] / with permission of Elsevier.

4.2 Synthesis Techniques for Hydrogels and Aerogels from Natural Fiber

4.2.1 Synthesis from Cotton

Cotton is an abundantly available natural polymer, which is primarily composed of cellulose (85–95%). The degree of polymerization of cotton-derived cellulose lies between 10,000 and 15,000. Cellulose is a homopolysaccharide having glucose as major monomer with β 1,4 linkage. In the formation of cellulose, glucose molecules initially combined to form a cellobiose molecule, and cellobiose molecules combine lengthwise with each other through strong oxygen bridging to form a giant cellulose molecule, which forms strong fibers and is highly cohesive in nature. The presence of hydroxyl groups in the cellulose structure is responsible for hydrogen bonding. It can be used as a sustainable raw material for the synthesis of hydrogels and aerogels (Figure 4.3).

Cellulose possesses very strong inter- and intramolecular hydrogen bonding, along with a high degree of polymerization, which results in a very slight dissolution of cellulose in water and organic solvents. Due to this characteristic, usually a two-step procedure is required to synthesize cellulose-based hydrogels. The first step involves the dissolution of cellulose, and the second step involves cross-linking (gelation). Different polar solvents (*N*-methylmorpholine oxide, triethylammonium chloride/dimethylsulfoxide, lithium chloride/dimethylacetamide, paraformaldehyde/

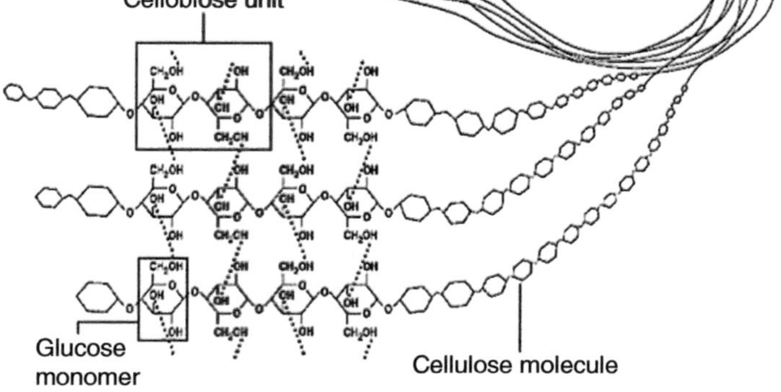

Figure 4.3 General structure of cellulose. *Source:* Adapted from https://www.onlinetextileacademy.com/ last accessed on 16 September 2025.

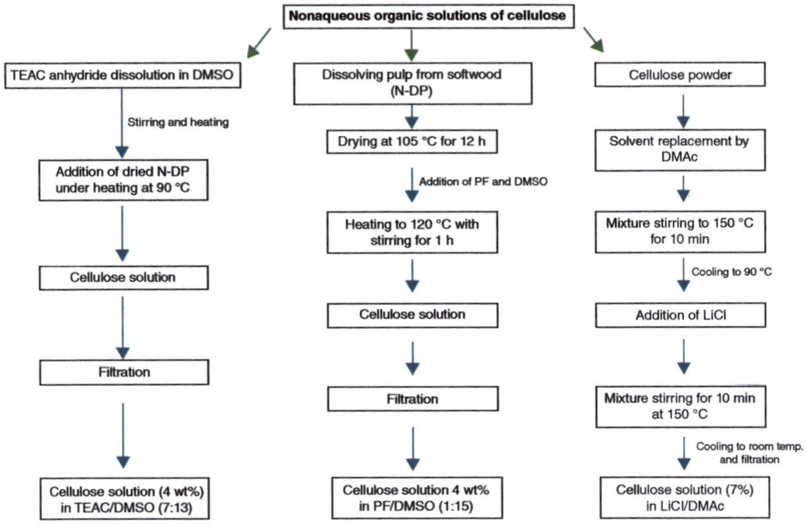

Figure 4.4 Synthesis of three different types of cellulose solutions.

dimethylsulfoxide, tetrabutylammonium fluoride/dimethylsulfoxide, triethylammonium chloride/dimethylsulfoxide, tetrabutylammonium fluoride/dimethylsulfoxide, etc.), ionic liquids (e.g. imidazolium, ammonium, pyridinium, or phosphonium cation coupled with a strong basic, hydrogen bond–accepting anion), deep eutectic solvents (e.g. combination of choline chloride with urea or thiourea), and alkali or alkali/(thio)urea aqueous systems have been developed for the dissolution of cellulose [9].

Cotton-based hydrogels can be synthesized by different cross-linking (chemical and physical) and processing techniques (solution casting, electrospinning, etc.). In a previous study, transparent cellulose-based hydrogel was synthesized by physical cross-linking. In the first step, cellulose solution was prepared using three different nonaqueous solvents [lithium chloride/dimethylacetamide (LiCl/DMAc), paraformaldehyde/dimethyl sulfoxide (PF/DMSO), and triethylammonium chloride/dimethyl sulfoxide (TEAC/DMSO)] and viscose for aqueous cellulose solution (9.5 wt%; NaOH, 6.0 wt%). Two different cellulose sources (cellulose powder and dissolved pulp from soft wood) were used for preparing cellulose solutions using nonaqueous solvents. The stepwise preparation of cellulose solution from three different nonaqueous solvents is presented below in Figure 4.4.

The synthesized cellulose solutions have been converted to hydrogels by casting on chromatographic plates. The procedure for hydrogel synthesis from aqueous and nonaqueous cellulose solutions is slightly different. The aqueous solution hydrogels were prepared by adding HCl in casting bath, along with water-miscible organic solvents, and the synthesized hydrogel was desulfurized and bleached before final washing (Figure 4.5).

Sol–gel technique can be applied to prepare cellulose hydrogel. In a previous study, cellulose hydrogel was prepared using 6% aqueous solution of microcrystalline

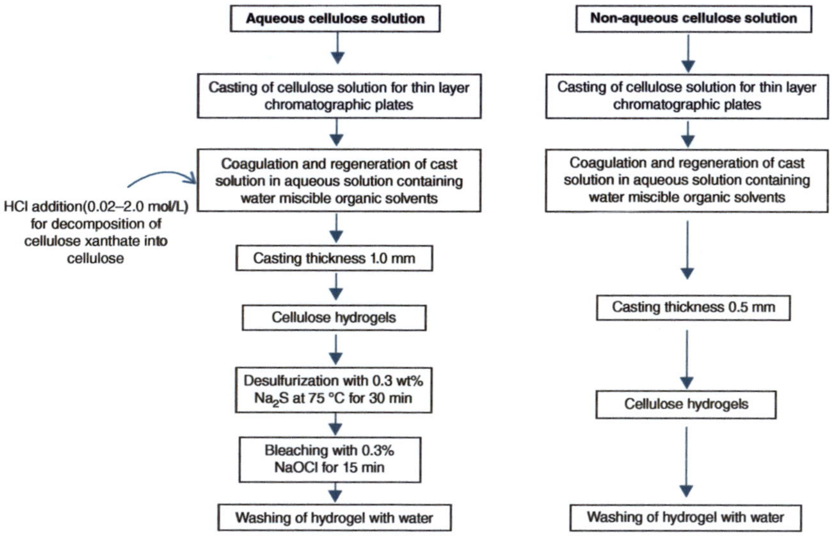

Figure 4.5 Synthesis of cellulose hydrogels using cellulose solutions (aqueous and nonaqueous).

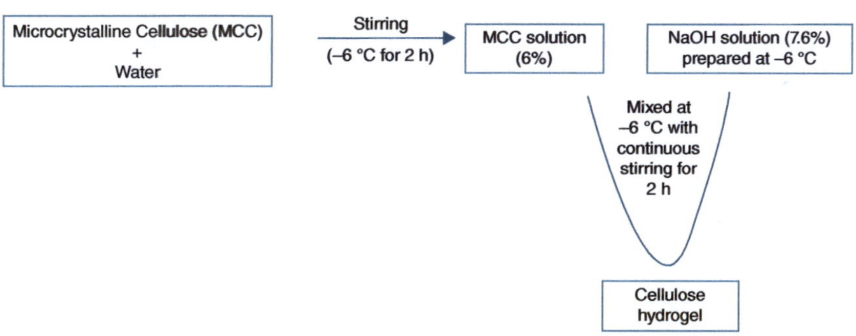

Figure 4.6 Synthesis of cellulose hydrogel from microcrystalline cellulose.

cellulose (MCC) solution and 7.6% solution of NaOH with continuous stirring for 2 hours at −6 °C (Figure 4.6). The prepared gel was applied on cotton nonwoven fabric by dipping for 1 hour, followed by oven drying at 50 °C for 2 hours [10].

An attempt was made by researchers to synthesize cotton-based hydrogels using succinic anhydride (SA) in the presence of 4-dimethylaminopyridine (DMAP) as the esterification catalyst (Figure 4.7). Lithium chloride (LiCl) and *N*-methyl-2-pyrrolidinone (NMP) were used as solvent, followed by NaOH neutralization. Precipitates were filtered, dried under reduced pressure, and sieved to get hydrogel in the form of off-white granules [11].

Chemical cross-linking can be achieved using different cross-linkers, for example, epichlorohydrin (ECH), citric acid, and genipin. In such a reported study, cellulose powder and sodium carboxymethylcellulose (CMC) were used for hydrogel

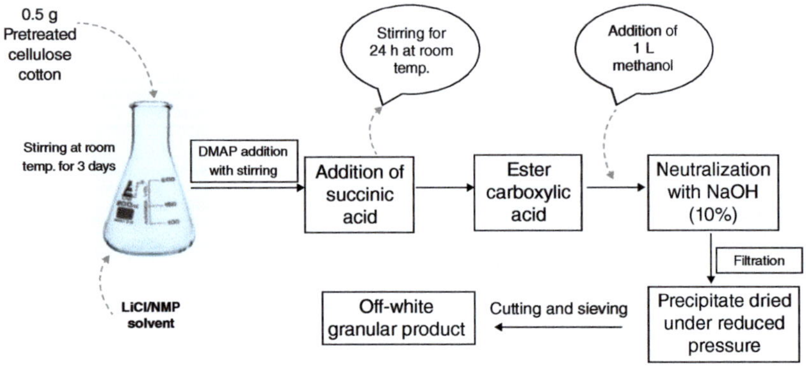

Figure 4.7 Synthesis of cotton-based hydrogel using succinic acid.

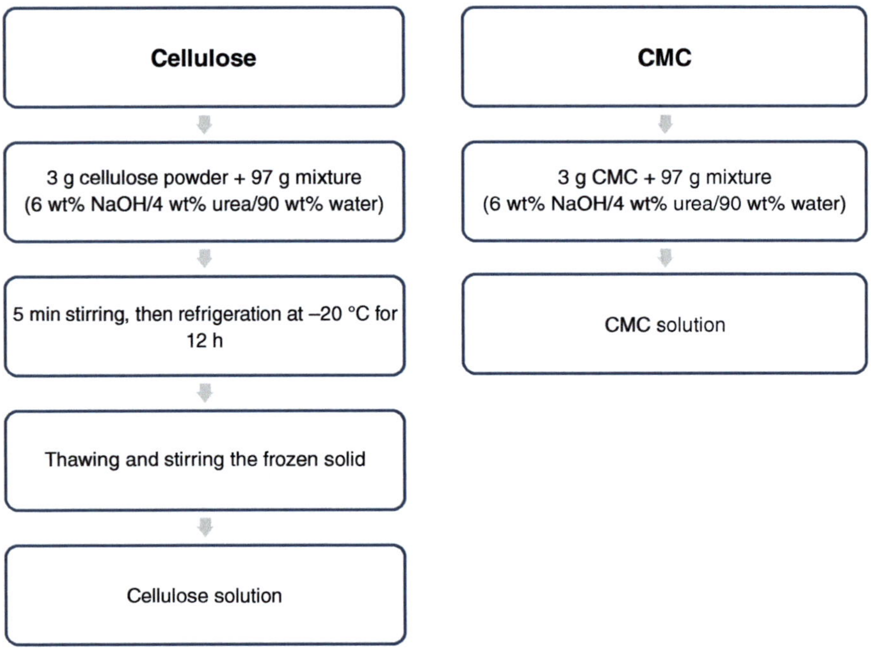

Figure 4.8 Preparation of cellulose and CMC solutions.

synthesis [12]. In the first step, 3 wt% cellulose solution was prepared by dispersing 3 g of cellulose powder in a 97 g mixture (6 wt% NaOH/4 wt% urea/90 wt% water). The mixture was stirred for 5 minutes and stored at −20 °C for 12 hours. The transparent cellulose solution was obtained by thawing and stirring the frozen solid at room temperature. A 3 wt% CMC solution was also obtained by dissolving CMC in the same solvent (Figure 4.8).

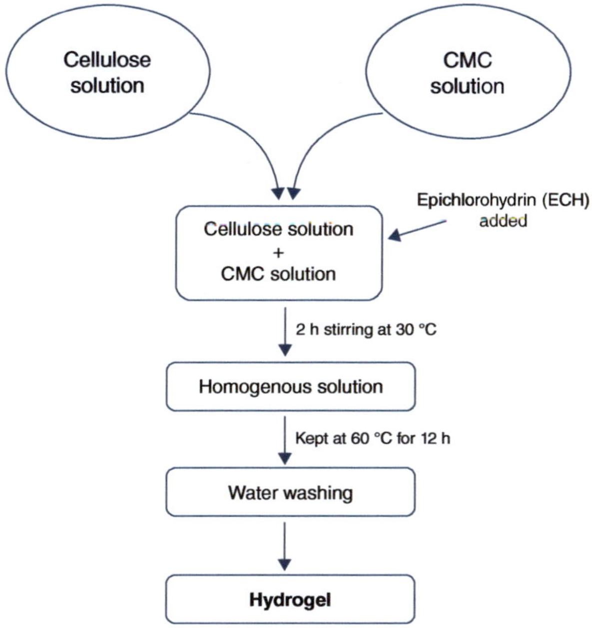

Figure 4.9 Synthesis of hydrogel from cellulose and CMC solutions.

Both the synthesized solutions (cellulose and CMC) were mixed in different ratios (5:5, 6:4, 7:3, 8:2, and 9:1) by weight. ECH was added in the mixture as a cross-linking agent, and the mixture was stirred for 2 hours at 30 °C to obtain a homogeneous solution (Figure 4.9). The hydrogels were obtained by keeping solutions at 60 °C for 12 hours followed by water washing [12].

For the preparation of aerogels, hydrogels were subjected to freeze-drying or supercritical drying. Supercritical drying involves replacing the liquid part of gel with a supercritical fluid (SCF) (e.g. supercritical carbon dioxide). The fluid is then brought to a gaseous state without causing the gel structure to collapse, while in the case of freeze-drying, the hydrogel is frozen, and then, the ice is removed by sublimation (transitioning directly from solid to gas) under vacuum.

Chemically, cellulose has a stable structure due to which there is a need to process the cellulose and convert it into elementary fibrils; these cellulosic nanofibrils can be converted into open porous filaments from which aerogels can be synthesized by freeze-drying and supercritical drying methods for textile applications. The first cellulose aerogel was synthesized by Jin et al. in 2004 in which cellulose powder was dispersed in a saturated solution of calcium thiocyanate (aq.$Ca[SCN]_2$). The solution was heated to 120–140 °C until a transparent solution was obtained, which was spread onto a glass plate to form a layer of 1.00 mm thickness. The resultant hydrogel was extracted using methanol and washed. The gel was subjected to drying using three different methods: regular freeze-drying using conventional freeze-dryer; solvent exchange drying in which water was first replaced by methanol, which was later replaced by *t*-butyl alcohol and subjected to regular freeze-drying; and

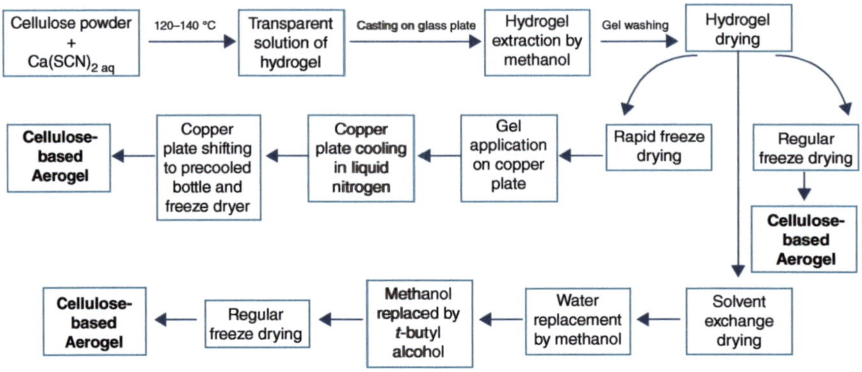

Figure 4.10 Preparation of cellulose-based aerogels.

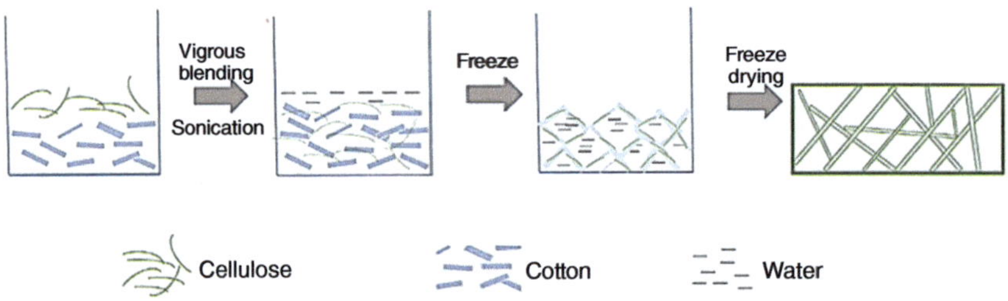

Figure 4.11 Synthesis of aerogel from pure cotton and cotton cellulose. *Source:* Cheng et al. 2017 / with permission of Elsevier.

rapid freeze-drying in which a copper plate was cooled in liquid nitrogen on which cellulosic gel was applied. The plate with the gel was transferred to a precooled glass bottle that was shifted to a freeze-dryer to get the aerogel (Figure 4.10). The synthesized aerogel was characterized, and best results were achieved from the solvent exchange drying method [13].

In another study, pure cotton and cotton cellulose are mixed and used for aerogel synthesis (Figure 4.11). For this purpose, cellulose fibers and small strands of cotton were mixed in deionized water (200 mL). The mixture was then homogenized for 15 minutes. After homogenization, polyamide-epichlorohydrin (PAE) solution (66.6 μL) was added, and the mixture was sonicated at 140 W for 5 minutes. The dispersion was kept at −18 °C for 24 hours and later vacuum dried at −98 °C for 96 hours to get the aerogel. Finally, the aerogel was cured for 3 hours at 120 °C [14].

4.2.2 Synthesis of Hydrogels and Aerogels from Silk

Silk is a potential natural fiber that is obtained from silkworms. It is mainly composed of silk fibroin (SF) [15]. SF depicts high strength due to its highly oriented crystalline phase. The amorphous phase of SF is responsible for energy absorption.

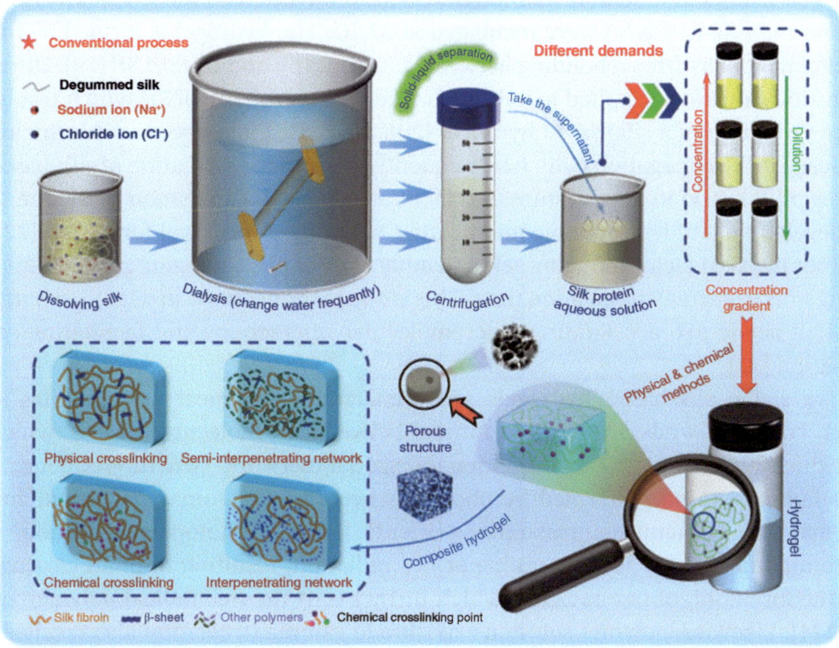

Figure 4.12 Conventional procedure for preparing silk protein hydrogel. *Source:* Liu et al. 2022 / Royal Society of Chemistry.

Hydrogels and aerogels can be made from SF with some attractive features for industrial applications such as high biocompatibility, safe nature, and coupling capability with other materials for desired applications [15]. SF is one of the least thermally conductive materials [16], which makes it a potential candidate for thermal management textiles. For thermal management, usually indoor heating systems are used, which consume approximately 47% of energy and also results in increased global warming due to greenhouse gas emissions [17].

Similar to cellulose-based hydrogels, silk-based hydrogels are also classified as physical and chemical gels, which depend on the nature of cross-linking in these gels. As a starting step, a protein solution is prepared for synthesizing silk-based hydrogels. For this purpose, conventionally, cocoons or silk is cut and degummed and then dissolved using different dissolution systems, for example, lithium bromide. The dissolved silk is then dialyzed with deionized water and concentrated to obtain a stock solution of silk protein aqueous solution [18] (Figure 4.12).

In physical gels, SFs are combined to form a hydrogel through physical interactions, for example, hydrogen bonding, electrostatic interaction, and chain winding. Unlike chemical gels, mild reaction conditions are required for the synthesis of physical hydrogels from silk. Change of physical factors such as pH, temperature, electric field, and the use of surfactants/organic solvents can transform the random coil structure of SF to a lower energy β-sheet conformation, which upon further aggregation transforms into silk-based hydrogels. One of the simplest physical methods of silk-based hydrogel synthesis is self-assembly. A group of researchers synthesized hydrogels using self-assembly method in which *Antheraea assama* SF (3% w/v) and *Bombyx mori* SF (3% w/v) were incubated at 37 °C. The hydrogel is formed by the development of hydrogen bonding between *A. assama* SF and *B. mori* SF [19]. Ultrasonication can also be applied to induce change in the structure of SF, which results in the formation of a silk-based hydrogel [20]. Similarly, high-speed vortex mixing also leads to SF aggregation, which subsequently leads to the formation of silk-based hydrogel [21]. Rise in temperature also facilitates the self-aggregation of SF due to the development of hydrophobic interactions, resulting in hydrogel formation [22]. Another physical factor inducing self-assembly of SF is pH. Changing pH by adding HCl results in decreasing gelation time. This is due to the decrease in negative charge on SF at acidic pH, accelerating macromolecular interactions and facilitating gel formation [23].

More stable silk-based hydrogels are prepared by chemical cross-linking in which covalent bonds are initially formed between SF molecular chains, which subsequently results in the formation of spatial network structure. Chemical cross-linking can be performed by photopolymerization, gamma irradiation, and enzymatic and chemical cross-linking agents [15]. In photopolymerization, a photoinitiator usually absorbs visible or UV radiation, resulting in bond cleavage and the formation of free radicals, which further start the free radical polymerization [24]. Likewise, polymer solution exposure to gamma irradiation also results in the formation of free radicals that reorganize to form covalent bonds, and further entanglement results in hydrogel formation [15]. The chemical cross-linking agents, for example, glutaraldehyde (GA), carbodiimide, and genipin can be used for hydrogel formation. These cross-linking agents react with different reactive groups (hydroxyl, amine, and carboxylic groups) present in silk protein molecules and subsequent cross-linking results in a network structure for hydrogel synthesis.

A group of researchers used SF for preparing silk-based hydrogels using three different physical methods, including electrogelation, vortex shearing, and cooling, and the synthesized gels were chemically cross-linked with GA using 20% w/w GA solution. The cross-linking reaction was completed in 20 minutes at room temperature [25]. The stepwise synthesis of silk-based hydrogel is presented below in Figure 4.13.

Yang and coworkers attempted to take advantage of the attractive properties of aerogel fibers in preparing wearable thermal materials. They first prepared hollow fibers from cellulose acetate (CA) and polyacrylic acid (PAA), followed by injection of SF aerogel in the hollow fibers, which were freeze-dried and characterized for application in wearable thermal textiles [16]. The synthesis steps for the preparation of CA/PAA hollow fibers can be seen below (Figure 4.14).

For the synthesis of SF aerogel, the first step is degumming of raw silk using 8 mol/L urea powder solution (liquor/silk ratio 30:1). The degummed silk is then reacted with a ternary solvent ($CaCl_2/C_2H_5OH/H_2O$) with molar ratio of 1:2:8 at 80 °C for 2 hours. The SF solution is then dialyzed for 72 hours to remove calcium chloride and ethanol. During dialysis process, the water was changed after every 4 hours, and the conductivity of dialysis water was tested to achieve conductance less than 0.8 μS/cm. The resultant SF solution was diluted to 1.4 wt% (Figure 4.15).

The next step was the synthesis of SF aerogel fibers (Figure 4.16).

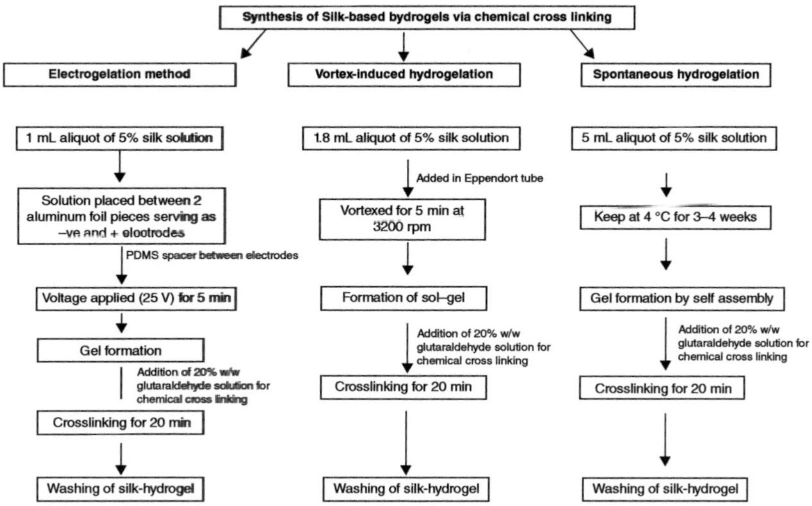

Figure 4.13 Synthesis of silk-based hydrogel by chemical cross-linking.

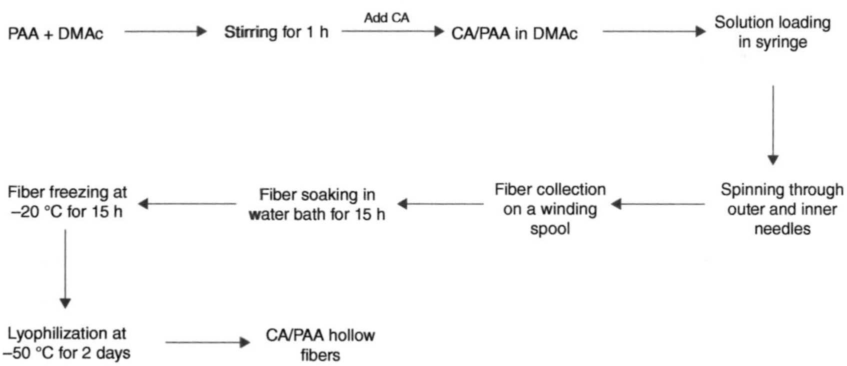

Figure 4.14 Stepwise procedure for the synthesis of CA/PAA hollow fibers.

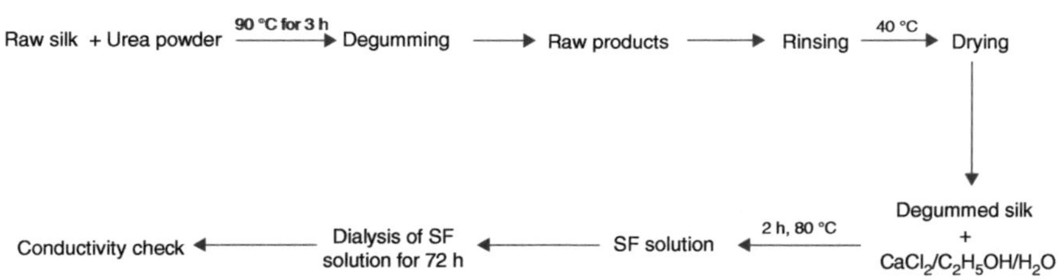

Figure 4.15 Stepwise synthesis of SF solution for injection into CA/PAA hollow fibers.

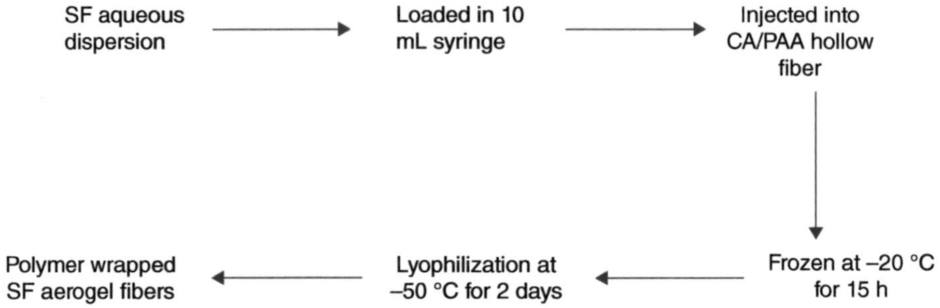

Figure 4.16 Synthesis of SF aerogel fibers.

4.2.3 Synthesis from Wool

Wool is a fibrous material mainly made up of three-dimensional mesh structure of keratin protein (95%) with potential textile applications (rugs, garments, etc.) [26]. Wool keratin possesses a complex structure with approximately 4–8 wt% sulfur. The structure and strength of wool fibers is mainly due to the presence of disulfide bonds in cystine amino acids. Keratin extraction from wool is a challenging step because of its stable structure and resistance to various chemical, biological, and physical treatments (Figure 4.17). It is usually performed using different chemical (reduction, oxidation, hydrolysis, and ionic liquid), enzymatic/microbial, and microwave irradiation [27] and supercritical water/steam treatments [28]. Reduction is usually performed using thiols, for example, mercaptoethanol, by breaking the disulfide bond (R-S-S-R) in cystine and the formation of cysteine (R–S–H). Although it is an efficient method that maintains the keratin chain structure, mercaptoethanol is an expensive chemical with harmful effects on the environment [29]. An economically viable chemical for keratin extraction from wool is sodium sulfide, which is responsible for sulfitolysis through the formation of cysteine (R–S–H) and cysteine-sulfonate

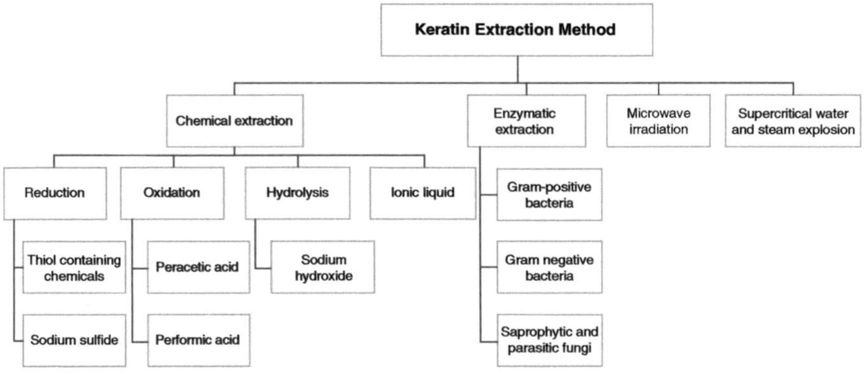

Figure 4.17 Different methods for keratin extraction from wool.

(R–S–S–O_3H), both of which are responsible for keratin solubilization. Oxidation is usually adopted for keratin extraction from wool by peracetic acid or performic acids, which form sulfonic acid (RSO_3H), but it is a time-consuming process. In hydrolysis, alkaline chemicals (NaOH, etc.) are usually used for keratin extraction, and acid neutralization is required in subsequent steps. This method damages the primary chain keratin structure [30]. Ionic liquids can also be used for keratin extraction, but the reaction is to be performed under nitrogen environment due to the requirement of precise temperature control [31]. The studies revealed that for keratin extraction from wool, chloride-containing ionic liquids are more efficient compared with the other tested anions (BF_4, PF_6, and Br), and dissolved keratin is separated using precipitation with water, methanol, or ethanol [32].

The microwave irradiation also assists keratin extraction from wool. It usually functions as a homogenous source of heating, which reduces process time, but this method usually results in significant cystine losses. In such a study, wool was subjected to microwave treatment for 90 minutes, and results revealed that the cystine concentration was reduced from 9.41 mol% in wool to approximately 0.5 mol% in the extracted keratin sample [26]. In steam explosion process, wool is treated with saturated steam (1:3 solid–liquid ratio) at 220 °C for 10 minutes with pressure. Although intense temperature conditions are applied in this method, the wool fragments remain still recognizable. Pure keratin can be obtained using centrifugation of the supernatant [28].

In a previous study, cellulose keratin composite hydrogel was synthesized by dissolving cotton pulp cellulose and wool keratin in [DBNH][Lev] which is a protic ionic liquid (PIL). The resultant solution was stirred for 1 hour at 120 °C (Figure 4.18). The homogenous solution was poured in petri dish and placed for 1 day to get the gel, which is immersed in ethanol to remove PIL [33].

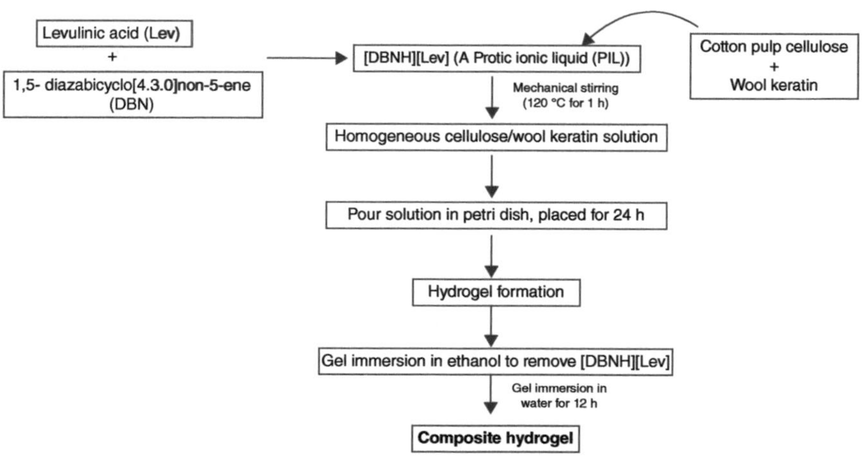

Figure 4.18 Synthesis of cellulose keratin hydrogel using PIL.

In a study, wool waste was used to synthesize aerogels. This first step was making a solution of polyvinyl alcohol (PVA) in water. Wool fibers were then carefully added in PVA solution ensuring homogenous spreading of fibers. The mixture was subjected to freezing at −18 °C for 8 hours, followed by freeze-drying at −50 °C for 48 hours at 0.1 mbar to get the aerogel [34].

4.3 Synthesis Techniques for Hydrogels and Aerogels from Synthetic Fibers

4.3.1 Synthesis from Polyester

Polyesters are synthetic polymers with versatile chemistry, which make them biocompatible and biodegradable. These characteristics provide an opportunity to exploit these polymers for hydrogel and aerogel synthesis. Polyester-based hydrogels combine the advantages of synthetic polyesters with the hydration capacity of hydrogels, offering a versatile platform for advanced applications in multiple fields [35], and can be synthesized using different methods. Generally, polyester-based precursors are cross-linked using cross-linking agents/irradiation/other techniques, which result in the formation of a three-dimensional network structure constituting hydrogels with tunable properties [35].

Among different types of polyesters available, polylactic acid (PLA), polyglycolic acid (PGA), and poly ε-caprolactone (PCL) can be used for hydrogel synthesis through chemical modification/copolymerization with hydrophilic monomers. Recently, a novel polyester was synthesized (polyoligoglycerol maleate), which is then converted into hydrogel [36]. For hydrogel synthesis, 1 g of polyoligoglycerol maleate was combined with citric acid using anhydrous tin (II) chloride as the catalyst (2 mol% of carboxylic groups of citric acid). The reaction was conducted in an oil bath at 150 °C for 3 hours under nitrogen atmosphere. The system pressure was then reduced to 100 mbar, and hydrogel was obtained with approximately 80% yield (Figure 4.19) [36].

PLA is also a polyester that can also be used for hydrogel synthesis using chemical and physical cross-linking methods. In chemical cross-linking, the PLA polymer is cross-linked with other polymer chains by the formation of covalent bonds. Different cross-linking agents have been reported for hydrogel synthesis from PLA using chemical cross-linkers, for example, polyethylene glycol (PEG), lithium diisopropyl amide (LDA), and *N*,*N*-methylene bisacrylamide [37]. Chemical cross-linking can also be achieved by free radical polymerization in which potassium persulphate ($K_2S_2O_8$) is usually used as an initiator and *N*,*N*-methylene bisacrylamide (MBS) is used as a cross-linker. In a study, dextron was linked with PLA [38]. Initially, 1 g of dextran is mixed in DMSO (40 mL) in a three-neck-round bottom flask placed in an oil bath under continuous magnetic stirring (Figure 4.20). The temperature was maintained at 80 °C. The reaction was conducted in nitrogen environment. KPS and MBA solutions were separately prepared in DMSO. To the dextran solution, KPS solution was first added, followed by MBA solution, under constant stirring

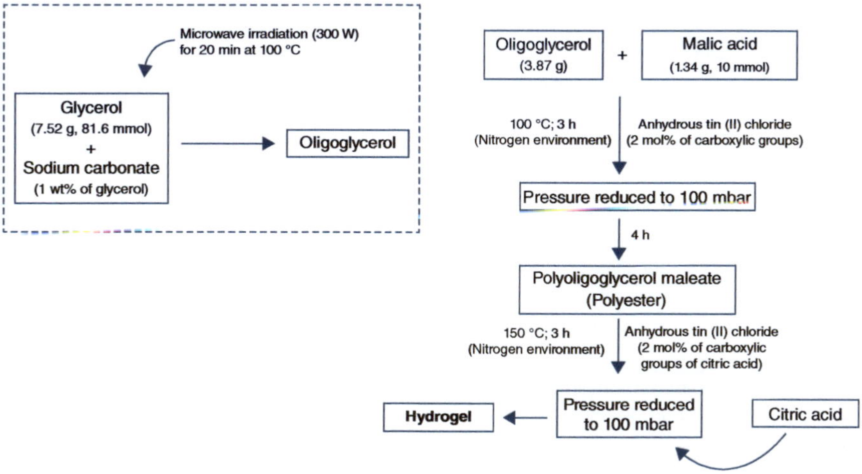

Figure 4.19 Polyester (polyoligoglycerol maleate) based–hydrogel synthesis scheme.

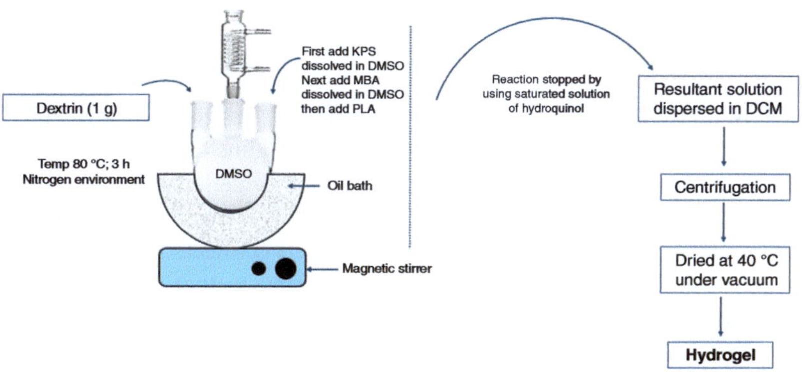

Figure 4.20 Stepwise synthesis of PLA–dextrin hydrogel.

(400 rpm). PLA was also dissolved in DMSO and added in the reaction mixture. The reaction conditions were kept same (80 °C, 400 rpm) for 3 hours. The reaction was topped by the addition of saturated solution of hydroquinol, and the final mixture was dispersed in DCM. The synthesized hydrogel was centrifuged and dried at 40 °C [38].

Similarly, cross-linking can also be achieved through photopolymerization, in which UV or visible light act as an initiator and convert light-sensitive monomers into free radicals, which subsequently result in cross-linked hydrogel formation.

The composites of PLA can be synthesized with other polymers to form hydrogels by grafting technique in which one polymer is grafted onto the other polymer. The resultant product properties are dictated by the nature of polymer grafted and method of grafting. In one such study, PLA was grafted onto chitosan (CS) using

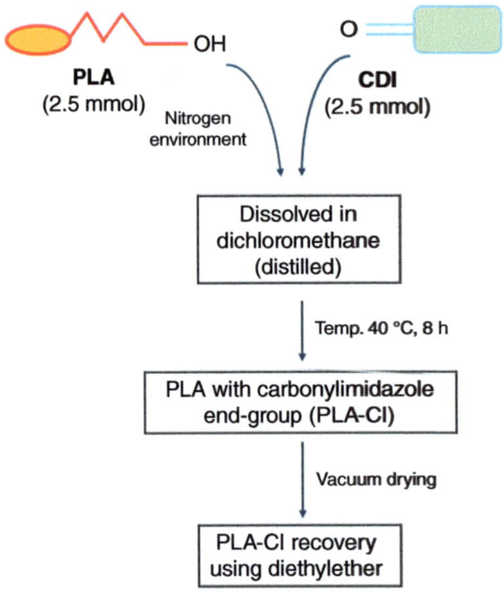

Figure 4.21 Synthesis of PLA-CI (precursor of PLA).

N,*N*-carbonyldiimidazole (CDI) as the coupling agent [39]. In the first step, PLA precursor (PLA-CI) was synthesized using PLA and CDI (Figure 4.21).

In the next step, CS (9 mmol), PLA-CI (0.3 mmol), and 4-dimethylaminopyridine (DMAP) (2.25 mmol) were added in a flask (25 mL) under nitrogen environment, followed by the addition of 15 mL of DMSO. The flask was connected to a Schlenk line to remove oxygen using the exhausting–refilling process that was repeated five times and stirred at 40 °C for 48 hours. After cooling to room temperature, the initial product was precipitated using isopropyl alcohol. The precipitate was vacuum dried, dialyzed in distilled water, and centrifuged to remove unreacted PLA precursor. The supernatant was then lyophilized to get the CS-g-PLA copolymer. Here, two types of grafted copolymers (CS-g-PLLA and CS-g-PDLA) were synthesized following the same procedure using PLLA (from L-lactide) and PDLA (using D-lactide). The gel synthesis was conducted by separately dissolving CS-g-PLLA and CS-g-PDLA copolymers in deionized water (30% solutions), followed by mixing under stirring. The gel formation was obtained after proceeding the reaction at 37 °C for 24 hours [39].

Aerogels can be synthesized using polyesters. In a study, an ultralight aerogel was synthesized using skin core structure polyester and natural fibers from kapok plant (Figure 4.22). The cost-effective aerogel synthesis was performed using physical bonding and freeze-drying process [40]. In the first step, homogeneous suspensions of kapok fibers and chopped polyester fibers were prepared in water at ambient

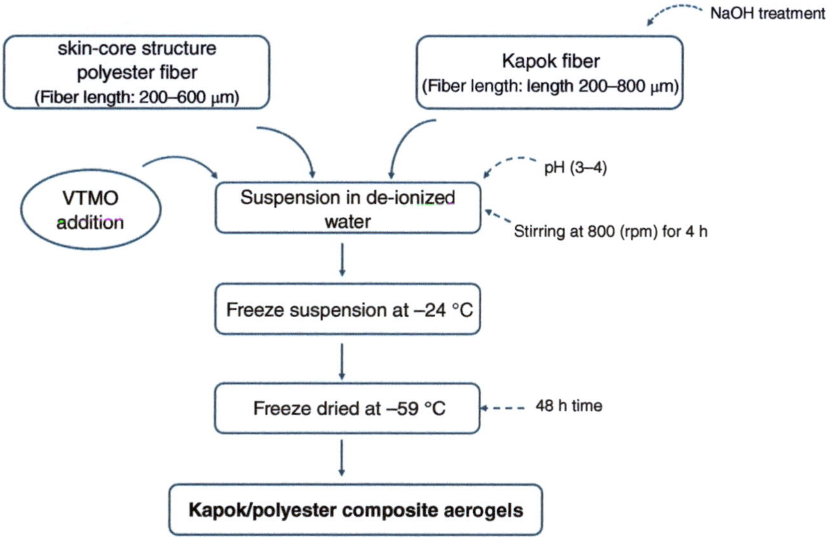

Figure 4.22 Synthesis of kapok/polyester composite aerogels.

temperature. The pH of the suspension was adjusted to 3–4 using acetic acid to hydrolyze vinyltrimethoxysilane (VTMO), which was added to the suspension as a silane modifier. The silanol groups formed by the hydrolysis of VTMO subsequently reacted with the kapok fiber. The suspension was frozen (−24 °C) and freeze-dried (−59 °C) for 48 hours to fabricate kapok/polyester composite aerogels, which were then oven-dried (170 °C) for 30 minutes [40].

In another attempt, waste PET plastic was used to synthesize aerogel with high fire-retardant properties. For this purpose, PET waste bottles were ground into powder. A solvent was prepared by mixing phenol (C_6H_5OH) and 1,1,2,2-perchloroethylene ($C_2H_2Cl_4$) (35 mL each), and PET waste plastic powder (3.17 g) was dissolved in the prepared solvent (50 mL). The mixture was sealed and heated at 68 °C for 2–3 hours. Aluminum isopropoxide ($C_9H_{21}AlO_3$) was used as a catalyst. Aluminum dihydrogen phosphate (AHP) was added into the reaction mixture. The mixture was sealed, and heating was continued at 95–97 °C for 7–9 hours with vigorous stirring. After completion of reaction, the mixture turned white, which was poured into a centrifugal tube (6 cm) for freeze-drying (−48 °C) to obtain the aerogel [41].

4.3.2 Acrylic Fiber-Based Hydrogels

Acrylic fiber is a synthetic fiber with 85% acrylonitrile component. Similar to wool, it is a warm fiber with a soft texture. It can be dyed vibrant colors and

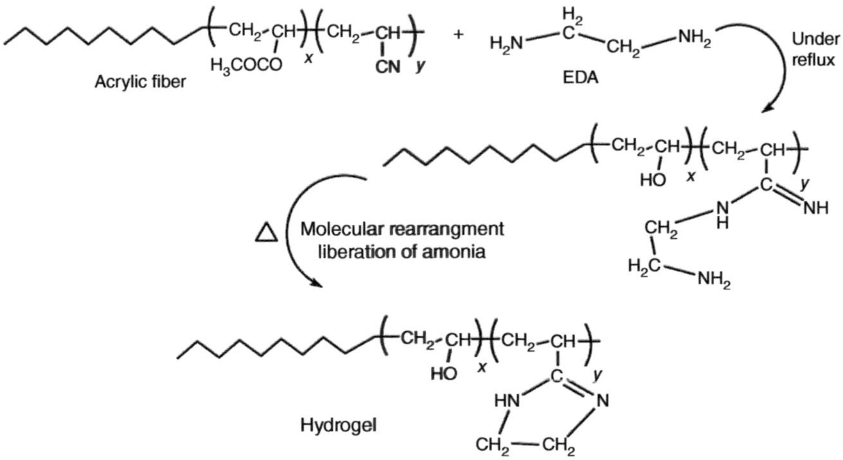

Figure 4.23 Reaction pathway showing the synthesis of Ac-EDA hydrogel. *Source:* Aniagor et al. 2022 / with permission of Elsevier.

is durable in nature. In a previous study, acrylic fiber was used to synthesize acrylic-based hydrogel by the reaction of acrylic fiber (5 g) with a binary mixture of 1,2-diaminoethane (EDA) and water with an equal ratio under highly alkaline pH conditions. The reactants were added in a round bottom flask (1 L) and refluxed for 8 hours at 70 °C (Figure 4.23). The contents were cooled and neutralized with methanol (acidified) to get pH 8.0 at ambient temperature with mechanical stirring. The synthesized hydrogel (Ac-EDA) was filtered off and oven-dried at 80 °C for 5 hours [42].

A composite aerogel was synthesized using polyurethane and polyacrylate through the free radical polymerization technique. In this study, two acrylate-terminated dendritic monomers were initially synthesized by the reaction of pentaerythritoltriacrylate (PETA) with two different triisocyanates [hexamethylene diisocyanate trimer (Desmodur N3300) and triphenylmethane-4,40,4 (TIPM)]. In this step, dibutyltindilaurate (DBTDL) was used as the catalyst. The resultant reaction mixture was stirred at room temperature for 30 minutes, and the solvent was removed using vacuum drying. In the next step, mixture contents were redissolved in dichloromethane, followed by the addition of hexane, which resulted in the formation of aL-Ac and aR-Ac monomers (Figure 4.24).

The synthesized monomers were polymerized via free radical polymerization and gel formation was induced by 2,2′-azobis(isobutyronitrile) (AIBN). The sol obtained was stirred at room temperature for 15 minutes under an Ar environment, which was later transferred to molds and kept for 24 hours at 60 °C for gelation and aging (Figure 4.25). The resultant gels were solvent exchanged with acetone and dried to aerogels using SCF CO_2 [43].

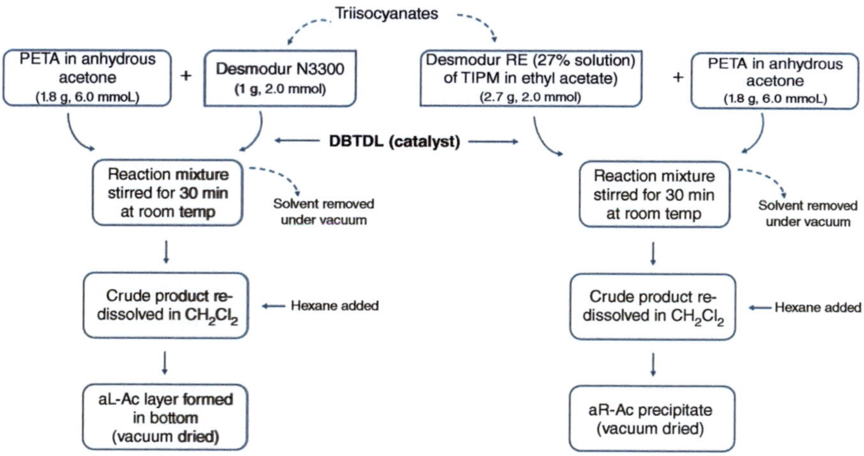

Figure 4.24 Synthesis of monomers from two different triisocyanates.

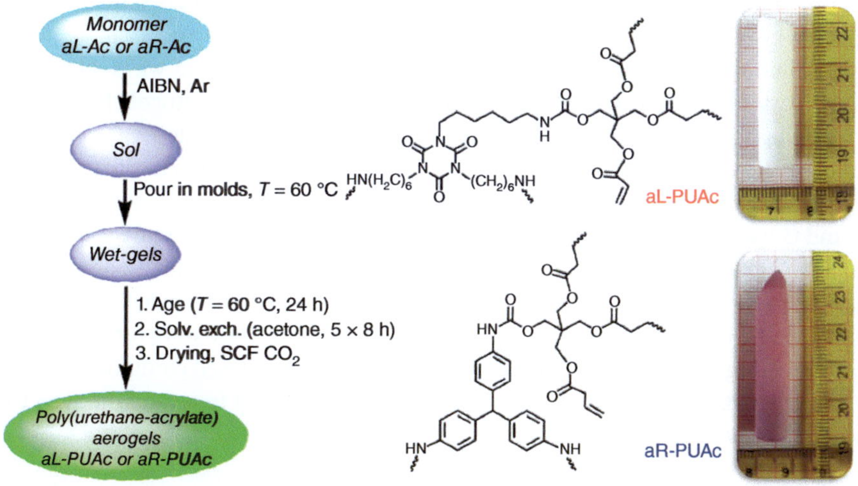

Figure 4.25 Synthesis of poly(urethane acrylate) aerogels. *Source:* Papastergiou et al. (2018) / MDPI / CC BY 4.0.

4.4 Blended-Fiber Hydrogels and Aerogels

Blended fibers are synthesized by combining two or more different types of fibers (natural, synthetic, or both), for example, wool–acrylic blends, polyester–cotton, and nylon–spandex. This is performed to get the required functionality for enhanced performance and cost-effectiveness. The blended fibers can also be used for the synthesis of aerogels and hydrogels, and the resultant products offer innovative solutions by combining high absorbency with lightweight properties, thus enhancing moisture management, thermal regulation, and even drug delivery in wearable fabrics.

In a previous study, two natural fibers (alginate and chitosan) were used for the synthesis of hydrogels. Chitosan was extracted from crab shell and treated with 2% acetic acid for protonation. Sodium alginate was dissolved in deionized water. The aqueous solutions of sodium alginate (anionic) and chitosan (cationic) were mixed in 1 N HCl, which resulted in the formation of precipitates. These precipitates containing gel-like complexes were obtained from the reaction mixture using centrifugation at 1500 rpm [44]. In another study, alginate cotton fibers have been tested for the synthesis of aerogels [45]. The cotton fibers (3 mm in length) were added to the aqueous solution of sodium alginate with shear mixing followed by wet spinning. The alginate fibers were extruded and subjected to cross-linking in a bath containing $CaCl_2$ solution (0.2 M). The fibers were extracted from the cross-linking bath after 15 minutes, rinsed with distilled water, followed by freezing for 12 hours, and subsequently subjected to vacuum freeze-drying (−80 °C) for 8 hours using 0.1 Torr pressure. The synthesized aerogel fibers were characterized and molded in different shapes [45].

In another attempt, textile waste fibers (2–20 mm) from crushed waste clothing (90% polyester and 10% cotton) were used for the synthesis of aerogels [46]. Two cross-linking agents were used in this study: PVA and GA. The first step was the treatment of textile waste fibers with NaOH at 80 °C for 3 hours with a material to liquor ratio of 1:100. The treated textile waste fibers were added to the aqueous solution of PVA and GA with sonication (400 W) at 80 °C for 20 minutes. The mixture was cured in an oven at 80 °C for 3 hours, followed by refrigeration at −18 °C for 12 hours and freeze-drying at −60 °C for 48 hours. These synthesized aerogels were hydrophilic in nature. For the transition from hydrophilic to hydrophobic, these were further modified by reacting with methyltrimethoxysilane (MTMS) in a closed container, which was heated at 70 °C for 3 hours. The product was then placed in a vacuum oven to remove excessive MTMS (Figure 4.26).

4.5 Conclusion

The synthesis of hydrogels and aerogels from natural, synthetic, and blended fibers offers a versatile platform for advanced textile applications. Natural polymers ensure biocompatibility and sustainability, while synthetic polymers improve the mechanical strength of the product. Blended fiber systems offer combined advantages with tailored functionalities. Various fabrication techniques (electrospinning, freeze-drying, sol–gel processes, cross-linking methods, etc.) play an important role in determining the structural and functional characteristics of synthesized hydrogels and aerogels. The synthesis route dictates important properties making them suitable for applications such as wound dressings, smart textiles, moisture management fabrics, and insulation materials. Future research should focus on

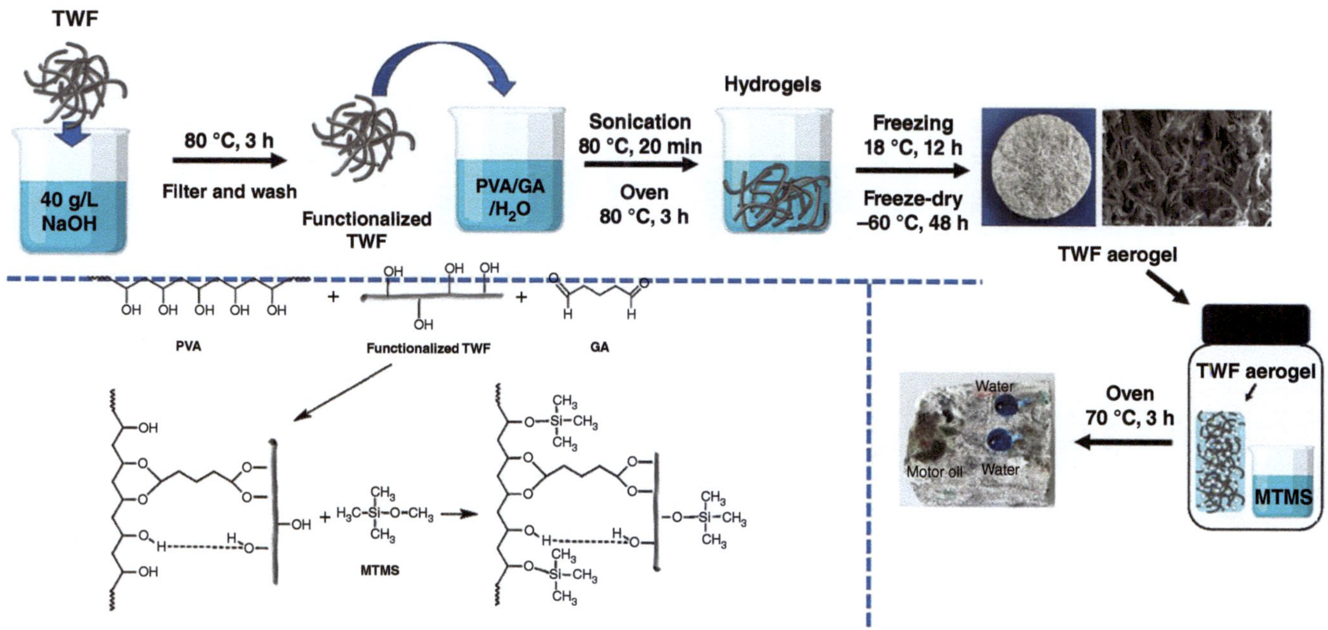

Figure 4.26 Stepwise illustration of aerogel synthesis from textile waste fibers. *Source:* Dong et al. 2022 / MDPI / CC BY 4.0.

cost-effectiveness, and environmental impact so that hydrogel- and aerogel-based textiles can pave the way for innovative, high-performance solutions in the evolving textile industry.

References

1 Ahmed, E.M. (2015). Hydrogel: preparation, characterization, and applications: a review. *Journal of Advanced Research* 6 (2): 105–121.

2 Mondal, M.I.H. (2024). *Smart Textiles from Natural Resources*. Elsevier.

3 Aegerter, M.A., Leventis, N., and Koebel, M.M. (2011). *Aerogels Handbook*. New York, NY: Springer New York.

4 García-González, C.A., Alnaief, M., and Smirnova, I. (2011). Polysaccharide-based aerogels—promising biodegradable carriers for drug delivery systems. *Carbohydrate Polymers* 86 (4): 1425–1438.

5 Maleki, H. (2016). Recent advances in aerogels for environmental remediation applications: a review. *Chemical Engineering Journal* 300: 98–118.

6 Petros, S., Tesfaye, T., and Ayele, M. (2020). A review on gelatin based hydrogels for medical textile applications. *Journal of Engineering* 2020 (1): 8866582.

7 Li, X., Wan, C., Tao, T. et al. (2024). An overview of the development status and applications of cellulose-based functional materials. *Cellulose* 31 (1): 61–99.

8 McNeil, S.J. and Gupta, H. (2022). Emerging applications of aerogels in textiles. *Polymer Testing* 106: 107426.

9 Shen, X., Shamshina, J.L., Berton, P. et al. (2016). Hydrogels based on cellulose and chitin: fabrication, properties, and applications. *Green Chemistry* 18 (1): 53–75.

10 Ahmad, F., Mushtaq, B., Butt, F.A. et al. (2021). Synthesis and characterization of nonwoven cotton-reinforced cellulose hydrogel for wound dressings. *Polymers* 13 (23): 4098.

11 Yoshimura, T., Matsuo, K., and Fujioka, R. (2006). Novel biodegradable superabsorbent hydrogels derived from cotton cellulose and succinic anhydride: synthesis and characterization. *Journal of Applied Polymer Science* 99 (6): 3251–3256.

12 Chang, C., Duan, B., Cai, J., and Zhang, L. (2010). Superabsorbent hydrogels based on cellulose for smart swelling and controllable delivery. *European Polymer Journal* 46 (1): 92–100.

13 Jin, H., Nishiyama, Y., Wada, M., and Kuga, S. (2004). Nanofibrillar cellulose aerogels. *Colloids and Surfaces A: Physicochemical and Engineering Aspects* 240 (1–3): 63–67.

14 Cheng, H., Gu, B., Pennefather, M.P. et al. (2017). Cotton aerogels and cotton-cellulose aerogels from environmental waste for oil spillage cleanup. *Materials & Design* 130: 452–458.

15 Zheng, H. and Zuo, B. (2021). Functional silk fibroin hydrogels: preparation, properties and applications. *Journal of Materials Chemistry B* 9 (5): 1238–1258.

16 Yang, H., Wang, Z., Liu, Z. et al. (2019). Continuous, strong, porous silk firoin-based aerogel fibers toward textile thermal insulation. *Polymers* 11 (11): 1899.

17 Wang, Z., Yang, H., Li, Y., and Zheng, X. (2020). Robust silk fibroin/graphene oxide aerogel fiber for radiative heating textiles. *ACS Applied Materials & Interfaces* 12 (13): 15726–15736.

18 Liu, J., Ge, X., Liu, L. et al. (2022). Challenges and opportunities of silk protein hydrogels in biomedical applications. *Materials Advances* 3 (5): 2291–2308.

19 Chouhan, D., Lohe, T.U., Samudrala, P.K., and Mandal, B.B. (2018). in situ forming injectable silk fibroin hydrogel promotes skin regeneration in full thickness burn wounds. *Advanced Healthcare Materials* 7 (24): e1801092.

20 Gorenkova, N., Osama, I., Seib, F.P., and Carswell, H.V.O. (2019). in vivo evaluation of engineered self-assembling silk fibroin hydrogels after intracerebral injection in a rat stroke model. *ACS Biomaterials Science & Engineering* 5 (2): 859–869.

21 Chen, D., Yin, Z., Wu, F. et al. (2017). Orientational behaviors of silk fibroin hydrogels. *Journal of Applied Polymer Science* 134 (32): 45050.

22 Kim, U.-J., Park, J., Li, C. et al. (2004). Structure and properties of silk hydrogels. *Biomacromolecules* 5 (3): 786–792.

23 Nagarkar, S., Nicolai, T., Chassenieux, C., and Lele, A. (2010). Structure and gelation mechanism of silk hydrogels. *Physical Chemistry Chemical Physics* 12 (15): 3834–3844.

24 Zhou, Y., Liang, K., Zhao, S. et al. (2018). Photopolymerized maleilated chitosan/methacrylated silk fibroin micro/nanocomposite hydrogels as potential scaffolds for cartilage tissue engineering. *International Journal of Biological Macromolecules* 108: 383–390.

25 Lin, Y., Xia, X., Shang, K. et al. (2013). Tuning chemical and physical cross-links in silk electrogels for morphological analysis and mechanical reinforcement. *Biomacromolecules* 14 (8): 2629–2635.

26 Shavandi, A., Silva, T.H., Bekhit, A.A., and Bekhit, A.E.D.A. (2017). Keratin: dissolution, extraction and biomedical application. *Biomaterials Science* 5 (9): 1699–1735.

27 Zoccola, M., Aluigi, A., Patrucco, A. et al. (2012). Microwave-assisted chemical-free hydrolysis of wool keratin. *Textile Research Journal* 82 (19): 2006–2018.

28 Tonin, C., Zoccola, M., Aluigi, A. et al. (2006). Study on the conversion of wool keratin by steam explosion. *Biomacromolecules* 7 (12): 3499–3504.

29 Yamauchi, K., Kusunoki, T., Kohda, A., and Konishi, Y. (1996). Preparation of stable aqueous solution of keratins, and physiochemical and biodegradational properties of films. *Journal of Biomedical Materials Research: An Official Journal* *of The Society for Biomaterials and The Japanese Society for Biomaterials* 31 (4): 439–444.

30 Tsuda, Y. and Nomura, Y. (2014). Properties of alkaline-hydrolyzed waterfowl feather keratin. *Animal Science Journal* 85 (2): 180–185.

31 Ghosh, A., Clerens, S., Deb-Choudhury, S., and Dyer, J.M. (2014). Thermal effects of ionic liquid dissolution on the structures and properties of regenerated wool keratin. *Polymer Degradation and Stability* 108: 108–115.

32 Xie, H., Li, S., and Zhang, S. (2005). Ionic liquids as novel solvents for the dissolution and blending of wool keratin fibers. *Green Chemistry* 7 (8): 606–608.

33 Deng, L., Wang, L., Zhang, L. et al. (2025). Cellulose–wool keratin composite hydrogels as selective support carriers for gold nanoparticles: synthesis and catalytic applications in the reduction of 4-nitrophenol in water. *Langmuir* 41 (18): 11526–11536.

34 Loh, J. et al. (2021). Advanced aerogels from wool waste fibers for oil spill cleaning applications. *Journal of Polymers and the Environment* 1–14.

35 Dhingra, S., Weisel, R.D., and Li, R.K. (2014). Synthesis of aliphatic polyester hydrogel for cardiac tissue engineering. *Cardiac Tissue Engineering: Methods and Protocols* 51–59.

36 Bitelo, B., Petzhold, C.L., Gamba, D., and Daitx, T.S. (2025). Synthesis of sustainable polyester-type hydrogels crosslinked with citric acid for the removal of crystal violet dye. *Journal of Molecular Liquids* 425: 127228.

37 Munim, S.A. and Raza, Z.A. (2019). Poly (lactic acid) based hydrogels: formation, characteristics and biomedical applications. *Journal of Porous Materials* 26 (3): 881–901.

38 Das, D., Das, R., Mandal, J. et al. (2014). Dextrin crosslinked with poly(lactic acid): a novel hydrogel for controlled drug release application. *Journal of Applied Polymer Science* 131 (7): 40039.

39 Hu, Y., Liu, Y., Qi, X. et al. (2012). Novel bioresorbable hydrogels prepared from chitosan-graft-polylactide copolymers. *Polymer International* 61 (1): 74–81.

40 Zhang, H., Chen, Y., Wang, J. et al. (2024). Natural fiber-based multifunctional aerogel with advanced oil absorption property and high thermal insulation performance. *Industrial Crops and Products* 216: 118724.

41 Gao, B., Sun, X., Yao, C., and Mao, L. (2022). A new strategy to chemically transform waste PET plastic into aerogel with high fire resistance and mechanical strength. *Polymer* 254: 125074.

42 Aniagor, C.O., Taha, G.M., Badawy, S.M. et al. (2022). Preparation of a novel acrylic fiber-based hydrogel and its utilization for the removal of aqueous lead ion. *Journal of Materials Research and Technology* 18: 1450–1459.

43 Papastergiou, M., Kanellou, A., Chriti, D. et al. (2018). Poly (urethane-acrylate) aerogels via radical polymerization of dendritic urethane-acrylate monomers. *Materials* 11 (11): 2249.

44 Brovko, O., Palamarchuk, I.A., Val'chuk, N.A. et al. (2017). Gels of sodium alginate–chitosan interpolyelectrolyte complexes. *Russian Journal of Physical Chemistry A* 91: 1580–1585.

45 Azam, F., Ahmad, F., Ahmad, S. et al. (2024). Alginate-cotton blended aerogel fibers: synthesis, characterization, and oil/water separation. *International journal of Environmental Science and Technology* 21 (5): 5065–5078.

46 Dong, C., Hu, Y., Zhu, Y. et al. (2022). Fabrication of textile waste fibers aerogels with excellent oil/organic solvent adsorption and thermal properties. *Gels* 8 (10): 684.

5

Functional Properties of Hydrogels and Aerogels in Textiles

Mohamed Assal[1,2], Ayoub Nadi[2], Abdeslam el Bouari[1], and Omar Cherkaoui[2]

[1]*HASSAN II University of Casablanca, Faculty of Sciences Ben M'Sik, Laboratory of Physical Chemistry, Materials and Catalyst (LCPMC) Chemistry Department, Bd Commandant Driss Al Harti Casablanca, 20670 Morocco*
[2]*Higher School of Textile and Clothing Industries, Laboratory Textile Materials Research (REMTEX), Km 8, El Jadida Road, Laymoune District Casablanca, 20202 Morocco*

5.1 Introduction

Incorporating hydrogels and aerogels into textiles is a remarkable step toward multifunctional materials engineering and is particularly important regarding textile enhancement in terms of functionality, versatility, and sustainability. Such materials provide textiles with a multifunctional set of advanced moisture management [1, 2] and temperature management capabilities, as well as antimicrobial [1, 2], flame-retardant [3–5], and biodegradable [6, 7] properties that stretch the usefulness of textiles.

The major feature of hydrogels is a three-dimensional hydrophilic polymer network that provides these materials with the capability to absorb and retain tremendous amounts of water [5, 7, 8]. This, in turn, is highly advantageous in terms of moisture management, which enhances comfort in the wearing of sportswear, medical textiles, and even clothes used in day-to-day activities. Aerogels, on the contrary, are well known for their ultra-lightweight and highly porous nanostructures that provide high thermal insulatory properties [9–12]. Their use in textile systems is important for the maintenance of body temperature in difficult ambient conditions, which is useful for the technical garments and protective clothing.

This chapter summarizes the review concerning the technological methods related to the incorporation of hydrogels and aerogels into textiles through coating, impregnation, and lamination, among other methods. It discusses the specific mechanical processes and the constituent parameters of temperature and pressure, which have to be controlled to achieve a permanent integration that does not restrict the flexibility and breathability of the textile. Moreover, the chapter describes the physicochemical properties of hydrogels and aerogels that impact their performance, such as the water absorption capacity, thermal conductivity, and stability of the structure.

It discusses how hydrogels allow effective moisture management through water absorption and release, aiding in performance apparel as well as in health care. It also analyzes the aerogels' contribution to the thermal regulation of the temperature by improving the thermal insulation of extreme cold weather and protective garments. Moreover, the chapter looks into other functions of hydrogels and aerogels in addition to thermal control such as being engineered with antimicrobial and antibacterial properties that minimize microbial contamination and enhance hygiene in medical and general textiles. It also discusses the flame-retardant properties that aerogels provide, contributing to the fire safety of protective garments and home textiles. Finally, the chapter looks at how biodegradable hydrogel and aerogel systems were developed in consideration of worldwide initiatives aimed at mitigating the ecological footprint of textile production and waste. By doing so, the chapter hopes to show the extent to which aerogels and hydrogels are revolutionizing the textile industry by providing multifunctional responsive systems to comfort, safety, and sustainability.

5.2 Moisture Management

Developed functional garments for the textile industry, frosted by functional nanomaterials, put modern clothing within the scope of devices capable of controlling heat and moisture for the users' comfort in a variety of conditions. Liu et al. [13] have developed innovative textiles featuring aerogels that function as aerogel-based passive thermal regulators (AT-PTR), which combat the exposure of humans to heat and cold stress in the outdoor environment. This system has a dual-structured design. The heating side, a carbon nanotube-modified cotton fabric serves as the heating side and provides enhanced solar energy harvesting of more than +23.2 °C warming in cold conditions. The cooling side, which is a poly(vinylidene-co-hexafluoropropene) nanofiber worked with silica aerogel (SA), provides subambient cooling of approximately −12.7 °C in the summer under strong sunlight. Field tests conducted across diverse climatic regions and seasons have demonstrated the AT-PTR's capacity for year-round thermal regulation. Notably, this textile maintains excellent wind resistance and moisture permeability, and users can easily switch between heating and cooling functions by simply flipping the material, thereby ensuring continuous thermal comfort in fluctuating outdoor environments.

Yet, as highlighted by Du and Kim [10], research on aerogels in textiles has largely centered on thermal insulation, often overlooking moisture management, a critical factor for thermo-physiological comfort. The intrinsic hydrophobicity and fragility of aerogels may restrict vapor transmission, risking local heat and humidity buildup. Furthermore, Liu et al. [14] reported the development of a temperature-responsive smart textile by grafting an interpenetrating network (IPN) hydrogel composed of sericin (SS), poly(*N*-isopropylacrylamide) (PNIPAAm), and silver nanoparticles (Ag NPs) onto polyester fabric. The modified textile exhibited a lower critical solution temperature (LCST) around 32.9 °C, close to average skin temperature, allowing it to reversibly switch between hydrophilic and hydrophobic states. This temperature sensitivity contributed to improved thermo-physiological comfort because the fabric adapted its moisture wicking behavior depending on ambient conditions.

Some studies directly investigated these issues. Shaid et al. [15] evaluated aerogel-coated fabrics, finding that appropriate coating methods and binders can mitigate the hydrophobic barrier, modestly improving air permeability and moisture transport. In additionally, by integrating phase change materials (PCMs) alongside aerogel coatings, they enhanced thermophysiological comfort, allowing excess body heat and moisture to be absorbed and later released, keeping the microclimate temperature closer to the comfort zone. This hybrid approach helped maintain wearer comfort without sacrificing thermal protection.

Despite these efforts, the authors note that human-wear and manikin tests focused on moisture and comfort remain scarce. Most evaluations rely on machine measurements of thermal and mechanical properties, leaving gaps in understanding real-world comfort and moisture management performance. Furthermore, encapsulation strategies meant to prevent aerogel dust contamination can reduce flexibility and breathability, indirectly impairing moisture regulation.

Unlike them, hydrogels provide dynamic moisture management due to their three-dimensional hydrophilic polymer networks, which can absorb and desorb substantial amounts of water. Liu et al.'s recent work [13] created the innovative aerogel-functionalized textile passive thermal regulator (AT-PTR), designed to provide protection against heat stress and hypothermia during exposure to dynamic outdoor environments. This AT-PTR contains a dual-sided structure: a heating side with carbon nanotube-modified cotton fabric, which has high solar absorption and can elevate local temperature by +23.2 °C, and a cooling side with subambient cooling SA (−12.7 °C) during high solar exposure along with P(VdF-HFP) cooling aerogel nanofibers. Field evaluations conducted during various climates and across seasons confirmed that it can maintain thermal comfort year-round. Moreover, the textile maintains good windproof and moisture-permeable properties, and the thermal effect can easily be reversed by flipping the garment, thus enhancing the practicality of the solution for outdoor wear requiring adaptable thermal regulation. In addition, Larusso et al. [8] proposed an efficient two-step method to graft hydrogel networks onto polyethylene terephthalate (PET) fabrics by using different acrylate-based monomers. Initially, the PET surface was functionalized through an aminolysis reaction with allylamine, introducing reactive double bonds. Subsequently, UV-induced radical photopolymerization in the presence of a cross-linker enabled the formation of grafted hydrogel networks, whose morphology and abrasion resistance varied according to the selected monomer. This approach demonstrates that UV photopolymerization offers a rapid, cost-effective technique to develop technical textiles with tunable surface properties and enhanced functional performance.

In conclusion, while aerogel composites effectively insulate and reduce garment weight, current research identifies the need for better balancing thermal insulation, flexibility, and moisture management to enhance overall wearer comfort. Future studies are encouraged to integrate more comprehensive comfort and moisture tests and explore composite solutions that preserve breathability.

Conversely, hydrogel-functionalized textiles introduce temperature-sensitive, moisture-responsive behavior that adapts to wearer conditions, enhancing wicking and skin comfort. Methods like UV-induced grafting expand the range of hydrogel structures that can be stably integrated into fabrics, improving durability and abrasion resistance.

5.3 Thermal Insulation Properties of Aerogel- and Hydrogel-Based Textiles

Aerogel-based textiles are widely recognized for their exceptional thermal insulation capabilities, primarily attributed to their ultralow thermal conductivity, typically in the range of 0.012–0.018 W/m·K. This remarkable performance results from their highly porous, three-dimensional network that effectively reduces heat transfer through solid–solid pathways and gas convection. Because the nanoporous structure restricts air movement, the effect of thermal convection becomes negligible, while the contribution of thermal radiation under normal environmental conditions is minimal. As a result, the primary factor influencing heat transfer in aerogels is gas conduction, which depends largely on pore size and the free path length of air molecules trapped inside the material.

In textile applications [16], aerogel materials are frequently incorporated into reinforced blanket structures, such as polyester fiber-reinforced SA blankets. These composite materials combine flexibility and hydrophobic properties with very low thermal conductivity, making them highly effective insulators while remaining lightweight and thin. Studies have shown that such aerogel blankets can achieve thermal performance several times higher than traditional insulating materials while also conserving space—an advantage for both clothing and building insulation. Typically, aerogel blankets maintain thermal conductivities ranging from 0.011 to 0.015 W/m·K over a wide temperature span (approximately 200–650 °C). In addition, Chang et al. [17] developed a SA/polyurethane composite aimed at enhancing thermal insulation for textile and flexible applications. SAs synthesized via a coprecursor method showed high porosity (97%) and notable hydrophobicity (contact angle $\approx$130°), outperforming those made by solvent exchange, as confirmed by Fourier Transform Infrared (Spectroscopy) (FTIR) through increased Si–CH_3 surface groups. Incorporating this aerogel into polyurethane matrices prepared with diisocyanate, poly(tetramethylene oxide), and 1,4-butanediol yielded uniform composite films by solvent casting. Among various formulations, the optimal blend (35/65 v/v) achieved low thermal conductivity ($\sim$0.13 W/m·K) while retaining hydrophobicity (contact angle $\approx$5.6°). Of importance, the composites overcame common limitations of pure aerogels, such as brittleness and hygroscopicity, thus offering practical potential as flexible, low conductivity insulating materials. Other studies such as those conducted by Sai et al. [18] had addressed the challenge of aerogels' intrinsic brittleness, which limits their direct use in textile applications such as thermal insulation fabrics. Using a novel approach, researchers developed fibrous silica–bacterial cellulose (BC) composite aerogels by diffusing silica sol into a reshaped BC hydrogel matrix, producing composite wet gel fibers with a nanoscale IPN. Thanks to the secondary shaping process, the quantity of BC nanofibers within the unit volume increased substantially, resulting in aerogel fibers with notable tensile strength (~5.4 MPa). These composite fibers also demonstrated a high specific surface area (~606.9 m^2/g), low density (<0.164 g/cm^3), and remarkable hydrophobicity. Of importance, they provided excellent thermal insulation under both elevated (≈210 °C) and subzero (≈ −72 °C) temperatures and exhibited enhanced thermal stability with decomposition temperatures approximately 330 °C—surpassing those of conventional natural polymer fibers. This innovative fabrication method offers promising prospects for producing flexible, high-performance aerogel fibers suitable for wearable thermal insulation applications. And He et al. [19] developed an environmentally friendly thermal insulation coating formulated by combining styrene-acrylic latex as the film-forming matrix with fillers, including SA, hollow glass beads, rutile titanium dioxide, and sericite powder. Prepared using water as the sole solvent—without additional organic solvents—the coating is described as nontoxic and safe for both environment and human health. Performance tests on the dried coating (≈4 mm thick) revealed excellent thermal insulation properties: when the heated surface temperature reached approximately 351 °C, the reverse side temperature remained approximately 188 °C, producing a temperature difference of approximately 163 °C. The incorporation of SA not only enhanced thermal insulation but also contributed to improved heat resistance, water resistance, corrosion resistance, and abrasion resistance. Of importance, the findings indicated that the beneficial effects of SA depended on its content; excessive amounts could diminish performance, suggesting that the SA concentration must be carefully optimized. Despite these advantages, conventional aerogels are naturally brittle and hydrophobic, which can affect flexibility and wearer comfort. Recent innovations have included reinforced composite aerogel fibers [13, 20, 21] and aerogel-coated nonwovens to balance insulation performance with mechanical durability [22–24].

Hydrogel-based textiles, by contrast, do not inherently possess the same level of passive thermal insulation as aerogels due to their higher water content and denser polymeric network, which can increase thermal conductivity (typically >0.2 W/m·K). However, hydrogels contribute to active thermal regulation through evaporative cooling: when environmental or body temperature rises, the absorbed water in the hydrogel gradually evaporates, drawing heat away from the wearer's skin. This process effectively cools the microclimate near the body, making it particularly useful in sportswear and summer garments [25]. Moreover, hydrogels with temperature-responsive properties—such as PNIPAAm-based hydrogels—undergo reversible phase transitions around physiological temperature (~32 °C). This enables dynamic adjustment of fabric moisture permeability and thermal buffering, improving thermophysiological comfort.

In summary, aerogels provide outstanding passive thermal insulation in textile systems due to their ultralight, nanoporous structures that dramatically limit heat transfer. In parallel, hydrogels contribute to active thermal regulation through their capacity for evaporative cooling and temperature-responsive phase transitions, supporting dynamic adjustment of the microclimate next to the skin. Looking forward, hybrid textile architectures that strategically combine the passive insulating performance of aerogels with the moisture-regulating functionality of hydrogels represent a promising pathway toward the development of smart, adaptive thermoregulatory fabrics capable of meeting the diverse demands of modern technical and wearable applications.

5.4 Antimicrobial and Antibacterial Properties

Incorporation of antibiotics and antibacterial properties to cloth has become an increasingly important step in improving hygiene, reducing odor, and limiting microbial colonization on the fabric. Hydrogels and aerogels have porous structures and tunable surface chemistry, which make them versatile platforms for loading antibacterial agents or obtaining intrinsic antimicrobial property. Hydrogels are networks of hydrophilic polymers, which can easily carry different agents, ranging from silver nanoparticles to quaternary ammonium compounds or natural and bioactive molecules with a controlled and sustained release on the textile. This method allows the antimicrobial effect to last longer than normal, even after washing. On the contrary, it is possible to functionalize silica and carbon aerogels with metal nanoparticles (Ag, Cu, and ZnO) or dopants, producing fabrics with broad-spectrum antimicrobial activity, excellent thermal insulation, and lightweight properties. Together, these materials support the development of multifunctional textiles that combine comfort, protection, and durability, aligning with growing demands in health care, outdoor, and technical textile markets.

Glazar et al. [6] developed cellulose-based hydrogel composites containing Ag nanoparticles, which demonstrated strong antimicrobial activity suitable for wound-healing and water purification applications. These materials combine the high porosity of aerogels with the moisture-retaining properties of hydrogels to support sustained antibacterial effects. In addition, Ortega-Portas et al. [26] confirmed that the antimicrobial effect seems to stem from direct interactions between the functionalized SiO_2 particles and bacterial membranes, ultimately disrupting membrane integrity and leading to cell death. These results highlight the promise of PU aerogels doped with modified silica for reducing Healthcare-Associated Infections (HAIs) in medical environments, although further studies are required to clarify the underlying mechanisms and assess long-term performance. Sharma et al. [27] highlight the rapid progress in designing cellulose-based materials, including membranes, hydrogels, and aerogels for water purification. These bio-derived materials benefit from inherent hydrophilicity, biodegradability, and nontoxicity, making them attractive for sustainable wastewater treatment. By incorporating nanomaterials or introducing functional groups, the adsorption and filtration capabilities of cellulose-based systems can be significantly enhanced, improving removal efficiencies for a wide range of pollutants such as heavy metals, dyes, salts, oils, and microorganisms. Advances in fabrication methods for regenerated cellulose, cellulose acetate, and nanocellulose structures have expanded their practical applications. In addition, studies have examined the mechanisms underlying contaminant removal and evaluated the environmental impacts through life cycle assessments, providing a broader perspective on the potential and limitations of these materials in addressing global water scarcity and pollution challenges. In textile, Jiang et al. [28] had developed a superhydrophobic cotton fabric modified with a 3D graphene–Ag/TiO_2 aerogel that achieved a water contact angle of approximately 161.8°, showed >99.99% antibacterial efficiency against *E. coli* and *S. aureus*, and maintained >98% oil–water separation under harsh conditions, demonstrating excellent durability and multifunctionality for technical textile applications.

The integration of antimicrobial and antibacterial properties into textiles using hydrogels and aerogels represents a promising strategy to meet the increasing demands for hygienic, odor-resistant, and protective fabrics. Hydrogels offer excellent potential as carriers for various antimicrobial agents, enabling controlled and sustained release that ensures long-lasting efficacy even after multiple washing cycles. Aerogels, due to their high porosity [22, 29, 30] and tunable surface functionalities, provide lightweight and thermally insulating substrates that can be functionalized with metal nanoparticles to deliver broad-spectrum antimicrobial action without compromising textile performance. The unique combination of hydrogel and aerogel properties encourage the development of multifunctional textiles for medical, sportswear, and technical applications. Continued research aimed at improving functionalization techniques, longevity, and safety of these materials will be the key to commercializing them into textile products.

5.5 Flame Retardancy

A challenge that still needs to be overcome is enhancing the flame retardancy of textiles, particularly for applications where strict fire standards are needed, such as protective clothing, upholstery, and technical fabrics. Flame-retardant treatments can hurt textile softness, breathability, or durability, creating demand for alternative, multifunctional solutions. In this, aerogels and hydrogels received much attention because of their intrinsic properties [3, 29, 31, 32]. Aerogels, with their highly porous networks and thermal stability, can act as effective thermal barriers that slow heat transfer and combustion. Conversely, hydrogels leverage their high water content and endothermic dehydration to delay ignition and suppress flame spread. By integrating these advanced materials, it becomes possible to develop lightweight, breathable textiles that combine flame resistance with other functionalities, aligning with current trends toward sustainable and high-performance protective fabrics. Assal et al. [32] demonstrated, using a vertical flame test (ISO 6940 [33]), that although raw and polyacrylate-coated cotton fabrics burned completely—showing high mass losses of approximately 92–96%—incorporating titanium xerogel into the polyacrylate matrix significantly improved flame retardancy. The PA/xerogel-coated fabrics exhibited notably reduced after-flame times and lower mass losses (approximately 40.8–51.1%), confirming the effectiveness of xerogel as a flame-retardant

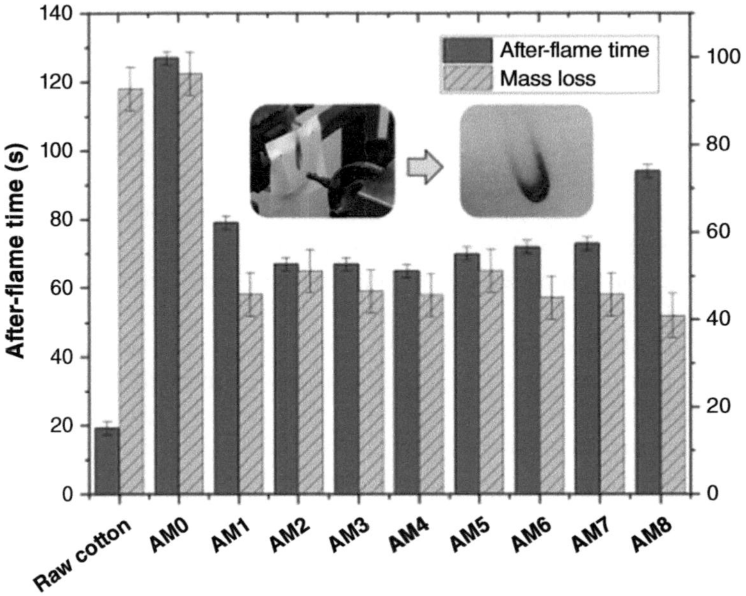

Figure 5.1 Vertical flammability test results. *Source:* Assal et al. [32] / Springer Nature.

filler that enhances fire resistance and slows combustion propagation, as illustrated in Figure 5.1.

Ghosh et al. [4] highlight how multifunctional textile coatings combining flame retardancy, antimicrobial activity, and self-cleaning properties are redefining protective fabrics for health-care, military, and industrial applications. Advances in coating methods such as sol–gel, dip-coating, spray-coating, and layer-by-layer assembly have enabled durable and thermally stable surfaces, while incorporating bioactive agents (e.g., chitosan, silver nanoparticles) and photocatalytic materials (e.g., TiO_2, silica) has improved pathogen resistance and sustainability. Challenges remain in balancing functional synergy, mechanical performance, and scalability, pointing to the need for biodegradable and smart-responsive systems to drive future innovation.

In addition, Jeong et al. [34] developed a self-assembled flame-retardant hydrogel based on poly(vinyl alcohol) and ammonium polyphosphate (PA hydrogel), which exhibited significant flame-retardant properties, including a limited oxygen index of 39.22% and a peak heat release rate near 158 kW/m^2. The hydrogel's dimensions were customizable up to 250 × 400 mm^2, and it demonstrated excellent mechanical flexibility, portability, and effective thermal protection across various substrates. These findings highlight its potential for use in applications such as thermal barriers and fire blankets. Özkan et al. [5] developed dynamically cross-linked cellulose nanocrystal (CNC) hydrogels incorporated with lignin, significantly enhancing flame-retardant properties by strengthening hydrogen bonding and forming a dense, elastic network that improves water retention and fuel dilution during combustion. This structure facilitates the formation of a compact glassy char layer that effectively impedes heat and oxygen transfer. Compared with neat

CNC films, the CNC–lignin–borax composites boosted the char yield up to 70.5% as well as improved their burn-through times by up to 71% on wood substrates, which indicates their potential use as a bio-based flame-retardant coating system.

Therefore, both aerogels and hydrogels represent appealing multifunctional solutions in improving flame retardancy of textiles while remaining comfortable and durable. These properties complement well in creating lightweight, breathable, and efficient fire-resistant fabrics that depend on aerogels (for thermal insulation) and hydrogels (for flame suppression driven by moisture). These inventions fit nicely with the increased need for environmentally safer, high-performance fibers in safety-sensitive devices.

5.6 Biodegradability

With rising environmental concerns shaping material choice in the textile sector, the biodegradability of functional additives such as hydrogels and aerogels has become a point of research focus. Although these substances provide excellent properties such as thermal insulation and fire retardance, their ecological footprint and end-of-life disposal need to be meticulously addressed to enable alignment with sustainability objectives. Biodegradable aerogels and hydrogels, commonly made from natural polymers or constructed from environmentally friendly chemistries, offer a compelling solution in that they reduce long-lived waste and enable biodegradation routes that are sustainable. Their degradation mechanisms, rates, and impact on textile performance need to be understood for the creation of next-generation functional and sustainable bio-based textiles.

Combining natural and biodegradable synthetic polymers can produce material with improved properties, but compatibility between the components must be ensured. In these blends, the better polymer can either enhance or negate the performance of the poorer polymer. The primary goal of polymer blending is to create new material with improved attributes. Natural polymers, particularly those derived from lignocellulosic sources, are abundant and highly valued for blending due to their flexibility and biocompatibility, making them well suited for advanced technological applications. Using natural polymer-based technologies promotes the use of environmentally friendly materials. Nevertheless, synthetic polymers remain important because they can mimic the behavior of natural polymers. Typically, polymer blending involves melting or dissolving the polymers to produce a homogeneous or heterogeneous mixture before regeneration. Various regenerated products, including hydrogels, aerogels, cryogels, and xerogels [7, 9, 35], are synthesized from blends of natural and inorganic components. While regeneration is commonly associated with hydrogels, it is less frequently applied to aerogels. Both hydrogels and aerogels have diverse applications, particularly in medical and agricultural fields, and are considered promising alternatives to nonbiodegradable synthetics due to their moldability, biocompatibility, and smart responsiveness to environmental stimuli. The strategic blending of natural and biodegradable synthetic polymers offers enhanced durability and versatility, especially given the range of products that can be developed. Composite materials derived from these blends show great potential for complex applications, thanks to their coherent structural integration. This chapter focuses on the blending of lignocellulose-based polymers (such as cellulose, hemicellulose, and lignin) with synthetic biodegradable polymers (including polyurethane, polylactic acid, polyvinyl alcohol, and polyethylene glycol) to produce hydrogels and aerogels tailored for medical and agricultural uses. Holocellulose nanofibrils/cellulose aerogel fibers (HCAFs) have been developed by Chen et al. [36] as biodegradable, flexible, and high-strength smart fibers with high surface area and porosity, enabling excellent self-cleaning and thermal insulation across a broad temperature range. These fibers exhibit multifunctional properties such as thermal management, Electromagnetic Interference (EMI) shielding, and responsive sensing of environmental stimuli such as temperature and humidity when combined with carbon nanotubes "Hybrid Carbon Composite Aerogel Fibers" (HCCAFs). HCCAF textiles allow for effective energy conversion and storage when used in conjunction with thermoelectric generators. Of interest, even after numerous washings, these biodegradable fibers retain their functionality and mechanical integrity, highlighting their appropriateness for applications in smart textiles and environmentally friendly wearable electronics. To address the environmental problems associated with discarded melt-blown masks, Liang et al. [37] developed a novel sodium alginate/nanosilver (SAg) composite aerogel as a biodegradable intermediate filter layer for face masks. The SAg air filter, which is made using a scalable pumping injection technique and softened with glycerol for flexibility, has outstanding antimicrobial activity through Coulomb adsorption of nanosilver particles, stable breathability, and superior air filtration. This multipurpose aerogel filter, which combines mechanical filtration and antimicrobial purification, provides a viable and efficient substitute for protective masks with encouraging prospects for widespread use. Despite cotton's inherent degradability, the rising demand for fashion and quick product turnover have led to a sharp rise in waste cotton textiles, creating resource and environmental problems.

Sustainable waste treatment methods are needed to address these problems. To enable the high-value reuse of textile waste, Huang et al. [38] concentrate on using solvent-based techniques to regenerate cellulose II aerogels from waste cotton. The structural properties, manufacturing processes, and uses of regenerated cellulose aerogels are summarized in detail in this review with the goal of promoting the long-term growth of the textile recycling sector and helping to achieve carbon neutrality. A key component of developing sustainable textile technologies is the biodegradability of hydrogels and aerogels. These materials can provide excellent performance while lessening their impact on the environment by using natural polymers and environmentally friendly chemistries. Designing bio-based textiles that strike a balance between environmental responsibility and functionality will be made possible by a thorough understanding of their degradation behaviors, which will aid in the transition to greener textile solutions.

5.7 Conclusion

A major turning point in the development of smart and functional textiles has been reached with the incorporation of hydrogels and aerogels into textile materials, which combine responsive moisture management, sophisticated thermal insulation, and lightweight structures into a single fabric system. As discussed in this chapter, hydrogels make a special contribution by absorbing, holding, and releasing moisture, which promotes wearer comfort and thermo-physiological balance in a variety of settings. Due to their nanoporous structure, aerogels provide unmatched thermal insulation, improving protection from heat and cold while preserving a low-weight profile that is essential for use in protective gear and performance apparel. Both hydrogels and aerogels can be further modified to add antimicrobial qualities beyond their primary uses, which will help prevent bacterial growth and enhance hygiene in everyday and medical textiles. The pursuit of biodegradable formulations responds to growing environmental demands for sustainable materials and less textile waste, while aerogels also aid in flame retardancy due to their inherent thermal stability. Despite these developments, it is still difficult to strike a balance between performance, robustness, adaptability, and mass production. According to recent studies, extensive testing in real-world settings is necessary to gain a better understanding of the practical effects on wearer comfort, particularly regarding mechanical resilience and moisture vapor transmission. To create textiles that can intelligently adjust to shifting physiological and environmental conditions, future research is probably going to investigate hybrid systems that combine the insulation and flame-retardant qualities of aerogels with the moisture-responsiveness of hydrogels. In conclusion, the combination of hydrogels and aerogels in textile science represents a change toward multipurpose, sustainable, and user-centered design, providing new avenues for satisfying the changing demands of environmental responsibility, comfort, and safety in contemporary textile applications.

References

1 Alhwaige, A.A., Ishida, H., and Qutubuddin, S. (2020). Chitosan/polybenzoxazine/clay mixed matrix composite aerogels: preparation, physical properties, and water absorbency. *Applied Clay Science* 184: 105403. https://doi.org/10.1016/j.clay.2019.105403.

2 Ng, S., Jelle, B.P., Sandberg, L.I.C. et al. (2015). Experimental investigations of aerogel-incorporated ultra-high performance concrete. *Construction and Building Materials* 77: 307–316. https://doi.org/10.1016/j.conbuildmat.2014.12.064.

3 Wen, O.Y., Tohir, M.Z.M., Yeaw, T.C.S. et al. (2023). Fire-resistant and flame-retardant surface finishing of polymers and textiles: a state-of-the-art review. *Progress in Organic Coating* 175: 107330. https://doi.org/10.1016/j.porgcoat.2022.107330.

4 Ghosh, J., Rupanty, N.S., Noor, T. et al. (2025). Functional coatings for textiles: advancements in flame resistance, antimicrobial defense, and self-cleaning performance. *RSC Advances* 15 (14): 10984–11022. https://doi.org/10.1039/d5ra01429h.

5 Özkan, A., Mıhlayanlar, E., Turhan, E.A. et al. (2025). Role of lignin on microstructure, mechanical properties and flame retardancy of nanocellulose-based composite hydrogels. *International Journal of Biological Macromolecules* 311: 144007. https://doi.org/10.1016/j.ijbiomac.2025.144007.

6 Glažar, D., Štular, D., Jerman, I. et al. (2025). Embedment of biosynthesised silver nanoparticles in polyNIPAAm/chitosan hydrogel for development of proactive smart textiles. *Nanomaterials* 15 (1): 1. https://doi.org/10.3390/nano15010010.

7 Salleh, K.M., Zainul Armir, N.A., Mazlan, N.S.N. et al. (2022). 12 - Hydrogel- and aerogel-based composites: biodegradable hydrogel and aerogel polymer blend-based composites. In: *Biodegradable Polymers, Blends and Composites*, Woodhead Publishing Series in Composites Science and Engineering (ed. S. Mavinkere Rangappa, J. Parameswaranpillai, S. Siengchin, and M. Ramesh), 355–388. Woodhead Publishing https://doi.org/10.1016/B978-0-12-823791-5.00019-3.

8 Lorusso, E., Ali, W., Hildebrandt, M. et al. (2019). Hydrogel functionalized polyester fabrics by UV-induced photopolymerization. *Polymers* 11: 1–10. https://doi.org/10.3390/polym11081329.

9 El-Naggar, M.E., Othman, S.I., Allam, A.A., and Morsy, O.M. (2020). Synthesis, drying process and medical application of polysaccharide-based aerogels. *International Journal of Biological Macromolecules* 145: 1115–1128. https://doi.org/10.1016/j.ijbiomac.2019.10.037.

10 Du, Y. and Kim, H.-E. (2022). Research trends of the application of aerogel materials in clothing. *Fashion and Textiles* 9 (1): 23. https://doi.org/10.1186/s40691-022-00298-5.

11 Aegerter, M.A., Leventis, N., Koebel, M., and Steiner, S.A. III, (ed.) (2022). *Springer Handbook of Aerogels*, 1e. Springer.

12 McNeil, S.J. and Gupta, H. (2022). Emerging applications of aerogels in textiles. *Polymer Testing* 106: 107426. https://doi.org/10.1016/j.polymertesting.2021.107426.

13 Liu, L., Yuan, D., Hu, X. et al. (2024). Wind-proof and moisture permeability aerogel-functionalized textile as all-seasonal passive thermal regulators. *Advanced Functional Materials* 34 (49): 2411551. https://doi.org/10.1002/adfm.202411551.

14 Liu, J., He, H., Yu, Z. et al. (2020). Investigation of temperature-responsive and thermo-physiological comfort of modified polyester fabric with Sericin/PNIPAAm/Ag NPs interpenetrating polymer network hydrogel. *Textile Research Journal* 90 (23–24): 2622–2638. https://doi.org/10.1177/0040517520931475.

15 Shaid, A., Furgusson, M., and Wang, L. (2014). Thermophysiological comfort analysis of aerogel nanoparticle incorporated fabric for fire fighter's protective clothing. *Chemical and Materials Engineering* 2 (2): 37–43. https://doi.org/10.13189/cme.2014.020203.

16 Azam, F., Ahmad, F., Ulker, Z. et al. (2022). The role and applications of aerogels in textiles. *Advances in Materials Science and Engineering* 2022, https://doi.org/10.1155/2022/2407769 (1): 1–22.

17 Chang, K.-J., Wang, Y.-Z., Peng, K.-C. et al. (2014). Preparation of silica aerogel/polyurethane composites for the application of thermal insulation. *Journal of Polymer Research* 21 (1): 338. https://doi.org/10.1007/s10965-013-0338-7.

18 Sai, H., Wang, M., Miao, C. et al. (2021). Robust silica-bacterial cellulose composite aerogel fibers for thermal insulation textile. *Gels* 7 (3): 145. https://doi.org/10.3390/gels7030145.

19 He, S., Wu, X., Zhang, X. et al. (2023). Preparation and properties of thermal insulation coating based on silica aerogel. *Energy and Buildings* 298: 113556. https://doi.org/10.1016/j.enbuild.2023.113556.

20 Tafreshi, O.A., Mosanenzadeh, S.G., Karamikamkar, S. et al. (2022). A review on multifunctional aerogel fibers: processing, fabrication, functionalization, and applications. *Materials Today Chemistry* 23: 100736. https://doi.org/10.1016/j.mtchem.2021.100736.

21 Boukind, S., Sair, S., Ait Ousaleh, H. et al. (2022). Ambient pressure drying as an advanced approach to the synthesis of silica aerogel composite for building thermal insulation. *Journal of Natural Fibers* 19 (15): 10142–10156. https://doi.org/10.1080/15440478.2021.1993486.

22 Geng, H. (2018). A facile approach to light weight, high porosity cellulose aerogels. *International Journal of Biological Macromolecules* 118: 921–931. https://doi.org/10.1016/j.ijbiomac.2018.06.167.

23 Vareda, J., Lamy-Mendes, A., and Durães, L. (2017). A reconsideration on the definition of the term aerogel based on current drying trends. *Microporous and Mesoporous Materials* 258: 211–216. https://doi.org/10.1016/j.micromeso.2017.09.016.

24 Du, A., Zhou, B., Zhang, Z., and Shen, J. (2013). A special material or a new state of matter: a review and reconsideration of the aerogel. *Materials* 6 (3): 941–968. https://doi.org/10.3390/ma6030941.

25 Liu, P., Gao, H., Chen, X. et al. (2020). In situ one-step construction of monolithic silica aerogel-based composite phase change materials for thermal protection. *Composites. Part B, Engineering* 195: 108072.

26 Ortega-Portas, C., Pinilla-Peñalver, E., Sánchez-Silva, L. et al. (2025). Polyurethane aerogels doped with modified SiO_2 particles: advanced materials with antibacterial properties for healthcare applications. *Colloids and Surfaces A: Physicochemical and Engineering Aspects* 721: 137204. https://doi.org/10.1016/j.colsurfa.2025.137204.

27 Sharma, K., Choudhary, P., Majeed, A. et al. (2025). Cellulose based membranes, hydrogels and aerogels for water treatment application. *Industrial Crops and Products* 225: 120474. https://doi.org/10.1016/j.indcrop.2025.120474.

28 Jiang, Q., Tang, P., Chen, W. et al. (2024). Fabrication and characterization of a novel superhydrophobic cotton fabric with integrated 3D graphene-Ag/TiO_2 aerogel for superior antibacterial performance, oil-water separation, and enhanced

durability. *Industrial Crops and Products* 222: 119882. https://doi.org/10.1016/j.indcrop.2024.119882.

29 Zhang, Y., Li, Y., Lei, Q. et al. (2023). A vertically aligned anisotropic boron nitride nanosheet (BNNSs)/polyvinyl alcohol (PVA) composite aerogel with ultra-low thermal conductivity prepared by the prefreeze-freeze-drying method for efficient thermal management. *Materials Today Communications* 36: 106742. https://doi.org/10.1016/j.mtcomm.2023.106742.

30 Timusk, M., Kangur, T., Locs, J. et al. (2021). Aerogel-like silica powders by combustion of sol-gel derived alcogels. *Microporous and Mesoporous Materials* 315: 110895. https://doi.org/10.1016/j.micromeso.2021.110895.

31 Yu, Z.-L., Yang, N., Apostolopoulou-Kalkavoura, V. et al. (2018). Fire-retardant and thermally insulating phenolic-silica aerogels. *Angewandte Chemie, International Edition* 57 (17): 4538–4542. https://doi.org/10.1002/anie.201711717.

32 Assal, M., El Wazna, M., Oumghar, K. et al. (2023). Multifunctional textile based on titanium xerogel: performance optimization through composition and microstructure. *Journal of Sol-Gel Science and Technology* 106 (3): 672–683. https://doi.org/10.1007/s10971-023-06102-0.

33 ISO 6940:2004 (2022). Textile fabrics—Burning behaviour—Determination of ease of ignition of vertically oriented specimens. ISO. Consulté le: 7 octobre 2022 [En ligne]. Disponible sur: https://www.iso.org/cms/render/live/en/sites/isoorg/contents/data/standard/03/56/35665.html

34 Jeong, S.Y., Song, H., Koo, K. et al. (2025). Facile and scalable preparation of poly(vinyl alcohol) hydrogels: insights into thermal kinetics, combustion behavior, and flame retardancy. *Advanced Industrial and Engineering Polymer Research* https://doi.org/10.1016/j.aiepr.2025.06.003.

35 Pietras-Ożga, D., Piątkowska-Sawczuk, K., Duro, G. et al. (2022). Chapter 9 - Sol-gel–derived silica xerogels: synthesis, properties, and their applicability for removal of hazardous pollutants*. In: *Advanced Materials for Sustainable Environmental Remediation* (ed. D. Giannakoudakis, L. Meili, and I. Anastopoulos), 261–277. Elsevier https://doi.org/10.1016/B978-0-323-90485-8.00008-4.

36 Chen, Y., Zhang, C., Tao, S. et al. (2023). High-performance smart cellulose nanohybrid aerogel fibers as a platform toward multifunctional textiles. *Chemical Engineering Journal* 466: 143153. https://doi.org/10.1016/j.cej.2023.143153.

37 Liang, J., Wang, Z., Ye, Q. et al. (2022). Pump-inject antimicrobial and biodegradable aerogel as mask intermediate filter layer for medical protection of air filtration. *Materials Today Sustainability* 19: 100211. https://doi.org/10.1016/j.mtsust.2022.100211.

38 Huang, Z., Zhang, Y., Xing, T. et al. (2024). Advances in regenerated cellulosic aerogel from waste cotton textile for emerging multidimensional applications. *International Journal of Biological Macromolecules* 270: 132462. https://doi.org/10.1016/j.ijbiomac.2024.132462.

6

Applications in Sustainable Fashion

Bushra Mushtaq[1], Nauman Ali Choudhry[2], and Abher Rasheed[1]

[1] *National Textile University, School of Engineering and Technology, Department of Textile Engineering, Sheikhpura Road 37610, Faisalabad, Pakistan*
[2] *RMIT University, School of Fashion and Textiles, WEFT Research Centre, 25 Dawson Street, Brunswick, VIC 3056, Australia*

6.1 Introduction

The fashion industry today is at a critical juncture of change and awakening, where sustainability, ecological awareness, and innovation in materials are coming to the forefront. Growing concerns over textile waste, pollution, microplastics, and high water and energy consumption drive both researchers and manufacturers toward greener, high-performance material options. Among such emerging responses, hydrogels and aerogels have come to the fore for their distinctive structural, thermal, and moisture-management capabilities. These materials not only embrace circular fashion but also allow for functionalities not possible with conventional textiles. The following section discusses the motive that defines sustainable fashion innovation and shows how both hydrogel- and aerogel-based technologies present promising pathways toward eco-efficient means of creating apparel, accessories, sportswear, and footwear.

6.1.1 The Need for Sustainable Fashion Innovation

The fashion sector is undergoing a transformative shift toward sustainability, driven by increasing environmental concern and consumer demand for eco-friendly products. Traditional textile manufacturing is resource-intensive and causes pollution, waste, and excess water use. For example, natural cotton cultivation requires huge quantities of water, approximately 2700 L to make a single T-shirt, putting huge pressure on freshwater supplies [1]. At the same time, synthetic fibers like polyester, which are used in the fast-fashion industry, are made from nonrenewable petroleum, release microplastics when washed, and pollute marine environments [2]. Textile finishing and dyeing also release chemicals into water, which contribute to environmental degradation [3].

As a response to these challenges, the industry is moving toward sustainable solutions that reduce ecological impact without compromising on performance and aesthetics. Hydrogels and aerogels are one such solution that presents functionalities and aligns with the principle of circular fashion. These products can save water and energy during production, increase biodegradability, and provide novel properties like self-regulated temperature and moisture management [4]. With the incorporation of these high-tech materials, the fashion industry can move toward more sustainable processes while meeting the customers' demand for eco-friendly products.

6.1.2 Role of Hydrogels and Aerogels in Sustainable Textiles

Hydrogels and aerogels are transforming textile engineering with functionality and sustainability. Hydrogels, composed of water-absorbing polymer networks, have exceptional moisture control and are suitable for active wear and daily wear. They not only absorb and release water according to environmental conditions but also regulate body temperature, minimizing the need for constant washing and extending the garment life cycle. Some hydrogels are designed with phase-change characteristics, enabling them to store and dissipate heat dynamically, improving wearer comfort across different climates [5, 6]. In addition, researchers are developing plant-based hydrogels as biodegradable alternatives to petroleum-based linings and coatings, lowering dependency on petroleum-based materials.

Aerogels, by contrast, are extremely lightweight, porous materials with high insulation and breathability qualities. In fashion design, aerogels ensure thermal regulation without the weight of conventional insulating materials such as down or synthetic fills. Aerogel jackets and aerogel footwear products ensure greater warmth while being thin and supple, fulfilling the requirements for both performance and everyday use [7].

6.1.3 Chapter Scope and Objectives

This chapter explores the potential of hydrogels and aerogels to transform sustainable fashion and their uses in four specific areas—sustainable clothing, biodegradable accessories, sportswear, and advanced footwear. The chapter begins by explaining how they develop clothes that minimize water usage, are stain-resistant, and control temperature, thereby reducing the environmental impact. Then, it investigates their application in biodegradable accessories, including plant-based leather handbags and adaptive jewelry, which reduce waste but retain beauty. The chapter then investigates their applications in sportswear, where water wicking and microplastic-free functionalities enhance performance with no environmental degradation. Last, it discusses developments in footwear, including self-cooling midsoles and recyclable shoe designs that incorporate hydrogel and aerogel technology.

6.2 Eco-friendly Clothing

The incorporation of hydrogels and aerogels into textile production is transforming the term green clothing by solving two fundamental sustainability issues: resource use across the garment life cycle and postconsumer waste disposal. Hydrogel-embedded textiles introduce a change in this respect through their water management attributes. These hydrophilic polymer systems are capable of absorbing as much as 90% of their weight in water without loss of structural integrity and facilitating the development of self-hydrating textiles. Advanced developments with temperature-sensitive polymers can dynamically change their water uptake as a function of ambient humidity and human body heat, producing a self-sustaining system that further improves wearer comfort.

Aerogel technology is used to support these developments by adding exceptional thermal performance to sustainable clothing. This innovation holds special application for outerwear, where conventional insulation tends to become bulky and hinder mobility. The use of bio-based aerogels produced from cellulose or chitosan enhances the sustainability of these clothes through their biodegradability. In addition to fundamental performance, these new materials allow novel self-cleaning processes to improve garment life. Photocatalytic aerogel coatings with titanium dioxide nanoparticles can degrade organic matter under light, essentially sanitizing the fabric surface without water or detergent, as shown in Figure 6.1. Hydrogel matrices with silver nanoparticles also offer sustained antimicrobial protection, limiting odor buildup and the need for frequent washing.

The improvements in durability by these materials also solve the fast fashion problem by increasing product lifetimes. Hydrogel-based fibers exhibited good mechanical resistance, whereas aerogel coatings can shield against UV degradation that usually leads to color fading and fiber degradation. All these features support new business models within the circular economy, where clothes are developed with

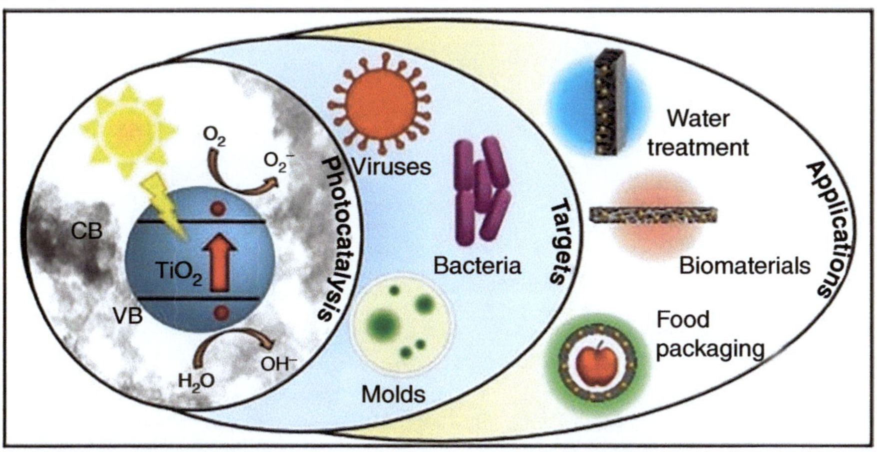

Figure 6.1 Photocatalytic titanium dioxide nanoparticles degrade organic matter under light. *Source:* [8] / MDPI / CC BY 4.0.

recyclable characteristics. Commercial applications have already developed jackets with panels of aerogel insulation that retain heat at 50% less thickness than regular insulation. Once manufacturing scales up, these technologies will revolutionize high-street fashion, providing consumers with truly sustainable choices without sacrificing style or function.

6.3 Biodegradable Fashion Accessories

6.3.1 The Sustainability Challenges in Fashion Accessories

The fashion accessories sector faces unique sustainability challenges due to its reliance on intricate material mixtures and small components, which complicate recycling processes. Conventional accessories tend to integrate metals, plastics, leather, and synthetic materials in configurations that make disassembly and recovery of the materials almost impossible. Fewer than 5% of fashion accessories are recycled today, as per the Ellen MacArthur Foundation reports, with the balance ending up in landfills where synthetic components may last centuries to break down. The manufacturing processes for these goods are just as concerning, with metal hardware involving energy-intensive mining and refining processes, while synthetic leather substitutes are based on petroleum-derived polyurethanes.

6.3.2 Hydrogel-Based Leather Alternatives

Recent breakthroughs in hydrogel technology are transforming sustainable alternatives to leather for accessories. Scientists have engineered plant-hydrogel composites that replicate the texture and appearance of real leather but with enhanced environmental advantages. These new materials generally consist of the following:

- Bacterial cellulose networks developed via fermentation processes, as shown in Figure 6.2 [9].

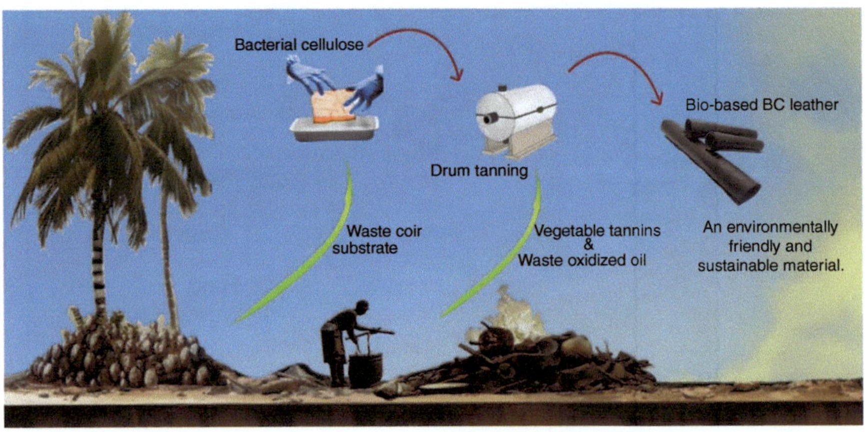

Figure 6.2 Synthesis of bio leather from discarded coconut. *Source:* [10] / with permission of Elsevier.

- Alginate hydrogels derived from seaweed, as shown in Figure 6.3 [11]. In this method, dialdehyde sodium alginate was used to react with dicyandiamide to form formaldehyde-free eco-leather with improved mechanical strength and good organoleptic characteristics.
- Natural cross-linking agents such as plant source tannins, as shown in Figure 6.4. In the method, tannins were isolated from *Tessmannia burttii* Harms (Caesalpiniaceae) stem. The findings reveal that *T. burttii* extracts provide an environmentally friendly alternative source of vegetable tannins, providing a green solution to reduce the dependency on chromium salts in the leather industry.

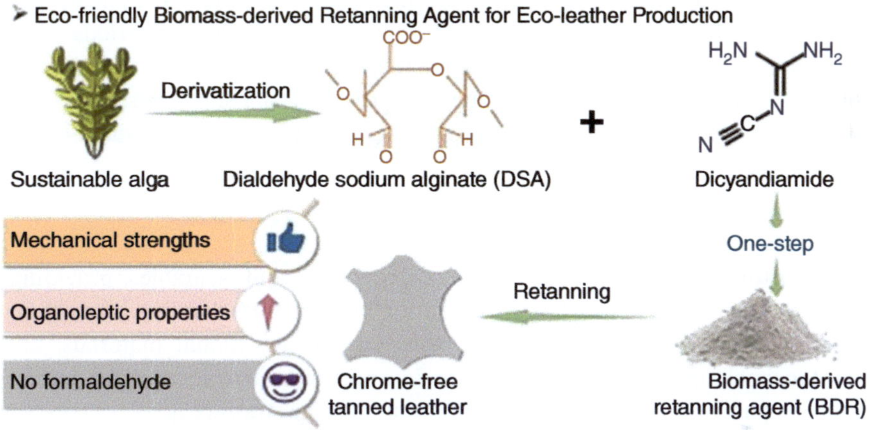

Figure 6.3 Dialdehyde sodium alginate-bonded dicyandiamide for formaldehyde-free leather production. *Source:* [12] / with permission of Elsevier.

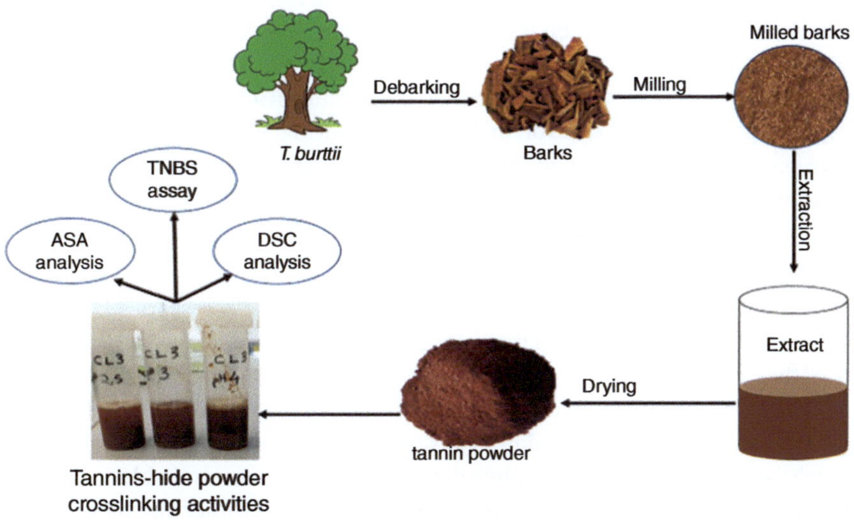

Figure 6.4 Tanning capacity of *Tessmannia burttii* extracts: the potential eco-friendly tanning agents for the leather industry. *Source:* [13] / Springer Nature / CC BY 4.0.

6.3.3 Aerogel-Based Functional Accessories

Aerogels are transforming fashion accessory design and functionality by their record-breaking combination of ultralight weight, mechanical strength, and multifunctionality.

6.3.3.1 Jewelry

Jewelry makers have started using silica aerogels that are reinforced with silk fibroin, combining high-tech functionality with aesthetics. The composite aerogels are transparent in appearance and reflect the appearance of frosted glass, but are up to 90% lighter. Practically speaking, a pendant or brooch that would otherwise be heavy can now become featherlight, so it is more comfortable to wear for longer periods.

Silk fibroin, a protein isolated from silk, increases the toughness and flexibility of the brittle silica aerogel matrix. Thus, such jewelry items are shock-absorbing and resistant to cracking or shattering, a frequent setback in conventional glass-based adornments [14].

6.3.3.2 Belts and Straps

Aerogel-elastomer composites are used in belts, straps, and wearables, providing lightweight, flexible, and tough substitutes for leather. Withstanding more than 10,000 flex cycles without degradation, they surpass traditional materials, are resistant to cracking, and offer green, nontoxic, and animal-free alternatives perfect for sports apparel, technological accessories, and luggage straps [15].

6.3.3.3 Eyewear

In the field of eyewear, the graphene-aerogel frame is a revolutionary development. The hybrid materials not only have very low density and high flexibility but also possess electrical conductivity, which is essential for smart glasses and AR (augmented reality) frames. The existence of graphene in the aerogel matrix makes it possible to integrate sensors, electrodes, and antennas into the frame itself, providing real-time data capture, gesture-based control, or health monitoring. Meanwhile, the aerogel structure keeps the frames light and thermally insulating, avoiding discomfort from prolonged use [16].

6.3.3.4 3D-Printed Aerogel Accessories

Scientists from MIT have developed a technique to 3D print accessories based on aerogel using direct ink writing (DIW). By extruding a silica aerogel ink through a nozzle, they can produce intricate geometries with adjustable densities ranging from 0.1 to 0.5 g/cm^3. This enables specific designs with adjustable mechanical strength, porosity, and biodegradability. Lower-density prints are suitable for lightweight accessories, whereas higher-density prints are more suitable for structural components. These sustainable products can be made to order and biodegrade in composting conditions in 3 months to 2 years, embracing the trend of circular fashion and sustainability [16, 17].

6.3.4 Smart and Responsive Biodegradable Accessories

The incorporation of responsive hydrogels is establishing a new field of "living accessories" that are engaged in their surroundings:.

- Color-changing components: pH-sensitive hydrogels composed of anthocyanin extracts produce accessories that change color based on skin chemistry or ambient conditions [18].
- Shape-memory devices: Heat-sensitive hydrogels enable accessories to change form, from a bracelet that contracts slightly in warmth to stay in place, for instance [19].
- Self-healing materials: Hydrogels containing dynamic covalent cross-links can recover from small scratches and cuts in the presence of water [20].

6.4 Sustainable Sportswear

6.4.1 The Environmental Impact of Conventional Sportswear

The international sportswear market is under increasing scrutiny for its massive environmental impact, with synthetic performance sports materials posing a particularly urgent sustainability issue. Polyester, nylon, and elastane, which together account for more than 60% of sportswear content, are sourced from nonrenewable oil deposits and make significant contributions to microfiber pollution. One synthetic sport garment can release as many as 1900 microplastic fibers when washed, eventually entering ocean ecosystems [21]. In addition, the finishing and dyeing of performance fabrics use enormous amounts of water (approximately 100–150 L/kg of material) and often contain perfluorinated chemicals (PFCs) that do not break down and remain in the environment forever [22]. Such concerns have necessitated an urgent need for novel material solutions that can provide the technical performance athletes require while accommodating sustainability needs.

6.4.2 Performance Enhancement by Hydrogel

Modern hydrogel technologies are transforming green sportswear through three fundamental mechanisms:

6.4.2.1 Moisture Management Systems

Cross-linked polyacrylamide hydrogels with customized pore size now provide moisture management in sportswear. These smart materials absorb 30 times their weight in sweat and are breathable, actively drawing water from the skin at rates of more than 500% relative to traditional wicking fabrics [23, 24].

6.4.2.2 Temperature Control

Phase-change hydrogels containing paraffin microcapsules exhibit active thermal control, absorbing surplus body heat under intense exercise and releasing it slowly during cooldown phases. Under Armour's Iso chill™ series applies this technology to keep skin temperature at optimal levels between 2 °C under all ambient conditions, lowering thermal stress by as much as 40% relative to conventional fabrics [25].

6.4.2.3 Biomechanical Support

Shear-thinning hydrogels are added to compression garments to offer adaptive support. These new-generation materials retain flexibility with regular movement but stiffen immediately when experiencing impact, providing joint protection without limiting mobility. Adidas offers 25% improved impact absorption than standard neoprene supports, using 60% less synthetic material [26].

6.4.3 Aerogel in Athletic Apparel

Aerogel technology is powering innovation in sustainable thermal management for sportswear.

6.4.3.1 Ultralight Insulation

Silica aerogel composites are now offering enhanced insulation with thicknesses as much as 80% lower than conventional materials. Nike Aeroloft™ jackets use aerogel particles incorporated into recycled polyester fibers and deliver a warmth equivalent to goose down at only 30% the weight, while being entirely waterproof [27].

6.4.3.2 Increased Breathability

New aerogel membranes with pore structure grading enable controlled moisture, vapor transmission, and the exclusion of liquid water. Patagonia's Nano-Air™ product line benefits from this technology to sustain breathability levels above 30,000 $g/m^2/24\,h$, while offering wind protection similar to conventional membranes [28, 29].

6.4.3.3 Impact Protection

Graphene-impregnated aerogel foams are transforming protective equipment. These foams dissipate as much as 90% of impact energy while adding no more than 70% of the weight of traditional foam padding. Specialized's latest range of cycling clothing proves 40% improved impact protection compared with EPS foam, with full recyclability [30].

6.4.4 Sustainable Manufacturing Breakthrough

The sportswear market is turning to new production techniques to limit its environmental footprint:

6.4.4.1 Waterless Dyeing

Supercritical CO_2 dyeing technologies that use hydrogel color carriers capture 95% of dye against 60–70% for traditional processes, making wastewater completely obsolete. Adidas has used this technique on 50% of its polyester range, conserving more than 100 million liters of water per year [31, 32].

6.4.4.2 Closed-Loop Recycling

Enzymatic hydrogel dissolution procedures now facilitate authentic fiber-to-fiber recycling of blended textiles. The H&M Conscious collection uses the technology to capture more than 85% of the material content from end-of-life apparel [33, 34].

6.4.4.3 3D Knitting

Additive manufacturing minimizes material loss by as much as 35% relative to cut-and-sew techniques. Nike's Flyknit innovation has saved almost 4 million pounds of waste since 2012, while enhancing performance via seamless construction [35].

6.5 Hydrogel and Aerogel-Infused Footwear

6.5.1 Environmental Impact of Conventional Footwear

The worldwide footwear market manufactures more than 24 billion pairs each year, which has a significant environmental effect along the entire product life cycle [36]. Conventional shoe production is highly dependent on petroleum-derived materials, with each pair having as many as 65 individual pieces constructed from plastics, synthetic rubber, and chemical-based adhesives that do not biodegrade. The carbon footprint is an average of 13.6 kg CO_2 equivalent per pair, with running shoes frequently being more than 20 kg due to multicomponent cushioning systems [37]. Postconsumer waste is a further urgent problem—nearly 90% of abandoned shoes find their way to landfills, where synthetic polymers can last hundreds of years. The water footprint of the industry is also problematic, with up to 8000 L needed per pair to tan leather and process materials [38]. These environmental costs have led to pressing innovation in eco-friendly footwear technology using hydrogels and aerogels.

6.5.2 Application of Hydrogel in Footwear

Hydrogel 3D polymeric networks, with the ability to hold large volumes of water, are transforming the face of footwear technology. Their superior mechanical flexibility, biocompatibility, and sustainability render them uniquely suited for incorporation into the next generation of shoes. The following is an investigation into two of the primary areas where hydrogels are having a lasting impact: high-performance cushioning systems and moisture management with hygiene improvement.

6.5.2.1 Advanced Cushioning System

Conventional cushioning materials used in footwear, including ethylene-vinyl acetate (EVA) foams, have ruled the industry long enough with their lightweight and shock absorption characteristics [39]. Hydrogel-based cushioning systems, though, are a technological advancement, providing superior energy dynamics, adaptive response, and ecological sustainability. Hydrogel midsoles provide better performance than EVA, with up to 75% energy return and gait-adaptive stiffness. Their self-healing ability increases product life, and water-based, recyclable compositions minimize waste. Light 3D-printed constructions maximize cushioning, support, and sustainability in sports shoes [40].

6.5.2.2 Moisture Management and Hygiene

Apart from mechanical benefits, hydrogels also enhance the microclimate within footwear by simulating natural thermal and moisture control systems. Hydrogels, which can be infused with phase-change materials, regulate optimum in-shoe temperatures (28–32 °C) by dissipating excess heat during exercise and releasing it during inactivity. Self-wicking channel-containing microstructured hydrogel insoles can transport moisture at rates of more than 50 $mL/h/cm^2$, keeping feet dry and preventing blisters or infection. Also, when loaded with antimicrobial agents such as silver nanoparticles or plant extracts, hydrogels inhibit bacterial growth. Nike's Phylon Shield™ shoes, for instance, recorded 99.7% bacterial inhibition, improving hygiene, odor, and product durability for extended usage [41].

6.5.3 Application of Aerogel in Sustainable Footwear

In insulation systems, aerogel-infused membranes outperform traditional materials like wool or foam. For instance, 0.5-mm thick aerogel films can match the insulation of 15 mm wool, providing warmth without compromising slim design. These membranes, typically made with silica or polymer aerogels in flexible carriers, retain performance even after 100,000 flex cycles, making them ideal for cold-weather athletic footwear. Innovations in biodegradable aerogels, derived from natural polymers such as cellulose or chitin, offer high thermal insulation and environmental sustainability, eliminating the microplastic issue of petroleum-based foams. The North Face's Aerogel Trail GTX® boots exemplify this by offering −40 °C protection with 60% weight reduction, increasing comfort and endurance in extreme conditions.

Structurally, aerogels enhance durability and reduce weight in components like shoe uppers, where aerogel composites offer triple the abrasion resistance of leather at just one-fifth the weight. They also enable breathable, waterproof membranes with MVTR (moisture vapor transmission rate) values exceeding 20,000 g/m^2/24 hours, surpassing materials like Gore-Tex®, ensuring dryness and comfort. Recyclable aerogel-based shank systems replace nonrenewable steel or carbon fiber while maintaining stiffness and support, contributing to circular footwear design.

Sustainable manufacturing practices now leverage hydrogels and aerogels to minimize environmental impact. Laser-activated hydrogel adhesives allow rapid, nontoxic bonding without VOCs (volatile organic compounds) or water waste, curing in 0.3 seconds and forming bonds 50% stronger than traditional glues. Meanwhile, ultrasonic welding of aerogel composites eliminates adhesives and stitching, reducing energy use by 70% and preserving structural integrity. These technologies represent a breakthrough in eco-conscious production, enabling high-performance, durable, and recyclable footwear aligned with circular economic principles.

6.6 Sustainable Manufacturing Breakthrough

In response to mounting environmental pressures and changing customer requirements, the footwear sector is adopting greener production techniques, with hydrogels and aerogels being at the forefront of facilitating sustainable innovation. A significant advancement is the use of waterless assembly technologies. Hydrogel adhesives, triggered by lasers and composed of light-sensitive cross-linkers, cure in 0.3 seconds under UV or infrared light and form bonds that are 50% stronger than conventional adhesives. These elastic, nontoxic adhesives are best suited to dynamic shoe areas, allowing for solvent-free, clean, and automatable production. Likewise, ultrasonic welding of aerogel composites uses high-frequency vibrations for bump-free bonding, cutting energy use by more than 70% over thermal lamination and maintaining thermal and structural integrity when stressed.

Equally transformative are closed-loop production systems. Ionic- or hydrogen-bonded hydrogels can be selectively dispersed in water, enabling more than 95% material recovery and reuse in closed-loop production—already tested by eco-friendly shoe companies. Silica- and biopolymer-based aerogels can be recycled using supercritical CO_2 extraction, a clean process that preserves their nanoporous architecture but facilitates additive removal and material customization. Modular design, based on reversible hydrogel connections and interlocking aerogel subunits, also facilitates circularity by enabling simple repair, component swapping, and effective recycling. Such advances extend product lifespan by as much as 35% and cut material waste by 20%, enabling a fully circular and low-impact footwear economy.

6.7 Future Perspectives and Challenges

The sustainable fashion of the future is on the cusp of being defined by revolutionary changes in material science, production technology, and circular economic systems. At the heart of this revolution are next-generation hydrogel and aerogel technologies that are not only environmentally friendly but also bring unprecedented functionalities. Yet, the path to mass market adaptation is fraught with roadblocks from the scalability of production to readiness for markets and regulatory adaptation. This section outlines the emerging material innovations, barriers to scale, policy gaps, and strategic pathways that will define the trajectory of sustainable fashion over the next two decades.

Future hydrogel and aerogel technologies are transforming green fashion by combining biotechnology, responsiveness, and adaptability. Bio-intelligent hydrogels, created at Harvard's Wyss Institute, include genetically programmed microorganisms that facilitate self-repair, photosynthetic pigmentation, and pollutant capture—earlier prototypes can heal 80% of scratches in 24 hours and capture 5 g of CO_2 per square meter per day. MIT's Media Lab holds the lead in 4D-printed aerogels infused with cellulose nanocrystals that change with temperature and humidity, allowing garments to alter insulation, breathability, and even self-assemble dynamically. Cambridge University's quantum dot–infused fabrics, on their part, allow for dye-free color change, ambient light energy harvesting as high as 15%, and real-time biometric sensing. Such advancements notwithstanding, industrial scale-up, prohibitive material costs, intricate manufacturing processes, and consumer skepticism over durability and care are major obstacles. Regulatory loopholes, weak standardization, and rising IP (intellectual property) conflicts also delay adoption. To bridge these gaps, hybrid materials, roll-to-roll manufacturing, digital deposition, AI-driven R&D, and circular business models like leasing and blockchain-supported tracking are being tested. Programmable apparel, atmospheric water-harvesting fabric, and neuromorphic garments could enter the mainstream by 2040. If stakeholders work together, these innovations have the potential to make fashion regenerative, lowering emissions, eradicating microplastic pollution, and establishing new green standards in living.

6.8 Conclusion

The convergence of hydrogel and aerogel technologies is a revolutionary step toward sustainable fashion. These next-generation materials minimize water consumption, maximize temperature control, and enable closed-loop recycling, promising a revolutionary redesign of clothing manufacturing and design. Unleashing their full potential demands a partnership between research, industry, policymaking, and consumer action. The 2030 road map describes a trajectory of mass adoption that makes fashion a regenerative force, not a burden on the environment. In addition to technical innovation, these materials redefine fashion as a sustainable art form, bringing human creativity into ecological consciousness and opening the door to a promising, responsible future.

References

1 Kazan, H., Akgul, D., and Kerc, A. (2020). Life cycle assessment of cotton woven shirts and alternative manufacturing techniques. *Clean Technologies and Environmental Policy* 22: 849–864.

2 Carney Almroth, B.M., Åström, L., Roslund, S. et al. (2018). Quantifying shedding of synthetic fibers from textiles; a source of microplastics released into the environment. *Environmental Science and Pollution Research* 25: 1191–1199.

3 Singha, K., Pandit, P., Maity, S., and Sharma, S.R. (2021). Harmful environmental effects for textile chemical dyeing practice. In: *Green Chemistry for Sustainable Textiles*, 153–164. Elsevier.

4 García-González, C.A., Blanco-Vales, M., Barros, J. et al. (2025). Review and perspectives on the sustainability of organic aerogels. *ACS Sustainable Chemistry & Engineering* 13 (18): 6469–6492.

5 Fang, Y., Chen, G., Bick, M., and Chen, J. (2021). Smart textiles for personalized thermoregulation. *Chemical Society Reviews* 50 (17): 9357–9374.

6 Chatterjee, S., Hui, P.C.-l., and Kan, C.-w. (2018). Thermoresponsive hydrogels and their biomedical applications: special insight into their applications in textile based transdermal therapy. *Polymers* 10 (5): 480.

7 Venkataraman, M., Mishra, R., Kotresh, T. et al. (2016). Aerogels for thermal insulation in high-performance textiles. *Textile Progress* 48 (2): 55–118.

8 De Pasquale, I., Porto, C.L., Dell'Edera, M. et al. (2020). Photocatalytic TiO_2-based nanostructured materials for microbial inactivation. *Catalysts* 10 (12): 1382.

9 Absharina, D., Padri, M., Veres, C., and Vágvölgyi, C. (2025). Bacterial cellulose: from biofabrication to applications in sustainable fashion and vegan leather. *Fermentation* 11 (1): 23.

10 Yang, F., Zhang, Z., Yuan, J. et al. (2024). Eco-friendly production of leather-like material from bacterial cellulose and waste resources. *Journal of Cleaner Production* 476: 143700.

11 Choi, S.M., Lee, D.H., Zo, S.M. et al. (2025). Gel-based approaches to vegan leather: opportunities and challenges in mimicking leather properties. *Gels* 11 (6): 395.

12 Ding, W., Wang, Y., Sun, J. et al. (2022). Dialdehyde sodium alginate bonded dicyandiamide for formaldehyde-free leather production with enhanced properties. *Carbohydrate Polymers* 295: 119838.

13 China, C.R., Nyandoro, S.S., Munissi, J.J. et al. (2021). Tanning capacity of *Tessmannia burttii* extracts: the potential eco-friendly tanning agents for the leather industry. *Journal of Leather Science and Engineering* 3: 1–9.

14 Michaloudis, I., Kanamori, K., Pappa, I., and Kehagias, N. (2023). U (rano) topia: spectral skies and rainbow holograms for silica aerogel artworks. *Journal of Sol-Gel Science and Technology* 106 (2): 319–330.

15 Du, Y. and Kim, H.-E. (2022). Research trends of the application of aerogel materials in clothing. *Fashion and Textiles* 9 (1): 23.

16 Tsegay, F., Ghannam, R., Daniel, N., and Butt, H. (2023). 3D printing smart eyeglass frames: a review. *ACS Applied Engineering Materials* 1 (4): 1142–1163.

17 Equbal, A., Equbal, M.A., Khan, Z.A., and Badruddin, I.A. (2025). A review on the rapid liquid printing (RLP): future 3D printing technology. *Progress in Additive Manufacturing* 10: 3157–3173.

18 Rosales-Murillo, S., Rosales-Murillo, S.S., Sánchez-Bodón, J. et al. (2024). Anthocyanin-loaded polymers as promising nature-based, responsive, and bioactive materials. *Polymers* 16 (1): 163.

19 Qian, D. (2024). *Thermal-Sensitive Nanocomposite Gel for Heat Shielding Application*. Hong Kong University of Science and Technology (Hong Kong).

20 Perera, M.M. and Ayres, N. (2020). Dynamic covalent bonds in self-healing, shape memory, and controllable stiffness hydrogels. *Polymer Chemistry* 11 (8): 1410–1423.

21 De Falco, F. (2018). *Microplastic Pollution from Synthetic Textiles: Quantitative Evaluation and Mitigation Strategies*. Università degli Studi di Napoli Federico II.

22 Raja, A., Arputharaj, A., Saxena, S., and Patil, P. (2019). Water requirement and sustainability of textile processing industries. In: *Water in Textiles and Fashion* (ed. S.S. Muthu), 155–173. Woodhead Publishing.

23 Govindan, N. (2018). Smart, waterproof, breathable sportswear: a review. *Journal of Textile and Apparel Technology and Management* 10 (3).

24 Nasrin, S. (2023). *Characterizing the Effects of Fabric Parameters on Drying Performance in Active Sportswear*. Oklahoma State University.

25 Pereira, T.M.P.A.N. (2021). *Under Armour Conquering the Sportswear Market*. Universidade NOVA de Lisboa (Portugal).

26 Hayes, S.G. and Venkatraman, P. (2015). *Materials and Technology for Sportswear and Performance Apparel*. Boca Raton, FL: CRC Press.

27 Childs, M. and Jin, B. (2018). Nike: an innovation journey. In: *Product Innovation in the Global Fashion Industry* (ed. B. Jin and E. Cedrola), 79–111. New York: Palgrave Macmillan US.

28 Angle, S. (2017). Hit the slopes. *Shape* 36 (5): 66–69.

29 Gharehaghaji, A.A. (2019). Nanotechnology in sport clothing. In: *Materials in Sports Equipment* (ed. A. Subic), 521–568. Elsevier.

30 Ye, H., Liu, Y., and Zhao, X. (2024). Recent research progress on graphene-based sound-absorbing materials. *Materials Today Chemistry* 38: 102139.

31 Periyasamy, A.P. (2024). Recent advances in the remediation of textile-dye-containing wastewater: prioritizing human health and sustainable wastewater treatment. *Sustainability* 16 (2): 495.

32 Zitkus, M., Lowe, B., Voss, B. et al. (2020). A case study of colorfastness of waterless dyed fabrics sustainable for the environment. In: *International Textile and Apparel Association Annual Conference Proceedings*, vol. 77(1). Iowa State University Digital Press.

33 Axentjärn, P. and Haddad, R. (2023). Circularity in the Fashion Industry: a study on recycling practices and end-of-life textile recycling among fashion retailers. Bachelor thesis for International Logistics Program. Chalmers University of Technology.

34 Egan, J., Wang, S., Shen, J. et al. (2023). Enzymatic textile fiber separation for sustainable waste processing. *Resources, Environment and Sustainability* 13: 100118.

35 Newman, Z. (2020). *A Comparison of Natural and Synthetic Yarns in Common Knit Footwear Structures*. North Carolina State University.

36 Wei, Y.D. (2009). China's shoe manufacturing and the Wenzhou model: perspectives on the world's leading producer and exporter of footwear. *Eurasian Geography and Economics* 50 (6): 720–739.

37 Bodoga, A., Nistorac, A., Loghin, M.C., and Isopescu, D.N. (2024). Environmental impact of footwear using life cycle assessment—case study of professional footwear. *Sustainability* 16 (14): 6094.

38 Humayra, S., Hossain, L., Hasan, S.R., and Khan, M.S. (2023). Water footprint calculation, effluent characteristics and pollution impact assessment of leather industry in Bangladesh. *Water* 15 (3): 378.

39 Swathy, K.K., Verma, R., and Kumar, L. (2023). Ethylene-vinyl acetate foam. In: *Polymeric Foams: Fundamentals and Types of Foams*, vol. Volume 1, 205–221. Washington, DC: American Chemical Society.

40 Kermen, E. (2020). *New Approach to Foot Orthotics Materials: Hydrogel Insoles.* University of British Columbia.

41 Galgali, A. (2022). *Prototyping of Injection EVA Foam Footwear Midsoles.* Massachusetts Institute of Technology.

7

Plant Fiber-Based Hydrogels and Aerogels for Biomedical Applications

Atike Köken[1] *and Ayşe Çelik Bedeloğlu*[2]

[1]*Bursa Technical University, Technology Transfer Office, Mimar Sinan Avenue, Eflak Street 177-4, 16310, Bursa, Türkiye*
[2]*Bursa Technical University, Polymer Materials Engineering Department, 16310, Bursa, Türkiye*

7.1 Introduction

Plant fibers have been frequently used in advanced sustainable biomedical materials in recent years due to their abundance, renewable nature, mechanical strength, and processability. Biopolymers found in plant-derived materials offer advantages in terms of biocompatibility, degradation, and sustainability. In biomedical hydrogels and aerogels, the most important and widely used biopolymer is cellulose, a β-1,4-linked linear glucan that serves as a structural component in plant cell walls. It exists in various hierarchical forms from macrostructures to nanoscale materials: cellulose nanofibrils (CNF), cellulose nanocrystals (CNC), bacterial cellulose (BC), etc. In addition, lignin and hemicelluloses are other plant-derived polymers [1, 2].

In addition to wood and agricultural residues (e.g. wheat straw), plant sources such as bamboo and grasses are also used for plant fibers [3, 4]. Many studies have reported that the plant source and plant species affect the properties of the extracted polymers (degree of polymerization, morphology, etc.). Extraction methods (mechanical, chemical, or enzymatic) and subsequent nanofibrillation or hydrolysis processes greatly affect the final material properties, including the availability of functional groups and surface chemistry, and the ability to be tailored for biomedical applications [5–7].

Plant fibers offer several properties that are critical to biomedical innovation, including mechanical strength, hydrophilicity and swelling, surface reactivity, biocompatibility and biodegradability, and renewability and sustainability [8]. These properties of plant fibers make them attractive for many biomedical applications, such as tissue engineering [4], wound dressings [9], biosensors and diagnostics [1], controlled drug delivery [3], and flexible bioelectronics and wearable devices [10, 11]. Compared to synthetic polymers, plant fibers are mostly hydrophilic, inherently biodegradable, and offer excellent surface functionality; these properties are difficult to mimic even by modifying synthetic polymers. However, synthetic polymers are better than cellulose, the most widely used biopolymer, in terms of chemical inertness and environmental durability [5]). Compared to animal-derived fibers (collagen, silk, etc.), plant fibers can be produced at larger scales, at lower cost, and without the risk of contamination by animal-derived pathogens [4]. Although animal-derived biomaterials often present stronger intrinsic bioactivities (e.g. signals for cell adhesion), plant fibers can overcome these disadvantages through methods such as composite preparation and functionalization.

7.1.1 Plant Fiber-Based Hydrogels and Aerogels: Synthesis Methods, Properties, and Applications

Hydrogels and aerogels derived from plant-based polymers such as nanocellulose and lignin have emerged as fundamental classes of materials that enable new approaches in biomedicine, flexible electronics, and sensing. The unique combination of mechanical flexibility, porosity, and biocompatibility results from the three-dimensional (3D) network architectures and intrinsic chemistry of plant-derived building blocks [12, 13]. These properties allow researchers to precisely design physical, chemical, and electrical functions by tailoring synthesis conditions and network assembly.

Hydrogel production begins with the extraction and purification of CNF, CNC, or other plant-derived polymers. Techniques include mechanical fibrillation, acid hydrolysis, and chemical oxidation such as TEMPO-mediated or periodate oxidation [14, 15]. These methods produce fibers with functional groups suitable for various cross-linking strategies [2]. The assembly of hydrogels utilizes a variety of basic cross-linking approaches, including physical cross-linking, dynamic covalent cross-linking, and double/multiplex networks, each of which imparts different properties to the final network. Plant-based hydrogels can be prepared by in situ gelation, shear thinning, and injectable hydrogels, as well as methods such as 3D printing [6, 7, 13, 16, 17].

Aerogels are obtained by carefully removing solvent from hydrogels using techniques such as freeze-drying (lyophilization), hydrogel drying, and supercritical CO_2 drying to preserve the network architecture and porosity. In addition, carbon aerogels with electrical conductivity, chemical stability, and more tunable porosity, which can be used in applications for energy storage and electrochemical converters, can be produced by different carbonization or pyrolysis methods [4, 18].

7.1.2 Properties and Application Areas of Plant-Based Hydrogels and Aerogels

Hydrogels exhibit high swelling rates, generally more than 1000%, thanks to their network structures and enable drug delivery and exudate management [17, 19]. Aerogels offer advantages in biomedical applications thanks to their properties such as low density, high surface area, and adjustable pore size (10 nm–10 μm) [10, 20]. Plant-based hydrogels exhibit a wide range of mechanical properties such as strength, compression, stiffness, toughness, flexibility, and fatigue resistance due to the design and density of their cross-linked networks. Polymeric materials are generally insulators, but different hydrogel and aerogel designs can provide biocompatibility and degradation properties, as well as tunable ionic or electrical conductivity properties [10, 13, 18]. In biomedical applications, flexible electronic conductors derived from flexible polymeric matrices with intrinsically conductive polymers or conductive fillers integrated are frequently used. In addition, flexible ionic conductors formed by embedding mobile ions in cross-linked polymer networks with intrinsic flexibility and electrical conductivity are also increasingly used in applications [21]. The main application areas of plant-based hydrogels and aerogels include tissue engineering scaffolds, wound dressings and healing materials, controlled drug delivery, sensing applications such as smart bandages and biosensor platforms, and wearable electronics.

7.2 Wound Dressings

Textile technology has advanced beyond traditional clothing and into high-performance applications, especially in medical fabrics. Wound dressings are one of the application areas of medical textiles. Wounds are classified as either acute or chronic based on the length of time and kind of healing process [22]. An acute wound is a skin injury that develops abruptly as a result of an accident or surgery. It normally takes 8–12 weeks to heal, though this depends on the size, depth, and severity of the injury to the skin's epidermis and dermis layers [23]. On the other hand, chronic wounds are unable to heal in a timely and organized way and do not advance through the typical stages of healing [24]. Burns and diabetic ulcers are examples of chronic wounds that provide significant clinical challenges due to their slower healing processes. The physiological process of wound healing is complex and includes hemostasis, inflammation, proliferation, and remodeling. To restore tissue integrity, these processes must be carefully coordinated [25]. Active intervention has replaced passive protection in the field of wound care in recent years. Smart and biofunctional materials are being developed to improve patient outcomes, decrease infection, and speed up healing. Particularly promising components of improved wound dressings include hydrogels and aerogels, which provide cutting-edge properties that greatly outperform those of conventional wound dressings like films, foams, or gauze. A moist wound environment that encourages tissue regeneration is produced by 3D hydrophilic networks known as hydrogels, which have a large water-holding capacity. Aerogels, on the other hand, are porous, lightweight materials with a wide surface area that offer improved thermal insulation and exudate management. A wound dressing material is expected to meet the following properties [26–28]:

- Absorbing the toxic substances and exudates from the wound surface
- Keeping the layer between the dressing and the wound extremely humid
- Shielding the wound from bacterial infiltration

- Being nontoxic
- Promoting gaseous exchange
- Offering thermal insulation
- Being readily removed without injuring the wound
- Being comfortable and sterilizable
- Having appropriate handling capabilities

When compared to conventional wound dressings, Hydrogel–textile composite wound dressings (HTCWD) and aerogels have many advantages, such as reducing discomfort, encouraging autolytic debridement, and acting as a delivery system for medications [29]. The other advantages of them are:

- Enhanced moisture control: Hydrogels maintain a moist wound environment, while aerogels can absorb exudate efficiently without causing maceration.
- Controlled drug delivery: Both materials can be loaded with therapeutic agents, enabling localized and sustained drug release.
- Biocompatibility and bioactivity: Natural polymer-based hydrogels and aerogels are generally nontoxic and can actively promote cell adhesion, proliferation, and tissue regeneration.
- Antimicrobial protection: Incorporation of antimicrobial agents (e.g. silver, antibiotics, or herbal extracts) into hydrogel or aerogel matrices provides infection control.
- Conformability and flexibility: Textile-based formats of these materials offer better adaptability to wound contours, enhancing patient comfort and reducing dressing displacement.

7.2.1 Hydrogel–Textile Composites for Wound Dressings and Applications

Hydrogels are one of the important materials used in many fields such as wound healing and have a 3D structure with hydrophilic polymer networks that can absorb and retain large amounts of water while maintaining their structural integrity. Hydrogels closely mimic the extracellular matrix (ECM) because to their high water content, flexibility, porous nature, and biocompatibility. They also provide a moist wound environment that is necessary for cell migration, proliferation, and tissue regeneration [30]. They are perfect for both acute and chronic wounds because of their ability to absorb exudate, dispense medication, and regulate temperature.

Hydrogel wound dressings consist of hydrophilic, inflatable, and insoluble materials, typically provided in gel and film forms. They have 70–90% water, enabling them to absorb several ulcers and rendering them ideal solely for the surface of wounds [31]. They are especially suitable for a range of wound types due to their adaptability, biocompatibility, and therapeutic loading capability. These substances create a moist wound environment, which is essential for encouraging fibroblast migration, autolytic debridement, and epithelialization. Hydrogels' therapeutic efficacy can also be increased by engineering them to release growth factors, antibacterial agents, or stem cells in a regulated way [32]. Hydrogels used in wound dressings are divided into two types: natural hydrogels and synthetic hydrogels. Natural biopolymers, including alginate, chitosan, gelatin, and hyaluronic acid, as well as synthetic polymers like polyvinyl alcohol (PVA), polyethylene glycol (PEG), and polyacrylamide (PAAm), are often utilized as hydrogel-forming polymers [33, 34].

Natural material-based hydrogels have a number of drawbacks, including poor mechanical qualities, high acquisition costs, low production, and challenging modification. Certain purposes, superior mechanical qualities, high output, affordability, and variety are all features of synthetic materials. Synthetic polymers, on the other hand, typically exhibit no biological or biodegradable activity and may result in harmful byproducts that cause tissue necrosis throughout the reaction [34]. The inherent mechanical fragility of many hydrogels might occasionally restrict their direct use, particularly in areas that need frequent handling or significant support. Hydrogel–textile composite wound dressings (HTCWD) combine the mechanical strength and conformability of textile substrates with the bioactive properties of hydrogels. Textile materials, including knitted, woven, and nonwoven fabrics, can be used in hydrogel matrices to get around these mechanical restrictions. Textiles give the composites better handling qualities, increased tensile strength, and structural integrity, which makes them simpler to apply, remove, and secure in place. Furthermore, textiles' fibrous nature can promote gas exchange, inhibit maceration, and provide adjustable porosity—all of which are essential for managing exudate and maximizing wound breathability. When it comes to fluid management, healing promotion, and patient comfort, these hybrid dressings are showing themselves to be superior to conventional materials [35].

HTCWDs for wound care are frequently designed using a variety of fabrication techniques, such as solvent casting, dip coating, electrospinning, and in situ polymerization, each of which has unique benefits for managing the architecture and functionality of the composite. The biocompatibility, biodegradability, and mechanical qualities of the composite are greatly influenced by the textile material selection, whether it be synthetic polymers like polyester or polyamide or natural fibers like cotton or silk. Furthermore, the hydrogel or textile component can be functionalized with bioactive substances such as growth factors, antimicrobials, or anti-inflammatory medications to give the composites tailored therapeutic properties that promote quicker healing and infection prevention [36].

Hydrogels must be integrated with textile substrates using specific fabrication methods that provide high interfacial adhesion, a controllable composite architecture, and the intended functional characteristics. Both the intended use and the particular hydrogel and textile materials often influence the technique selection [37]. The techniques used for these interactions are detailed as follows: (1) in situ polymerization, (2) sol–gel method, (3) solvent casting, (4) dip coating, and (5) electrospinning.

1) In situ polymerization: This method promotes superior interfacial adhesion and homogeneous composite formation by producing the hydrogel concurrently with its interaction with the textile scaffold, which sets it apart from post-fabrication integration techniques (such as lamination or coating). In situ polymerization

involves the application of initiators, cross-linkers, and monomers to the cloth, which acts as a template to direct the creation of hydrogels. The composite's stability and mechanical strength are improved by these interactions. The fabric is prepared, then soaked in a precursor solution and polymerized by enzymatic, thermal, UV, or redox initiation techniques. For applications like wound dressings, the resultant HTCWDs are next cleaned to get rid of unreacted components and dried, commonly by air or freeze drying, to have the right structure and biocompatibility [38].

2) Sol–gel method: Sol–gel process involves chemically converting molecular precursors into a gel network that is directly merged with a textile substrate. Pre-incorporating therapeutic ingredients, like antimicrobial nanoparticles, into the sol for even dispersion is possible with this technique. A crucial step is to fully impregnate a biocompatible fabric with the sol solution that has been created, enabling deep penetration. Next comes the important gelation stage, in which the sol transforms into a stable hydrogel inside the fabric. After gelation, thorough washing eliminates unreacted components to guarantee biocompatibility. Finally, the wet gel undergoes a controlled drying process to become the final wound dressing. Through the production of multipurpose dressings with specialized mechanical properties, high absorbency, and regulated therapeutic release, this meticulously controlled sol–gel approach enables a significant advancement in wound treatment [39].
3) Solvent casting: Hydrogel–textile wound dressings can be easily made using the solvent casting technique. A preformed hydrogel polymer is physically deposited onto a fabric in this process, and a solid composite is created by evaporating the solvent. A hydrophilic polymer (such as PVA, alginate, cellulose derivatives, or gelatin) is first dissolved in a volatile solvent, usually water, to produce a "dope" solution. This solution is subsequently applied or impregnated onto a prepared textile substrate using methods such as blade casting. The dissolved polymer solidifies by further controlled solvent evaporation, forming a cohesive hydrogel sheet that is closely woven into the textile to create a synergistic material [40].
4) Dip coating: The dip coating process, which applies thin, homogeneous polymeric films onto porous fiber substrates, is a flexible and scalable way to create hydrogel–textile wound dressings. A textile is submerged in a prepared solution of hydrogel-forming polymers (such as PVA, alginate, and chitosan) dissolved in a solvent, usually water, as part of the process. Antimicrobials and growth factors are examples of therapeutic substances that can be added to this solution. The hydrogel is solidified by further drying, forming a composite that is closely linked to the fiber structure of the fabric. Dip coating hydrogel–textile composites, which combine the advantages of hydrogel with textile support, show promise for wound treatment. Because the textile reinforcement keeps the hydrogel from tearing, removal is painless and trauma is reduced. Despite being easy, affordable, and versatile for large-scale manufacturing, difficulties include controlling material waste and deep hydrogel penetration in thick textiles [41].
5) Electrospinning: Electrospinning is a ground-breaking and highly adaptable method that provides unique structural and functional advantages that are

challenging to do with conventional methods. A fine jet emerges from a polymer melt or solution after it is exposed to a high voltage. Hydrogel-forming polymers, such as PVA, chitosan, alginate, gelatin, collagen, and hyaluronic acid, are prepared as electrospinning-suitable solutions for hydrogel–textile composites. A major benefit in active wound management is the ability to directly incorporate therapeutic agents, such as growth factors, antibiotics, and anti-inflammatory compounds, into the spinning solution, enabling their controlled and sustained release at the wound site. In order to maintain a moist and sterile healing environment, electrospun mats' high surface area-to-volume ratio and interconnected porous structure enable optimal gas exchange, efficiently manage wound exudate, and act as a barrier against external pathogens. By adding antimicrobial agents like silver nanoparticles or plant extracts to the fibers, these dressings can be designed to have active functionalities like antibacterial qualities in addition to their structural resemblance to the ECM. By encouraging angiogenesis, collagen synthesis, and re-epithelialization, electrospun hydrogel–textile composites have demonstrated promise in the treatment of a variety of wound types, including diabetic foot ulcers, burns, and chronic wounds [42, 43].

7.2.2 Aerogels for Wound Dressings and Applications

Aerogels are the lightest solid materials in the world. High porosity, low heat conductivity, and wide surface area are some of their special physicochemical properties that make them perfect for wound-healing applications, especially in drug administration and moisture regulation. Recent studies emphasize the usage of aerogels based on alginate, chitosan, cellulose, and their combinations for wound dressing applications. Bioactive substances like antibiotics or silver nanoparticles can be added to these aerogels to functionalize them and provide antibacterial action while keeping the wound surface dry.

Aerogel-textile wound dressings, which have a nanostructured network and offer a large specific surface area, are highly porous solid materials that are generally distinguished by their exceptionally high porosity (sometimes exceeding 90%) and ultra-low density [44, 45]. These inherent qualities are quite beneficial for applications involving wound dressings. Aerogel–textile composites' high porosity guarantees superior gas permeability, which is essential for avoiding hypoxia at the wound site and promoting the cellular respiration required for healing. At the same time, their porous shape enables better fluid absorption, which efficiently controls wound exudate and preserves the ideal moist environment necessary for encouraging cell migration, re-epithelialization, and generally faster wound healing [44, 46]. Additionally, these dressings' ultra-lightweight design greatly improves patient comfort and lessens the physical strain that larger, traditional bandages frequently cause.

The two main types of aerogels used in wound dressing applications are organic (polymeric) aerogels and those made from natural biopolymers; the latter are frequently used in composite forms to improve mechanical integrity. Although there are composite forms of inorganic aerogels, such as silica aerogels, which are known for their insulating qualities, their brittleness and potential for toxicity during preparation may restrict their direct use in medicine. Because of their intrinsic biocompatibility, biodegradability, and reduced immunogenicity, organic aerogels are preferred more frequently [47].

Cellulose aerogels, which are made from the most prevalent natural polymer, have exceptional absorption, flexibility, and adjustable qualities that make them appropriate for a range of wound applications and drug loading [48]. These aerogels offer the perfect scaffold for cell adhesion, proliferation, and differentiation because of their high porosity and large specific surface area. For the manufacture of aerogels, cellulose can be made by bacteria (BC) or obtained from plants (wood, agricultural waste, etc.) [49]. Particularly, BC has exceptional mechanical strength and water retention properties due to its high purity, crystallinity, and distinct nanofiber networks. They are great options for direct wound contact because of their innate biodegradability and biocompatibility, which reduces negative reactions. It is simple to functionalize cellulose aerogels by adding active substances during production or by chemically altering their plentiful hydroxyl groups. This enables them to demonstrate improved antimicrobial qualities, regulate the long-term release of medicinal medications (such as antibiotics and anti-inflammatory drugs like ibuprofen), and encourage tissue regeneration by giving fibroblasts and other wound-healing cells a favorable environment, thus boosting their capacity to heal wounds [50].

Alginate aerogels, which are made from brown algae, are very biocompatible and biodegradable, which makes them great options for absorbing exudates from wounds and delivering medications that have an anti-inflammatory impact.

Alginates are anionic polysaccharides composed of repeating blocks of β-D-mannuronic acid (M) and α-L-guluronic acid (G) units [51, 52]. Their most significant functional property is their ability to form hydrogels through ionotropic gelation. When alginate-based textile dressings come into contact with wound exudate (which contains sodium ions), an ion exchange occurs: calcium ions from the dressing are replaced by sodium ions from the fluid. This causes the insoluble calcium alginate fibers to swell and transform into a soft, cohesive gel that effectively absorbs large volumes of exudate. The management of extremely exudative wounds, prevention of maceration of the surrounding skin, and preservation of the ideal moist environment necessary for faster healing all depend on this high absorption capacity [53, 54]. Alginate aerogels have important therapeutic benefits beyond fluid management. In bleeding wounds, the dressing's released calcium ions can actively contribute to the clotting cascade and aid in hemostasis. Alginate has also been demonstrated to promote cell migration and proliferation, which may directly contribute to improving the healing process of wounds. Alginate-based dressings have been shown in studies to enhance collagen deposition and promote angiogenesis or the creation of new blood vessels, resulting in the production of more ordered and mature scars [55]. They perform even better when certain ions, such as zinc and silver, are added, offering broad-spectrum antibacterial action and immune-modulatory effects, respectively [52]. In addition to being useful for a variety of biomedical applications, such as bone and cartilage tissue engineering, their porous structure acts as a scaffold for cellular infiltration and tissue regeneration [56]. It is possible to precisely manipulate the porosity architectures of these biopolymer-based aerogels during synthesis, enabling customized features that promote ideal wound microenvironments.

Specialized fabrication processes are needed to integrate aerogels with textile substrates to create wound dressings. These procedures must maintain the aerogel's delicate porous structure while guaranteeing strong adhesion and functional integration with the textile. A number of techniques have been created to achieve this goal: (1) particle integration/embedding, (2) in situ sol–gel synthesis and textile impregnation, (3) coating technique, and (4) fiber-based aerogels and integration.

1) Particle integration/embedding: This method entails directly integrating pre-made aerogel granules or particles into the textile structure. Techniques include thermal bonding, which uses heat to fuse textile fibers around aerogel particles; resin bonding, which uses a polymeric resin as a binder to physically entrap the aerogel within nonwoven fabrics; and needle punching, which mechanically embeds aerogel particles into fibrous mats. This technique, which is frequently used for larger-scale production, enables the dispersed presence of aerogel properties throughout the fabric.
2) In situ sol–gel synthesis and textile impregnation: The classic sol–gel synthesis of aerogel is utilized in this technique, in which the cloth is impregnated with the liquid precursor solution (sol). A thorough drying procedure [such as supercritical drying (SCD) or freeze-drying (lyophilization)] is then used to extract the solvent while maintaining the porous aerogel structure after the gelation process takes place in situ within the textile's fibrous network. A more intimate bond between the textile and the aerogel may be created with this method, creating a material that is really composite. To create cellulose–silica aerogels, for instance, cellulose layers have been directly impregnated into silica hydrosols or alcosols [57].
3) Coating techniques: Several coating techniques can be used to apply liquid aerogel precursors or coating paste containing aerogel to textile surfaces. These include spray-coating, which permits targeted and controlled application; and dip-coating, in which the fabric is submerged in the aerogel dispersion or knife coating methods [58, 59].
4) Fiber-based aerogels and integration: Aerogels are first converted into fibers in certain sophisticated methods, and these fibers are subsequently incorporated into textile constructions. Utilizing methods like wet spinning or solution blow spinning (SBS) of cellulose or chitosan solutions, followed by freeze-drying or supercritical CO_2 drying, aerogel microfibers or nanofibers can be created [49]. To create the final wound dressing, these aerogel fibers can subsequently be knitted, woven, or incorporated into nonwoven mats. Prior to being included in the broader textile structure, this method provides improved shapeability and mechanical qualities for the aerogel component itself.

Aerogel–textile wound dressings have substantial therapeutic potential due to their ability to promote tissue regeneration and medication release, in addition to their structural advantages. Aerogels have significant loading capabilities for a variety of bioactive substances, such as growth hormones, antibiotics, anti-inflammatory medications, and even stem cells, thanks to their large internal surface area and adjustable pore network. By optimizing local medication concentrations and reducing systemic adverse effects, this enables the regulated and prolonged release of these therapeutic substances directly at the wound bed, thereby fostering targeted healing responses. Furthermore, the aerogel–textile composite's 3D porous scaffold provides a biomimetic environment that promotes cell adhesion, proliferation, and differentiation. This facilitates the formation of granulation tissue and, in the end, promotes skin regeneration with less scarring. These properties make aerogel–textile dressings more than just passive barriers; they make them active players in the healing cascade.

As a summary of the wound dressings, the mechanical strength and usefulness of textile structures are further improved by the incorporation of these smart materials, opening the door for the development of next-generation wound dressings that blend therapeutic functionality with an intuitive design. With an emphasis on their composition, modes of action, therapeutic potential, and prospects for individualized wound care in the future, this review seeks to critically analyze the function of hydrogel and aerogel fabrics in wound healing

7.3 Drug Delivery Systems

Advanced biomaterials and smart textiles have created new opportunities in personalized medicine, especially in the area of medication delivery. Problems with dosage management, site-specific targeting, and patient compliance are common in traditional medication delivery. By using the natural flexibility, breathability, and comfort of textiles, textile-based drug delivery systems present a strong alternative. These systems can administer therapeutic drugs systemically through transdermal pathways or directly to the target site. Hydrogels and aerogels stand out in this cutting-edge industry as materials with special qualities that support regulated drug release [60].

Hydrogels can provide properties including moisture control, wound healing, and prolonged medication release when incorporated into fabrics. They are great options for encapsulating and releasing a variety of therapeutic compounds, including proteins, cells, and small-molecule medications, due to their porous architectures, biocompatibility, and adjustable swelling kinetics [61].

In contrast, aerogels are extremely light and porous materials made from gels by substituting air for the liquid component, usually during supercritical drying. They are suited for high-capacity drug loading and precise control over release patterns because of their tunable pore size distribution, wide surface area, and extraordinarily high porosity. Aerogels can be incorporated into textile substrates to produce innovative long-term drug delivery systems, especially for uses that call for stability and effective payload protection. The creation, description, drug loading processes, and release kinetics of textile-based hydrogel and aerogel composites will be covered in detail in this chapter, along with their unique benefits and difficulties in contemporary drug delivery applications [62].

7.3.1 Hydrogels for Wound Dressings and Applications

Several methods are used in the creation of hydrogel–textile composites, with the main goal being to maintain the mechanical integrity and wearer comfort of the textile while establishing a stable interface between the polymeric gel and the textile substrate. In situ polymerization, coating, and the integration of preformed hydrogel particles are examples of common techniques (see Section 7.2.1).

Hydrogel–textile composites can be loaded with drugs either during the synthesis process (e.g. by adding the drug to the monomer solution prior to polymerization) or after the synthesis is complete (e.g. by submerging the hydrogel–textile in a drug solution, which permits absorption into the swollen network). The characteristics of the drug and the affinity of the hydrogel determine the loading method to be used [63].

The main factors controlling drug release from hydrogel–textile composites are the hydrogel matrix's swelling, diffusion, and breakdown.

- Diffusion-controlled release: A concentration gradient can cause small drug molecules to diffuse out of the hydrogel network. The hydrogel's pore size, cross-linking density, and drug molecular size all affect the rate of diffusion [30].
- Swelling-controlled release: The expansion of the polymer network, which raises the effective diffusion channel, can be used to control the release of encapsulated pharmaceuticals in hydrogels that expand considerably when hydrated. On-demand release is made possible by the ability of environmental cues such as pH, temperature, or ionic strength to cause swelling [64].
- Degradation-controlled release: Biodegradable hydrogels, which are frequently composed of synthetic polymers with hydrolyzable linkages (like polyesters) or natural polymers (like chitosan and alginate), release encapsulated medications when the polymer matrix breaks down. This method works especially well for delivering biologics that need to be protected until they are released or for long-term drug delivery [63].

7.3.2 Aerogels for Wound Dressings and Applications

Aerogels offer a unique drug delivery platform because of their ultralow density and distinctive hierarchical porosity structure, particularly when paired with textile substrates. Aerogels have the potential for multimodal drug release and extraordinarily high drug loading capacities due to their high surface area and adjustable pore volume [65]. Aerogels are excellent for loading drugs because of their large internal surface area; they frequently achieve noticeably larger payloads than hydrogels. Usually, loading is accomplished by adsorption, in which drug molecules are incorporated into the pore network of the aerogel by submersion in a drug solution and then solvent extraction [66].

Aerogel–textile systems have potential for long-term, stable drug delivery applications, like the delivery of anti-inflammatory medications in orthopedic supports. They enable sustained release of antibiotics in chronic wound care or serve as reservoirs for volatile therapeutic agents because of their insulating qualities. Although the majority of drug release from aerogel–textile composites is controlled by diffusion, the unique pore structure provides the following benefits:

- Hierarchical porosity: The hierarchical pore structure (macropores, mesopores, and micropores) that aerogels frequently have can be used to regulate release rates. Deeper, smaller pores may discharge continuously after an initial burst from the surface pores.
- Matrix dissolution/degradation: Just like biodegradable hydrogels, certain forms of aerogels (such as cellulose aerogels and biopolymer-based aerogels) may have prolonged drug release due to the gradual dissolution or biodegradation of the aerogel matrix.
- Surface modification: By functionalizing the internal surfaces of aerogels, certain interactions with drug molecules (such as hydrophobic or electrostatic contacts) can be introduced, further adjusting the release kinetics and avoiding the burst release of highly soluble medicines.

The integration of aerogels into textiles often focuses on maintaining the aerogel's structural integrity while ensuring robust attachment to the fabric. Particle integration/embedding, in situ sol–gel synthesis and textile impregnation, coating techniques, and fiber-based aerogels and integration methods are used to have drug delivery textile composites (see Section 7.2.2).

Hydrogel and aerogel drug delivery systems based on textiles are a revolutionary development in customized medicine and smart biomaterials. Aerogels' special porous structure allows for regulated release and remarkable drug loading capacity, while hydrogels' ability to modulate release through swelling and breakdown makes them versatile. Even though there are still issues with mechanical stability, scalability, and regulatory concerns, continuous advancements in material science and engineering are laying the groundwork for these sophisticated textile composites to transform drug delivery by providing better patient compliance, increased therapeutic efficacy, and a new standard for wearable medical technology.

7.4 Smart Bandages

Smart bandages, in addition to covering and protecting wounds, also actively interact with the wound environment to provide real-time information or controlled treatment in response to physiological cues [8, 9]. Smart bandages can often integrate sensing (via electrical, chemical, or optical signals) and/or sometimes controlled drug delivery [12] in addition to their traditional functions such as moisture management, barrier properties, and antimicrobial protection. As a smart material, it functions by sensing, detecting, and responding appropriately to external stimuli. In the wound condition, it acts as a biomarker for detecting pH, temperature, glucose, various metabolites, infection, exudate level, or healing progress, and it reveals its responses/reactions in response to the detected changes and also ensures that the environmental conditions remain functional and hygienic to perform all these functions [13]. Equipped with advanced sensors, these systems enable personalized

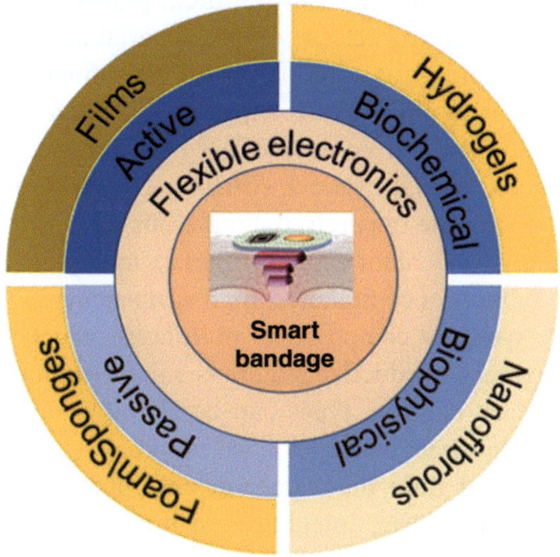

Figure 7.1 Schematic illustration of the smart bandage structure levels. *Source:* Levin et al. 2023 / MDPI / CC BY 4.0.

medical applications by providing wireless data transmission and patient-specific treatment support.

Traditional bandages provide general protection but lack the ability to respond to or report on the dynamic biochemical conditions of a healing wound, which can lead to delayed intervention and increased risk in critical wounds. Smart bandages are especially needed for chronic wounds such as diabetic ulcers or pressure sores, where continuous monitoring, precise, and adaptive care are required to prevent infection and accelerate healing. The use of bandages that can provide real-time feedback and programmable therapy supports personalized, precise wound care [9].

Because they are made of biocompatible materials, plant-based hydrogels are non-toxic and naturally compatible with skin and tissue. Due to their high absorbency, CNF/CNC hydrogels can manage abundant exudates, shorten dressing change intervals, and reduce the risk of infection [9]. Due to their mechanically versatile properties, hydrogel stiffness, flexibility, and fatigue resistance through cross-linking and control of nanostructure can meet the needs of delicate skin dressings or durable wound pads. Their engineerable electrical and ionic conductivity allows the integration of conductive pathways required for enhanced biosignal transmission [10]. Sustainably sourced cellulose-based smart bandages are renewable, low-carbon, and biodegradable depending on the formulation [12] (Figure 7.1).

7.4.1 Materials Used in Smart Bandages and Mechanisms

Hydrogels and aerogels obtained from natural polymers such as cellulose, lignin, pectin, chitosan, and starch are widely used in smart bandage systems due to their biodegradability, biocompatibility, and environmental friendliness. The high water retention capacity of hydrogels accelerates healing by keeping the wound environment moist while providing water-rich environments and the hosting of active molecules (e.g. enzymes and dyes) for enhanced sensing and treatment capabilities. CNF/CNC form the moist structure, forming robust, flexible, and swelling networks. The high surface area, porosity, and low density of aerogels facilitate the storage and release of antibacterial agents while supporting breathability and rapid fluid utilization [10, 18]. Lignin fractions are bonded or blended with CNF to provide antioxidant, anti-UV, and additional mechanical properties. Oxidized and functionalized cellulose is used for improved sensory response and a highly porous, special network structure [6, 7]. Plant fiber-based hydrogels and aerogels have become the preferred materials for smart bandages due to their high swelling ability (critical for exudate uptake), mechanical adjustability and strength for comfortable skin compliance, high surface area, and biocompatibility. Smart bandages are typically composed of multilayer structures and may include biocompatible polymers, conductive materials, micro-/nanosensors, and drug delivery systems.

Smart bandage types can be classified into two groups as passive and active. Passive smart bandages have minimal sensing or reactive functions but can offer advanced fluid management, antimicrobial effect, or structural support. Active smart bandages integrate sensing devices (electrochemical, colorimetric, and mechanical), provide feedback, and adapt the function in real time. Active smart bandages use hydrogels/aerogels with integrated ionic or electrical conductivity to convert chemical or mechanical signals (pH, temperature, oxygen, moisture, mechanical, electrical, etc.) [68] into readable outputs.

Some cellulose-based hydrogels change volume (swell/shrink) or show color changes in response to pH, ionic strength, or specific metabolites, providing a simple visual indicator of wound status. This can help nonexpert users detect infection or high exudate levels. In terms of capacitive/resistance measurement, changes in network conductance or permeability due to swelling or biochemical interactions are converted into electrical signals. In terms of amperometric sensing, enzymes immobilized on oxidized cellulose hydrogel (e.g. dialdehyde CNF) catalyze the conversion of glucose or lactate, generating a measurable electric current. In colorimetric response, plant-derived hydrogels are conjugated with pH or metabolite-sensitive dyes for visual readout [13].

Hierarchical porous hydrogels and aerogels made from plant nanocellulose can rapidly absorb exudate (swelling ratio > 1000%), retain moisture to optimize healing, and simultaneously maintain breathability, which is important for patient comfort and prevention of maceration. Freeze-drying and ice-molding techniques provide directional porosity, enabling directional fluid transport and effective exudate clearance.

7.4.2 Applications

Smart bandage applications consisting of plant fiber-based hydrogels and aerogels have focused on topics such as pressure and biosignal sensing, integration with sensing, and feedback devices. Carbon aerogel developed from cellulose nanofibers and lignin derived from wood has a lightweight, highly compressible (up to 95% strain)/flexible structure with high sensitivity over a wide pressure range (–16.89 kPa). The aerogel has been used as a flexible, fully solid-state pressure sensor that can accurately sense human biosignals, demonstrating its direct application in smart, responsive wound care and monitoring [10].

The hydrogel designed for an ultrasensitive pressure sensor with low-molecular-weight cellulose and rough surface structure has high piezocapacitive sensitivity (from 0.04 to 89.81 kPa^{-1}), enabling precise monitoring of deformations and physiological pressures [14]. Ion-exchange cellulose hydrogels exhibit robust mechanical and sensory properties, which can be applied to detect weak pressure changes in conformal contact with the skin [11]. Lignin and cellulose derivative-based hydrogels present both asymmetric adhesion and significant mechanical and electrical properties and are directly applied as bioelectrodes and self-powered sensors for wearable and wound care devices [6, 7]. Mostafalu et al. [69] studied a wireless, 3D-printed, flexible sensor smart bandage for continuous monitoring of wound oxygenation.

7.4.3 Future Perspective

Smart bandages are of great importance in terms of continuous wound monitoring and intervention, controlling drug release, early diagnosis of infection, and providing treatment-specific feedback. Recent studies indicate the high swelling, mechanical adaptability, functionalization capacity with sensing modules (enzymes, dyes, and nanoparticles), and robust integration into biomedical sensing platforms of smart bandages containing plant-based hydrogels and aerogels. Being biocompatible, biodegradable, and environmentally friendly, being relatively cheap and easy to produce from sustainable sources, and also being multifunctional (stimulable, antibacterial, release-controlled, etc.) greatly expand their areas of use. However, especially in humid environments, insufficient mechanical strength, limited electrical or ionic conductivity, difficulties in functional sensor integration, long-term accuracy and stability of the sensor, scalability, and reliability are issues that need further study and attention regarding smart bandages. In addition, various water retention strategies are being investigated with different additives for dehydration and stability problems that hydrogels encounter over time. Increasing the cross-link density for mechanical strength may reduce flexibility or sensitivity. Multilayered, multifunctional, and wireless communication integrated forms of plant-based smart bandage systems are expected to become widespread in the near future. Some advanced hydrogels utilize reversible (dynamic covalent or ionic) cross-linking for self-healing capacity, extending the functional life of smart bandages exposed to repeated mechanical stress. In addition, smart wound management will be possible with patient-specific biosensor integration in personalized treatment scenarios.

7.5 Wearable Electronics and Health Monitors

In the IoE era, with the rapid development of soft electronics, flexible electrical and ionic conductive materials, which are indispensable components of soft electronics, have come to the fore and have also faced various challenges. In wearable and stretchable electronics, some of the most important features of hydrogels and aerogels, especially those incorporated into the sensor structure, are thermal stability, high sensitivity, self-healing, and transparency.

Wearable electronics are an expanding class of portable, lightweight, and flexible devices designed for continuous, noninvasive monitoring or manipulation of physiological, behavioral, or environmental variables directly on the human body. These include pressure/strain sensors, smart textiles, energy harvesters, and energy storage elements, as well as integrated biosignal monitors for health diagnostics, rehabilitation, sports, and human–machine interfaces. The necessity of wearable electronics is driven by the increasing demands in healthcare (continuous patient monitoring, prosthetics, and smart wound care), security (motion and fatigue detection), personalized fitness tracking, and advanced human–computer interaction (virtual reality, robotics, and motion control) [1, 19, 70].

7.5.1 Materials Used and Mechanism

The materials used in wearable electronics and health trackers, in addition to their high-performance functionality, must be biocompatible, lightweight, robust, and adaptable to skin or dynamic movement. On the other hand, traditional sensors made of metal or inorganic semiconductors are generally rigid and brittle, so they are not compatible with the human body. Therefore, industry and academia have begun to study this issue in depth and improve many of its properties, including flexibility, durability, lightness, and biocompatibility, so that they can be used in various parts of the human body.

Plant-derived hydrogels and aerogels, which offer sustainable alternatives to synthetic or inorganic materials, have emerged as leading candidates for environmentally friendly, high-performance wearable electronics. Materials to be used for wearable electronics provide high flexibility, stretchability, and toughness against repeated bending or stretching. They provide compatible and comfortable contact with the skin or tissue and, when appropriate, sufficient ionic and/or electronic conductivity for reliable signal transmission or energy management.

Biocompatibility, minimum toxicity, and environmental degradability are important for single-use devices. In this context, plant fiber-based hydrogels and aerogels formed from CNF, CNC, lignin, and their dynamic or chemically functionalized networks have sufficient properties to be used in wearable electronics. Their mechanical properties are tunable (modulus from kPa to MPa; >50–95% strain tolerance), and they offer intrinsic hydrophilicity and swelling properties for ionic mobility. Ionic hydrogels (especially oxidized CNF) provide conductivity up to 10^{-1} S/cm in the swollen state, which is sufficient for electrophysiological and electrochemical signaling. This allows tuning of conductivity, mechanical strength, and signal sensitivity through chemical and processing routes. Many CNF-based sensors can withstand compressive strains up to 95% and stretch/flexure for over 50,000 cycles without significant structural or performance degradation. They are designed for conformal skin contact (modulus of 10–50 kPa) or robust device support (up to MPa scale) when needed, thus providing sustainable, nontoxic alternatives to petrochemical-based or dense inorganic composites [8].

7.5.2 Applications

Wearable electronics using plant fiber-derived hydrogels and aerogels focus on pressure and strain sensors, biosignal monitors, energy devices, such as supercapacitors and batteries, and smart textiles and integrated circuits. Plant fiber-derived hydrogels and aerogels are widely used as active layers in pressure- and strain-sensitive wearable sensors. These "electronic skin" devices can monitor slight changes in pressure, torsion, or tactile contact by piezoresistive or piezocapacitive transduction or iontronic sensing, which is important for biomedical applications, health monitoring, gesture recognition, and robotic feedback systems. Integration of plant-derived networks into wearable biosignal electrodes and health monitors enables detection of electrophysiological signals (e.g. ECG and EMG), sweat metabolites, or other physiological cues.

Carbon aerogels (after pyrolysis) derived from plant polymers are used as lightweight, flexible electrodes in wearable supercapacitors and rechargeable batteries. By combining plant-based hydrogels/aerogels with textiles or substrates, it is possible to create composite materials with integrated sensory or energy storage capabilities.

7.6 Sensor Applications

Hydrogels and aerogels can be used in pressure, tactile, and gesture recognition applications of sensors. They change resistance or capacitance when compressed or stretched due to changes in network geometry or ion mobility. In wood-derived CNF–lignin carbon aerogel sensors, these aerogels exhibit high compressibility (up to 95% strain) and fatigue resistance, achieving high sensitivity (sensing range of 0–16.89 kPa). The devices can accurately detect biosignals from humans and are suitable for wearable and flexible pressure sensor arrays [10].

CNF anisotropic carbon aerogels provide high shape retention and sensitivity, functioning reliably for thousands of cycles and over a wide pressure range [18]. Pressure sensors fabricated from the bacterial nanocellulose/wood CNF aerogel composite have been shown to have adjustable sensitivity (0–100 kPa), fast response, and mechanical durability and have been validated for motion detection, footstep recognition, and smartphone-based communication applications [13]. Cellulose ion-gel pressure sensors offer applications for human–machine interaction and medical monitoring via the piezocapacitive effect [14].

Supramolecular hydrogels exhibit tunable ionic conductivity under pressure or deformation, which can result in highly sensitive outputs. By developing cellulose-based supramolecular biopolymer gels with high ionic conductivity and exceptional hardness, ultrahigh iontronic sensitivity (1,475,000 kPa^{-1} ultrahigh sensitivity pressure sensor) has been used to distinguish objects via tactile feedback [70]. Wu et al. [71] significantly improved the thermal sensitivity of hydrogels (thermal sensitivity 13.1%/°C) by adding ethylene glycol/glycerol (Gly) as a solvent to hydrogels. The obtained Gly–organohydrogel exhibited improved anti-drying and anti-freezing properties, ultrahigh flexibility (1103% strain), self-healing ability, and high transparency by preventing drying and freezing at 70 and −18 °C, respectively. In another study, researchers [72] studied the fabrication of hydrogels with high conductivity and anti-swelling ability for precise temperature and strain detection over a wide temperature range. Among many types of salts, the multifunctional transparent sensors fabricated with LiBr showed the best performance for hydrogels,

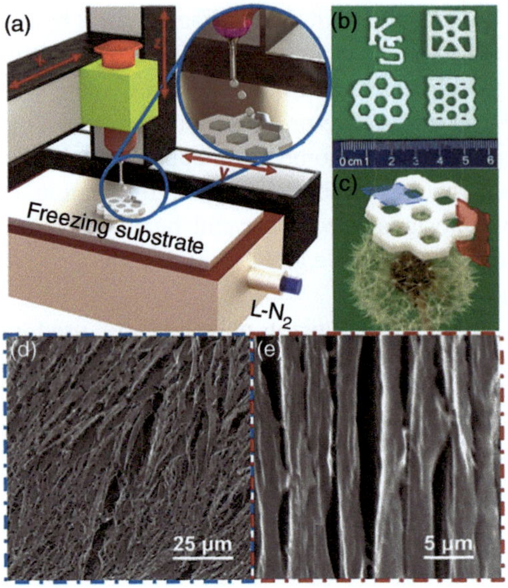

Figure 7.2 (a) Schematics of the 3D freeze printing process. (b) Photographs of the 3D freeze-printed cellulose aerogels. (c) 3D freeze-printed cellulose aerogel with a honeycomb shape standing on a dandelion. (d) SEM image showing the top surface of the 3D freeze-printed CNC aerogel. (e) SEM image showing the cross-sectional surface of the 3D freeze-printed CNC aerogel. *Source:* Reproduced from Tetik et al. (2021) / with permission of Elsevier.

exhibiting high thermal sensitivity (2.54%/°C), wide temperature sensing range (−78.5 to 97 °C), low detection limit (0.1% strain), and low hysteresis in cyclic strain sensing. The fabricated epidermal sensors can monitor various physiological signals such as facial expressions, word pronunciation, and knee bending in real time.

Cao et al. [20] developed a sensitive pressure sensor with a three-part microstructured carbon nanotube (CNT)/cellulose aerogel. The sensor components are a cellulose/poly(vinyl alcohol)/CNT aerogel-based sensing layer and a top and bottom thermoplastic polyurethane elastomer (TPU)/silver nanowire (Ag NW) nanofiber electrode, which were produced by freeze-drying and electrospinning. The combined sensors, which utilize microcone and micropore structures on the nanofiber electrode, showed a high sensitivity of 66.4 kPa^{-1} and an excellent response speed of 10 ms.

Tetik et al. [16] produced cellulose-based aerogels (Figure 7.2) with micropores aligned along the freezing direction by using a drop-based material deposition approach and a 3D freeze-printing method to investigate the mechanical properties and stability of these anisotropic materials. The stress reduction values obtained after 100 cyclic compression up to 25% strain were ~30% and ~16% in the axial and radial directions, respectively. Aerogels were functionalized with PEDOT:PSS, and their performance in piezoresistive sensing applications was evaluated.

7.7 Wearable Electronics and Health Monitors

Aerogel materials that provide both flexibility and high electrochemical performance (high specific capacitance [150–250 F/g], fast charge/discharge, and robust cycle life) can be developed for use in wearable electronics. Carbon aerogels generated from plant sources have been shown to endure extraordinarily high compressive stress (95%) for at least 50 cycles, or 50,000 cycles, with an 84% height retention at 50% strain. Carbon aerogels' mechanical capabilities, stable 3D lamellar architecture, and high conductivity have proven quick current response and good sensitivity over a wide pressure range (0.002–7 kPa) [57]. Flexible, fully solid-state symmetric supercapacitors fabricated using wood-derived CNF–lignin carbon aerogels exhibit high capacitance, good rate capability, and robust mechanical durability under repeated strain [10]. Zn-ion batteries with sorbitol-modified cellulose (CNF) hydrogel electrolyte derived from wheat straw operate efficiently and are resistant to environmental changes [15]. Thus, materials integrated into compatible, fully solid-state, and wearable devices for energy storage can be developed. Wang et al. [6, 7] produced a new bilayer hydrogel with asymmetric adhesion, durability, and electrical properties using lignin and cellulose-based sources. In the lower part that contacts the skin, a soft and skin-compatible lower hydrogel layer mediated by lignosulfonate sodium (LS)–borax (young modulus: ~14.2 kPa, adhesion strength: ~18.7 kPa) was designed, and in the upper part, a nonadhesive hydrogel layer mediated by quaternary hydroxyethyl cellulose (QHEC) (young modulus: ~101.3 kPa, adhesion strength: ~2.2 kPa).

7.8 Future Perspective

Plant-based hydrogels and aerogels can be easily and cost-effectively integrated into thin film, patch, textile, or wearable module designs. Biocompatible structures that can be made entirely from sustainable plants minimize the potential degradation after use and the toxicological and environmental burden. Tunable cross-linking, surface modification, and design allow for controllable mechanical and electrical outputs. On the other hand, the risk of fluid loss can affect comfort and ion/electrical performance. Long-term validation, reliability, scalability, sensitivity, durability balance, and reproducibility are still issues that need to be worked on. In particular, explaining structure–property relationships with advanced characterization and computational modeling techniques, developing new generation material designs, manufacturing with innovative techniques, and making measurement processes based on international standardization will overcome these difficulties.

References

1 Cui, C., Fu, Q., Meng, L. et al. (2021). Recent progress in natural biopolymers conductive hydrogels for flexible wearable sensors and energy devices: materials, structures, and performance. *ACS Applied Bio Materials* 4 (1): 85–121. https://doi.org/10.1021/acsabm.0c00807.

2 Du, Z., Wang, N., and Du, J. (2025). Recent progress in cellulose-based conductive hydrogels. *Polymers* 17 (8): 1089. https://doi.org/10.3390/polym17081089.

3 John, A. and Zhong, W. (2019). Nanocellulose-based hydrogels for biomedical applications. *Current Nanoscience* 15 (4): 371–381. https://doi.org/10.2174/1573413714666180723145038.

4 Khalil, H.P.S.A., Adnan, A.S., Yahya, E.B. et al. (2020). A review on plant cellulose nanofibre-based aerogels for biomedical applications. *Polymers* 12 (8): 1759. https://doi.org/10.3390/polym12081759.

5 Saleh, A.K., El-Sayed, M.H., El-Sakhawy, M.A. et al. (2025). Cellulose-based conductive materials for bioelectronics. *ChemSusChem* 18 (6). https://doi.org/10.1002/cssc.202401762: e202401762.

6 Wang, C., Bai, J., Tian, P. et al. (2021a). The application status of nanoscale cellulose-based hydrogels in tissue engineering and regenerative biomedicine. *Frontiers in Bioengineering and Biotechnology* 9. https://doi.org/10.3389/fbioe.2021.732513.

7 Wang, Q., Pan, X., Guo, J. et al. (2021b). Lignin and cellulose derivatives-induced hydrogel with asymmetrical adhesion, strength, and electriferous properties for wearable bioelectrodes and self-powered sensors. *Chemical Engineering Journal* 414: 128903. https://doi.org/10.1016/j.cej.2021.128903.

8 Zou, P., Yao, J., Cui, Y. et al. (2022). Advances in cellulose-based hydrogels for biomedical engineering: a review summary. *Gels* 8 (6): 364. https://doi.org/10.3390/gels8060364.

9 Rumon, M.M.H. (2025). Advances in cellulose-based hydrogels: tunable swelling dynamics and their versatile real-time applications. *RSC Advances* 15 (15): 11688–11729. https://doi.org/10.1039/d5ra00521c.

10 Chen, Z., Zhuo, H., Hu, Y. et al. (2020). Wood-derived lightweight and elastic carbon aerogel for pressure sensing and energy storage. *Advanced Functional Materials* 30 (17). https://doi.org/10.1002/adfm.201910292.

11 Zhou, S., Guo, K., Bukhvalov, D. et al. (2020). Cellulose hydrogels by reversible ion-exchange as flexible pressure sensors. *Advanced Materials Technologies* 5 (9). https://doi.org/10.1002/admt.202000358.

12 Jiang, G., Xu, G., Xia, Q. et al. (2025). Cellulose gels: functional design and promising smart applications. *Small*. https://doi.org/10.1002/smll.202412538, 21, e2412538.

13 Zhang, M., Xu, T., Liu, K. et al. (2025). Modulation and mechanisms of cellulose-based hydrogels for flexible sensors. *SusMat* 5 (1). https://doi.org/10.1002/sus2.255.

14 Chen, M., Wan, H., Hu, Y. et al. (2023). Rationally designed cellulose hydrogel for an ultrasensitive pressure sensor. *Materials Horizons* 10 (10): 4510–4520. https://doi.org/10.1039/d3mh01051a.

15 Quan, Y., Zhou, W., Wu, T. et al. (2022). Sorbitol-modified cellulose hydrogel electrolyte derived from wheat straws towards high-performance environmentally adaptive flexible zinc-ion batteries. *Chemical Engineering Journal* 446: 137056. https://doi.org/10.1016/j.cej.2022.137056.

16 Tetik, H., Zhao, K., Shah, N., and Lin, D. (2021). 3d freeze-printed cellulose-based aerogels: obtaining truly 3d shapes, and functionalization with cross-linking and conductive additives. *Journal of Manufacturing Processes* 68: 445–453. https://doi.org/10.1016/j.jmapro.2021.05.051.

17 Yang, L., Yuan, Q., Lou, C. et al. (2025). Modified nanocellulose hydrogels and applications in sensing fields. *Gels* 11 (2): 140. https://doi.org/10.3390/gels11020140.

18 Lai, H., Zhuo, H., Hu, Y. et al. (2021). … & Zhang, M. *Anisotropic carbon aerogel from cellulose nanofibers featuring highly effective compression stress transfer and pressure sensing ACS Sustainable Chemistry & Engineering* 9 (29): 9761–9769. https://doi.org/10.1021/acssuschemeng.1c02051.

19 Li, Y., Guo, J., Li, M. et al. (2021). Recent application of cellulose gel in flexible sensing-a review. *ES Food & Agroforestry.* https://doi.org/10.30919/esfaf466.

20 Cao, J., Sun, G., Wang, P., and Meng, C. (2024). Microstructured CNTs/cellulose aerogel for a highly sensitive pressure sensor. *ACS Applied Materials & Interfaces* 16 (40): 54652–54662. https://doi.org/10.1021/acsami.4c12566.

21 Niu, W. and Liu, X. (2022). Stretchable ionic conductors for soft electronics. *Macromolecular Rapid Communications* 43 (23). https://doi.org/10.1002/marc.202200512: e2200512.

22 Jahani, K., Pourjahed, A., and Abbasi, H. (2024). Ziziphora clinopodioides extract-loaded chitosan/polyvinylpyrrolidone casting films for potential applications in wound dressings. *Journal of Biomaterials Science, Polymer Edition* 35 (4): 559–578. https://doi.org/10.1080/09205063.2024.2301815.

23 Dhivya, S., Padma, V.V., and Santhini, E. (2015). Wound dressings - a review. *Biomedicine* 5 (4): 22. doi: 10.7603/s40681-015-0022-9. Epub 2015 Nov 28. PMID: 26615539; PMCID: PMC4662938.

24 Lazarus, G.S., Cooper, D.M., Knighton, D.R. et al. (1994). Definitions and guidelines for assessment of wounds and evaluation of healing. *Wound Repair and Regeneration* 2 (3): 165–170. https://doi.org/10.1001/archderm.1994.01690040093015.

25 Velnar, T., Bailey, T., and Smrkolj, V. (2009). The wound healing process: an overview of the cellular and molecular mechanisms. *Journal of International Medical Research* 37 (5): 1528–1542. https://doi.org/10.1177/147323000903700531.

26 Gupta, B.S. and Edwards, J.V. (2019). Textile materials and structures for topical management of wounds. *Advanced Textiles for Wound Care* 55–104. https://doi.org/10.1016/b978-0-08-102192-7.00003-5.

27 Köken, A. and Kanık, M. (2022). Aerojellerle Isı Yalıtımı ve Tekstil Uygulamaları [Thermal insulation and textile applications with aerojels]. *Tekstil ve Mühendis* 29 (128): 249–260. https://doi.org/10.7216/teksmuh.1222488.

28 Shiekh, P.A., Singh, A., and Kumar, A. (2020). Exosome laden oxygen releasing antioxidant and antibacterial cryogel wound dressing OxOBand alleviate diabetic and infectious wound healing. *Biomaterials* 249. https://doi.org/10.1016/j.biomaterials.2020.120020.

29 Triplett, T.C. and Richlen, B. (2025). The benefits of hydrogel wound dressings. https://blog.wcei.net/wound-care-dressings-hydrogels (accessed 30 November 2025).

30 Peppas, N.A., Bures, P., Leobandung, W.S., and Ichikawa, H. (2000). Hydrogels in pharmaceutical formulations. *European Journal of Pharmaceutics and Biopharmaceutics* 50 (1): 27–46. https://doi.org/10.1016/S0939-6411(00)00090-4.

31 Rezvani Ghomi, E., Khalili, S., Nouri Khorasani, S. et al. (2019). Wound dressings: current advances and future directions. *Journal of Applied Polymer Science* 136 (27): 47738. https://doi.org/10.1002/app.47738.

32 Zhang, X., Qin, M., Xu, M. et al. (2021). The fabrication of antibacterial hydrogels for wound healing. *European Polymer Journal* 146: 110268. https://doi.org/10.1016/j.eurpolymj.2021.110268.

33 Chaudhari, A.A., Vig, K., Baganizi, D.R. et al. (2016). Future prospects for scaffolding methods and biomaterials in skin tissue engineering: a review. *International Journal of Molecular Sciences* 17 (12): 1974. https://doi.org/10.3390/ijms17121974.

34 Su, J., Li, J., Liang, J. et al. (2021). Hydrogel preparation methods and biomaterials for wound dressing. *Life* 11 (10): 1016. https://doi.org/10.3390/life11101016.

35 Ahmad, F., Mushtaq, B., Ahmad, S. et al. (2023). A novel composite of hemp fiber and alginate hydrogel for wound dressings. *Journal of Polymers and the Environment* 31: 2294–2305. https://doi.org/10.1007/s10924-023-02756-7.

36 Haidari, H. and Kopecki, Z. (2024). Chapter One - Advanced hydrogel for management of bacterial wound infections. In: *Methods in Microbiology*, vol. 54 (ed. P.A. Ramsland, A. Elbourne, and V. Gurtler), 1–38. Academic Press ISSN 0580-9517, ISBN 9780443191190. https://doi.org/10.1016/bs.mim.2023.10.001.

37 Guo, W., Li, Z., Liu, B. et al. (2024). Hydrogel-based textile composites. *Progress in Chemistry* 36 (6): 914–927. https://doi.org/10.7536/PC231002.

38 Amante, C., Falcone, G., Aquino, R.P. et al. (2024). In situ hydrogel formulation for advanced wound dressing: influence of co-solvents and functional excipient on tailored alginate–pectin–chitosan blend gelation kinetics, adhesiveness, and performance. *Gels* 10 (1): 3. https://doi.org/10.3390/gels10010003.

39 Demeter, M., Scărişoreanu, A., and Călina, I. (2023). State of the art of hydrogel wound dressings developed by ionizing radiation. *Gels* 9 (1): 55. https://doi.org/10.3390/gels9010055.

40 Borbolla-Jiménez, F.V., Peña-Corona, S.I., Farah, S.J. et al. (2023). Films for wound healing fabricated using a solvent casting technique. *Pharmaceutics* 15 (7): 1914. https://doi.org/10.3390/pharmaceutics15071914.

41 Yuan, X., Zhang, J., Shi, J. et al. (2024). Cotton fabric-reinforced hydrogels with excellent mechanical and broad-spectrum photothermal antibacterial properties. *Polymers* 16 (10): 1346. https://doi.org/10.3390/polym16101346.

42 Kou, J., Li, Y., Zhou, C. et al. (2025 Mar). Electrospinning in promoting chronic wound healing: materials, process, and applications. *Frontiers in Bioengineering and Biotechnology* 6 (13): 1550553. https://doi.org/10.3389/fbioe.2025.1550553.

43 Moazzami Goudarzi, Z., Zaszczyńska, A., Kowalczyk, T., and Sajkiewicz, P. (2024). Electrospun antimicrobial drug delivery systems and hydrogels used for wound dressings. *Pharmaceutics* 16 (1): 93. https://doi.org/10.3390/pharmaceutics16010093.

44 Bernardes, B.G., Del Gaudio, P., Alves, P. et al. (2021). Bioaerogels: promising nanostructured materials in fluid management, healing and regeneration of wounds. *Molecules* 26 (13): 3834. https://doi.org/10.3390/molecules26133834.

45 McNeil, S.J. and Gupta, H. (2022). Emerging applications of aerogels in textiles. *Polymer Testing* 106: 107426. https://doi.org/10.1016/j.polymertesting.2021.107426.

46 Smirnova, I. and Gurikov, P. (2018). Aerogel production: current status, research directions, and future opportunities. *The Journal of Supercritical Fluids, Volume* 134: 228–233. ISSN 0896-8446, https://doi.org/10.1016/j.supflu.2017.12.037.

47 Blanco Parte, F.G., Santoso, S.P., Chou, C.C. et al. (2020 May). Current progress on the production, modification, and applications of bacterial cellulose. *Critical Reviews in Biotechnology* 40 (3): 397–414. https://doi.org/10.1080/07388551.2020.1713721.

48 Khalil, H.A., Yahya, E.B., Jummaat, F. et al. (2023). Biopolymers based aerogels: a review on revolutionary solutions for smart therapeutics delivery. *Progress in Materials Science* 131: 101014. https://doi.org/10.1016/j.pmatsci.2022.101014.

49 Rostamitabar, M., Subrahmanyam, R., Gurikov, P. et al. (2021). Cellulose aerogel micro fibers for drug delivery applications. *Materials Science and Engineering: C* 127: 112196. https://doi.org/10.1016/j.msec.2021.112196.

50 Yahya, E.B., Alzalouk, M.M., Alfallous, K.A., and Abogmaza, A.F. (2020). Antibacterial cellulose-based aerogels for wound healing application: a review. *Biomedical Research and Therapy* 7 (10): 4032–4040. https://doi.org/10.15419/bmrat.v7i10.637.

51 Jaiswal, S., Yadav, P., and Kumar, R. (2024). Biomedical applications of aerogels: therapeutic potential, safety, and future research directions. *Polymers for Advanced Technologies* 35 (12): e70019. https://doi.org/10.1002/pat.70019.

52 Keil, C., Hübner, C., Richter, C. et al. (2020). Ca-Zn-Ag alginate aerogels for wound healing applications: swelling behavior in simulated human body fluids and effect on macrophages. *Polymers* 12 (11): 2741. https://doi.org/10.3390/polym12112741.

53 Karamikamkar, S., Yalcintas, E.P., Haghniaz, R. et al. (2023). Aerogel-based biomaterials for biomedical applications: from fabrication methods to disease-targeting applications. *Advanced Science* 10 (23): 2204681. https://doi.org/10.1002/advs.202204681.

54 Sinclair, C. and Yorke, M. (2015). Gelling Alginate Dressings and Their Contribution to Wound Management. *Wounds UK* 11 (2).

55 Koga, A.Y., Felix, J.C., Silvestre, R.G.M. et al. (2020). Evaluation of wound healing effect of alginate film containing Aloe vera gel and cross-linked with zinc chloride. *Acta Cirúrgica Brasileira* 35: e202000507. https://doi.org/10.1590/s0102-865020200050000007.

56 Lázár, I., Čelko, L., and Menelaou, M. (2023). Aerogel-based materials in bone and cartilage tissue engineering—a review with future implications. *Gels* 9 (9): 746. https://doi.org/10.3390/gels9090746.

57 Cai, J., Liu, S., Feng, J. et al. (2012). Cellulose-silica nanocomposite aerogels by in-situ formation of silica in cellulose gel. *Angewandte Chemie, International Edition* 51 (9): 2076–2079. https://doi.org/10.1002/anie.201105730.

58 Köken, A. and Kanık, M. (2023). Investigation of thermal and solar properties of aerogel powder coated textiles. *Materials Science* 29 (3): 382–390. https://doi.org/10.5755/j02.ms.32869.

59 Mekonnen, B.T., Ding, W., Liu, H. et al. (2021). Preparation of aerogel and its application progress in coatings: a mini overview. *Journal of Leather Science and Engineering* 3: 1–16. https://doi.org/10.1186/s42825-021-00067-y.

60 Parın, F.N. (2024). Mini-review of the bi-component nanofibrous scaffolds and drug delivery applications. Journal of Innovative Engineering and Natural. *Science* 4 (2): 391–413. https://doi.org/10.61112/jiens.1417524.

61 Shoukat, H., Buksh, K., Noreen, S. et al. (2021). Hydrogels as potential drug-delivery systems: network design and applications. *Therapeutic Delivery* 12 (5): 375–396. https://doi.org/10.4155/tde-2020-0114.

62 Bakhori, N.M., Ismail, Z., Hassan, M.Z., and Dolah, R. (2023). Emerging trends in nanotechnology: aerogel-based materials for biomedical applications. *Nanomaterials* 13 (6): 1063. https://doi.org/10.3390/nano13061063.

63 Thang, N.H., Chien, T.B., and Cuong, D.X. (2023). Polymer-based hydrogels applied in drug delivery: an overview. *Gels* 9 (7): 523. https://doi.org/10.3390/gels9070523.

64 Rodríguez-Rodríguez, R., Espinosa-Andrews, H., and García-Carvajal, Z.Y. (2022). Stimuli-responsive hydrogels in drug delivery. In: *Functional Biomaterials: Drug Delivery and Biomedical Applications*, 75–103. Singapore: Springer Singapore. https://doi.org/10.1007/978-981-16-7152-4_3.

65 Ulker, Z. and Erkey, C. (2014). An emerging platform for drug delivery: aerogel based systems. *Journal of Controlled Release* 177: 51–63. https://doi.org/10.1016/j.jconrel.2013.12.033.

66 Veronovski, A., Novak, Z., and Željko, K. (2011). Organic biodegradable aerogels used in controlled drug release. *MRS Online Proceedings Library* 1306 (1): 702. https://doi.org/10.1557/opl.2011.315.

67 Levin, A., Gong, S., and Cheng, W. (2023). Wearable smart bandage-based bio-sensors. *Biosensors* 13 (4): 462. https://doi.org/10.3390/bios13040462.

68 Derakhshandeh, H., Kashaf, S.S., Aghabaglou, F. et al. (2018). Smart bandages: the future of wound care. *Trends in Biotechnology* 36 (12): 1259–1274. https://doi.org/10.1016/j.tibtech.2018.07.007.

69 Mostafalu, P., Lenk, W., Dokmeci, M.R. et al. (2015). Wireless flexible smart bandage for continuous monitoring of wound oxygenation. *IEEE Transactions on Biomedical Circuits and Systems* 9 (5): 670–677. https://doi.org/10.1109/tbcas.2015.2488582.

70 Çiçek, M.Ö., Durukan, M.B., Yıldız, B. et al. (2023). Ultra-sensitive bio-polymer tactile feedback. *Advanced Materials Technologies* 8 (16). https://doi.org/10.1002/admt.202300322.

71 Wu, J., Wu, Z., Wei, Y. et al. (2020). … & Xie, X. *Ultrasensitive and stretchable temperature sensors based on thermally stable and self-healing organohydrogels ACS Applied Materials & Interfaces* 12 (16): 19069–19079. https://doi.org/10.1021/acsami.0c04359.

72 Wu, Z., Shi, W., Ding, H. et al. (2021). Ultrastable, stretchable, highly conductive and transparent hydrogels enabled by salt-percolation for high-performance temperature and strain sensing. *Journal of Materials Chemistry C* 9 (39): 13668–13679. https://doi.org/10.1039/d1tc02506f.

8

Animal Fiber-Based Hydrogels and Aerogels for Biomedical Applications

Bushra Mushtaq[1], *Muhammad Amber Fareed*[2,3], *and Muhammad Sohail Zafar*[2,3,4]

[1] *National Textile University, School of Engineering and Technology, Department of Textile Technology, Sheikhupura Road, 37610, Faisalabad, Pakistan*
[2] *Ajman University, College of Dentistry, Department of Clinical Sciences, 346, Ajman, United Arab Emirates*
[3] *Ajman University, Centre of Medical and Bio-allied Health Sciences Research, 346, Ajman, United Arab Emirates*
[4] *University of Jordan, School of Dentistry, Queen Rania Street, 11942, Amman, Jordan*

8.1 Introduction

Biomaterials have evolved from inert materials to biocompatible systems that mimic human tissue. Animal fiber-based hydrogels and aerogels are the most promising ones with their very high biocompatibility, biodegradability, and mechanical flexibility. These products bridge the gap between synthetic polymers (lacking biological activity) and natural matrices with natural bioactivity [1, 2].

Hydrogels are three-dimensional polymeric material networks loaded with water, while aerogels are highly porous structures obtained by evaporating liquid from gels. They are protein based and obtained from collagen, gelatin, keratin, silk fibroin, and chitosan. They impart biological properties typical of the extracellular matrix (ECM) to allow tissue regeneration, drug delivery, and biosensing [3–5].

8.1.1 Definition of Hydrogel and Aerogel

Hydrogels are three-dimensional (3D) hydrophilic polymer networks capable of absorbing and retaining significant amounts of water (up to 90–99%) without dissolving due to physical or chemical cross-linking. Their high water content mimics the soft tissues in the human body, making them suitable for cell encapsulation and tissue regeneration [6].

Aerogels are highly light materials produced by removing the liquid phase of a gel using supercritical drying or freeze-drying to develop a highly porous, low-density network of solids. Animal fiber-based aerogel has a fibrous microarchitecture with good gaseous exchange, thermal insulation, and drug loading properties [7, 8] (Table 8.1).

Table 8.1 Comparison of the characteristics of hydrogel and aerogel.

Characteristics	Hydrogel	Aerogel
Water content	7–99%	<10%
Porosity	Moderate	>90%
Density	~1 g/cm^3	0.003–0.5 g/cm^3
Biomedical use	Cell delivery, wound healing	Bone scaffold, biosensors

Source: Adapted from References [9–12].

Table 8.2 Animal fiber-derived biopolymers and their advantages.

Biopolymer	Source	Properties and biomedical benefits
Collagen	Skin, tendon	Native ECM promotes cell adhesion and migration
Gelatin	Collagen derivatives	Injectable, biodegradable, thermosensitive
Chitosan	Crustacean shells	Antimicrobial, mucoadhesive sensitive
Silk fibroin	*Bombyx mori* cocoons	Beta sheet domain for strength, tunable degradation
Keratin	Wool, hair	Cell adhesion motifs (LDV [Leucine-Aspartic Acid-Valine], RGD [Arginine-Glycine-Aspartic acid]), disulfide cross-linking

Source: Adapted from References [13–19].

8.1.2 Animal Fiber-Derived Biopolymers and Their Advantages

Animal fiber-derived biopolymers offer unique advantages over plant based or synthetic alternatives for biomedical applications (Table 8.2).

Compared with synthetic polymers (PEG [polyethylene glycol]) and PVA (polyvinyl alcohol), these biopolymers have superior cell material interactions, making them ideal for clinical applications.

8.1.3 Historical Development

The biomedical use of fiber from animals has a long and rich history, and the practice of using silk sutures in surgeries is one of the earliest examples. But it is in recent decades that these natural fibers have been considerably developed in terms of their application and performance. In the 1950s–1970s, collagen wound dressings initiated contemporary clinical uses, paving the way for biocompatible and bioactive scaffolds in regenerative medicine [20, 21]. The 1990s witnessed a significant breakthrough with the development of recombinant spider silk production, which provided improved mechanical strength and reproducibility for biomedical applications [22]. In the 2010s, the integration of animal fibers into 3D bioprinting technologies gained momentum, with silk fibroin and keratin-based bioinks enabling the fabrication of tissue-engineered constructs with high fidelity and biological relevance [23, 24]. Most recently, the 2020s have witnessed the emergence of smart aerogels derived from animal fibers, capable of stimuli-responsive drug delivery and real-time therapeutic response [25, 26].

This rekindled interest and innovation have been driven by a range of converging trends. Green chemistry strategies have made it more environmentally friendly to extract and process fibers. Concurrently, breakthroughs in nanotechnology have enabled precise control of the structural and functional properties of these biomaterials. Furthermore, the increased worldwide focus on sustainable and naturally sourced health-care products has created animal fiber-based hydrogels and aerogels as prime targets for next-generation biomedical technologies.

8.1.4 Current Market Landscape and Clinical Impact

The world biomaterials market is rapidly growing and is anticipated to reach approximately $245.6 billion in 2027, driven by technological advancements and increasing demand for biologically compatible products in health care [27]. Within this general market, animal fiber-based hydrogels and aerogels are major industries due to their sustainability, biocompatibility, and functional diversity. One of the most significant areas of application is in wound care, one of which is a present market value of $7.45 billion with projected growth to $12.11 billion by 2032. Keratin-based dressings lead to this growth with improved healing, antimicrobial activities, and minimized scarring [28]. In orthopedics, biomaterials derived from silk have been very promising, especially in cartilage repair devices, where their mechanical properties and bioactivity develop a perfect matrix for tissue growth [29, 30]. Moreover, the drug delivery market is also growing, with protein-based systems, most of them derived from animal fibers, estimated to grow at a CAGR (Compound Annual Growth Rate) of 37%. These trends emphasize the increasing clinical and commercial relevance of biomaterials derived from animal fibers and pinpoint their increasing role in the future of regenerative and personalized medicine [31].

8.1.5 Multidisciplinary Convergence

Advanced animal fiber-based therapeutic systems depend on the multidisciplinary intersection of several important fields, each bringing essential expertise to the table. Materials science underpins the entire field by allowing precise control of the nanostructure, critical for optimizing the physical and chemical properties of the materials. Biology informs cell–matrix interactions, ensuring biocompatibility and maximizing therapeutic impact. Engineering provides solutions to scalable manufacturing issues, enabling laboratory research to scale up industrial production. Medicine completes the cycle by ensuring these innovations pass regulatory approval and can be properly used in actual medical treatments. This chapter examines how the convergence of these fields enables the development of next-generation therapeutic systems based on animal fibers, emphasizing the cooperative efforts required in advancing this leading-edge field.

8.2 Preparation and Fabrication Methods

Animal fiber-derived hydrogels and aerogels are obtained from natural proteinaceous or polysaccharide sources such as collagen, silk fibroin, keratin, and chitosan that inherently exhibit biocompatibility, biodegradability, and ECM-like structural resemblance [32, 33]. Preparation of these biomaterials is typically a multistep process including extraction, purification, solubilization, gelation or cross-linking, and drying (for aerogels) [34]. Every technique applied throughout these steps is important in determining the final material's properties and its suitability for use in biomedical applications.

8.2.1 Extraction and Purification

See Table 8.3.

Table 8.3 Extraction and purification.

Material	Extraction method	Innovation
Silk fibroin	Degumming with Na_2CO_3 or enzymatically	Nanofibrillation via homogenization enhances strength [35]
Keratin	Reductive (thioglycolate) or oxidative method	Plasma-assisted extraction improves yield and preserves the S–S bond [36]
Collagen	Acid + pepsin digestion	Electrodialysis removes endotoxins for implant-grade materials [37]
Chitosan	Deacetylation of chitin	Required acid for dissolution, e.g. acetic acid [38]

8.2.1.1 Silk Fibroin Extraction (*Bombyx mori* and Recombinant Spider Silk)

Conventional degumming processes for silk include boiling the fibers in a 0.02 M Na_2CO_3 solution for 30–60 minutes to eliminate sericin, as shown in Figure 8.1. This process, however, may break down the molecular weight of fibroin, which could undermine its structural integrity [35].

To mitigate these shortfalls, new alternatives have been developed. One is enzymatic degumming with protease XIV, efficiently removing sericin while maintaining fibroin's molecular structure [39]. Another novel approach is microwave-assisted extraction, which cuts processing time and energy utilization by as much as 40% compared with traditional methods [40]. Moreover, nanofibrillation via high-pressure homogenization yields silk nanofibrils of 3–10 nm width, significantly improving the mechanical characteristics of silk hydrogels [41]. These new approaches provide more effective and eco-friendly means of processing silk without sacrificing or even enhancing its material performance [41].

8.2.1.2 Keratin Extraction (Wool/Human Hair)

Keratin, a cysteine-rich structural protein with disulfide bonds, is readily isolated from human hair and wool for biomedical purposes such as wound dressings

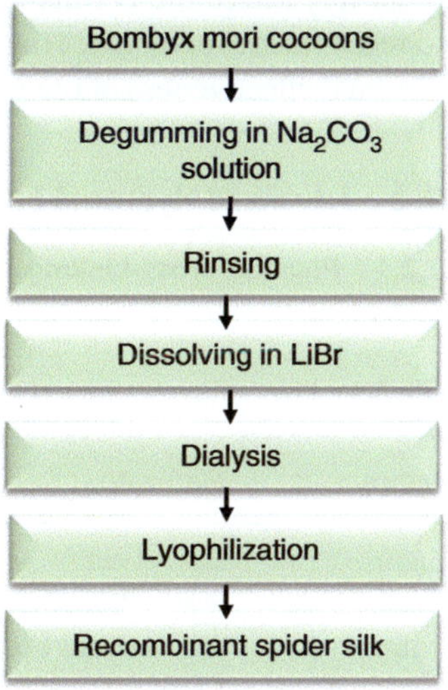

Figure 8.1 Conventional degumming process for silk.

and tissue scaffolds. Two general chemical procedures have been used to isolate keratin: oxidative and reductive procedures. For the reductive procedure, a thioglycolate–urea–urea system is used with 2.5 M thioglycolic acid and 8 M urea at 50 °C. This method reduces the disulfide bonds of keratin, resulting in a high yield of 85–90% soluble keratin. Nevertheless, a significant limitation of this process is partial loss of cysteine residues, which are critical for recreating cross-links during product manufacturing, which may affect the end mechanical property of keratin-based products [42].

Conversely, the oxidative process uses peracetic acid to act as an oxidizing agent for keratin extraction while keeping the cysteine residues in an oxidized state. While this method holds intact reactive sites critical for further cross-linking, the yield is typically compromised at 60–70%. More recently, plasma-assisted extraction has emerged as a new method that yields up to 82% keratin with the additional benefit of preventing chemical breakdown of disulfide bonds. This process provides a cleaner and more effective method by using physical plasma activation in conjunction with minimal chemical intervention, yielding structurally whole and functionally active keratin [43].

8.2.1.3 Collagen Extraction (Type I from Bovine Tendon)

Type I collagen isolation from bovine tendon usually consists of acid solubilization, followed by purification to meet biomedical-grade standards. The process starts with the extraction of collagen, where tendon tissue is digested in 0.5 M acetic acid and pepsin to break down collagen without depolymerizing the triple-helix structure. An important consideration is keeping the temperature less than 25 °C during digestion to avoid thermal denaturation [44].

After solubilization, the extract is purified through electrodialysis, a very effective purification technique for eliminating contaminants. This removes 99.8% of the endotoxins to <0.5 EU/mg, complying with FDA (Food and Drug Administration) standards for implantable biomaterials. The high-purity collagen that is obtained is structurally intact and is thus ideal for use in tissue engineering, wound dressings, and regenerative medicine. This improved protocol guarantees biocompatibility and functional effectiveness of the purified collagen [45].

8.2.2 Advanced Hydrogel Fabrication

8.2.2.1 Physical Cross-Linking

See Table 8.4.

Table 8.4 Physical cross-linking method and its mechanism.

Method	Mechanism	Example	Advantages
Thermal	Beta-sheet crystallization	37 °C incubation of 6% silk solution	No chemical residues
Ionic	Ca^{2+} bridges between COO^-	100 mM $CaCl_2$ for alginate silk hybrids	Rapid gelation (<2 min)
Self-assembly	pH-induced fibrilization	Keratin at pH 4.2 near pl	Nanofibrous architecture

Source: Adapted from References [46–49].

8.2.2.2 Chemical Cross-Linking

When choosing cross-linking techniques for silk biomaterials, genipin and EDC (1-ethyl-3-(3-dimethylaminopropyl) carbodiimide)/NHS (*N*-hydroxysuccinimide) have various pros and cons. Genipin is a natural cross-linker that has much lower cytotoxicity with an $IC_{50} > 500\,\mu M$ against that of glutaraldehyde, as low as 50 μM, and thus is more biocompatible for biomedical use. On the contrary, it necessitates a longer reaction time of 12–24 hours [50].

Conversely, the EDC/NHS system, based on carbodiimide cross-links, 90% in only 1 hour, but requires rigorous pH control for effectiveness. For fast gelation, click chemistry methods such as thiol-ene click between PEG-diacrylate, and modified keratins facilitate the formation of gel in a mere 30 seconds under ultraviolet illumination (370 nm) [51].

Silk composites show superior performance in 3D printing and bioprinting applications. Mixing 8% silk fibroin with 2% gelatin methacryloyl (GelMA) has shown improved printability along with high cell viability (>95%). The constructions have a compression modulus of 15–25 kPa, proving to be optimal for soft tissue engineering [23]. These developments show the versatility of materials based on silk in biofabrication, with optimum mechanical properties, biocompatibility, and processing efficiency [23, 52].

8.2.3 Advanced Aerogel Fabrications

8.2.3.1 Sol–Gel Transition Control

Silk fibroin aerogels need a critical concentration of 2–4% w/v to create a stable nanofibrillar network for maximum structural integrity. Traditional gelation takes as much as 72 hours, but this can be dramatically shortened by sonication-assisted assembly, whereby the gelation time can be confined to 2 hours. This is achieved through a fast assembly that not only increases production efficiency but also maintains the porous, light structure needed for uses such as tissue engineering, drug delivery, and thermal insulation. The synergy between controlled fibroin concentration and modern processing methods provides tunable mechanical and morphological properties in silk-based aerogels [53].

8.2.3.2 Drying Technology Comparison

See Table 8.5 [54–56].

Table 8.5 Drying techniques for aerogel.

Method	Pore size (μm)	Shrinkage (%)	Surface area (m^2/g)
Supercritical CO_2	0.1–10	<5	400–600
Freeze-drying −80 °C	50–300	15–30	100–200
Ambient pressure drying	1–50	40–60	50–150

Hydrogels can be transformed into aerogels by removing water without collapsing their porous structure. It entails the following.

8.2.3.3 Freeze-Drying (Lyophilization)

Freeze-drying, or lyophilization, is a widely used technique for creating highly porous and lightweight biomaterials. The process involves freezing a hydrogel followed by the sublimation of ice under vacuum, which preserves the material's structure while removing water. This method typically results in materials with aligned or anisotropic pore architectures, making them particularly useful for tissue engineering applications. For instance, freeze-dried collagen and silk fibroin aerogels have been successfully used as bone scaffolds, where their interconnected porosity promotes cell infiltration, nutrient diffusion, and mechanical support. The ability to tailor pore size and orientation through controlled freezing conditions further enhances their suitability for regenerative medicine and other biomedical applications [54].

8.2.3.4 Supercritical Drying

Supercritical drying is a cutting-edge method used to produce highly porous aerogels with intact, delicate nanostructures. Here, water in the hydrogel is initially replaced with a solvent such as ethanol, which is subsequently extracted under supercritical conditions (above its critical pressure and temperature). This technique avoids surface tension-induced pore collapse, providing better preservation of nanostructural integrity compared with freeze-drying. Owing to its accuracy, supercritical drying is used extensively in the generation of silk fibroin and chitosan aerogels, which are worthwhile for biomedical technologies such as drug delivery and tissue engineering [55].

Table 8.6 Summary of key methods and applications.

Step	Method	Material	Application
Extraction	Enzymatic digestion	Collagen	Skin and cartilage scaffold
Solubilization	Acid/alkali/ionic solvent	Chitosan/silk	Drug delivery matrices
Cross-linking	Physical (Beta-sheet), chemical (EDC), and enzymatic	Silk, collagen, keratin	Injectable hydrogel, implants
Drying	Freeze-drying/ supercritical CO_2	Silk, chitosan, keratin	Aerogel for bone scaffolds
Advanced fabrication	3D printing, electrospinning	Collagen, keratin	Tissue engineering scaffolds

8.2.3.5 Ambient Pressure Drying

While supercritical drying uses less energy input than other drying techniques, it entails supplementary operations such as solvent exchange and surface treatment (e.g. salinization) to avoid structural collapse during processing. These additional operations and issues about possible chemical residues from solvents or modifying reagents have restricted their use in large-scale biomedical applications. Consequently, although the method excels at maintaining nanostructures in materials such as aerogels, researchers tend to use competing methods such as freeze-drying for biomaterials, where residue-free processing and biocompatibility are paramount [56] (Table 8.6).

8.3 Properties and Characterization

Knowledge of the physicochemical, structural, mechanical, thermal, and biological properties of hydrogels and aerogels based on animal fibers is important to maximize their biomedical functionalities. These physical properties represent their drug delivery, wound healing, tissue engineering, and biosensing applications. The subsequent subsections highlight the most important physicochemical, structural, mechanical, thermal, and biological properties and analytical methods used to determine them.

8.3.1 Physicochemical Properties

8.3.1.1 Swelling Behavior

Hydrogels, which are hydrophilic amino acid or carbohydrate networks, soak up water without sacrificing structural integrity, making them perfect for drug delivery, cell culture, and wound hydration. Chitosan hydrogels exhibit pH-sensitive swelling based on their amino groups [57], while keratin hydrogels realize moderate swelling (∼500%) through hydrophilic peptide chains [58]. Swelling behavior is most assessed by placing samples in PBS and conducting gravimetric analysis. This attribute directly has an impact on their functionality in biomedical applications.

8.3.1.2 Porosity and Pore Size

Porosity of biomaterials is essential to define cell infiltration, diffusion of nutrients, and drug permeability, and aerogels usually have porosity above 90% with pores from 50 to 500 μm, making them suitable for tissue scaffolding. Freeze-dried collagen aerogels, for example, have pore diameters of 100–300 μm, which are ideal for bone cell migration. To appropriately evaluate these structural attributes, characterization methods such as scanning electron microscopy (SEM), mercury intrusion porosimetry, and micro-CT imaging are used to evaluate pore morphology, distribution, and connectivity to ensure that the material is suitable for the desired biomedical application [59].

8.3.2 Mechanical Properties

Mechanical stability is essential for load-carrying biomedical applications (e.g. cartilage, bone). Fiber-based materials from animals display adjustable mechanical properties depending on polymer blends, cross-linking density, and fiber content [60]. The mechanical properties of biomaterials are greatly different based on their structure and composition. For example, gelatin/chitosan hydrogels have a compressive modulus of 20–50 kPa that closely replicates the mechanics of soft tissues. Silk fibroin hydrogels, on the contrary, may be nanofilled with graphene oxide for higher tensile strength, greater than 1 MPa [61].

The aerogels, especially cross-linked silk aerogels, exhibit even more mechanical strength, with compressive moduli of over 2 MPa. This could be further enhanced via enzymatic cross-linking or by the addition of inorganic fillers (e.g. hydroxyapatite in bone scaffolds) for enhanced load-bearing applications. To properly analyze these mechanical properties, standardized test methods are used, such as uniaxial tensile/compression tests, rheological tests, and dynamic mechanical analysis (DMA), to provide thorough characterization for both biomedical and engineering uses [62].

8.3.3 Thermal Stability

Thermal analysis is significant in analyzing the stability and degradation characteristics of biomaterials under sterilization and processing. Thermogravimetric analysis (TGA) indicates that chitosan aerogels exhibit two-stage degradation: initial loss of water at 50–150 °C and polymer decomposition at 250–300 °C [63].

Simultaneously, differential scanning calorimetry (DSC) identifies denaturation peaks of silk fibroin and collagen between 200 and 220 °C, which give us information on their thermal transitions. The most important instrumentation for such analyses is TGA, DSC, and differential thermal analysis (DTA), which together assist in characterizing thermal stability, decomposition trends, and phase transition, key to optimizing the performance of the material in biomedical applications.

8.3.4 Biodegradability and Enzymatic Degradation

Biomaterials' biodegradability is a significant characteristic that makes them get absorbed or replaced by the native tissue over time, with the degradation being affected by cross-linking techniques, fiber content, and processing conditions. For example, collagen and gelatin hydrogels degrade quickly by collagenase in approximately 1–2 weeks, and thus, they can be used for short-term applications [64]. Conversely, keratin hydrogels break down more slowly because they contain stable disulfide bridges, offering long-term structural stability for extended usage [65]. To determine biodegradation, standard tests involve putting the material in an enzyme solution (e.g. lysozyme or protease) and monitoring weight loss over time, as well as using SEM to detect morphological changes. These tests guarantee that biomaterials satisfy target degradation patterns for targeted biomedical devices.

8.3.5 Surface Morphology and Microstructure

These results emphasize the need to tailor surface properties to maximize scaffold performance in tissue engineering and regenerative medicine.

The morphology of the surface of biomaterial scaffolds is central to controlling cell adhesion, growth, and drug interaction because it needs to closely mimic the in vivo ECM to provide biocompatibility. SEM identifies the porous, fibrous structure of aerogels and freeze-dried hydrogels, and atomic force microscopy (AFM) delivers quantitative data on surface roughness, a determinant in controlling cell–material interaction [66].

Research has shown that silk fibroin scaffolds with a controlled surface roughness (Ra) of 30–80 nm support greatly enhanced osteoblast attachment and bone regeneration [67]. In a similar vein, nanoporous chitosan-silica hybrid aerogels with their large surface area (~300 m^2/g) show improved cell adhesion because of their improved topographical characteristics [68].

8.3.6 Biocompatibility and Cytotoxicity

Biocompatibility and cytotoxicity are key parameters for biomaterial clinical utility, ensuring that they are nontoxic and do not provoke adverse reactions while promoting cell viability and tissue integration. In vitro tests, for example, MTT and live/dead assays are performed with human dermal fibroblasts or osteoblasts to measure cell survival and growth. For example, chitosan hydrogels are very biocompatible, retaining more than 90% viability in NIH 3T3 fibroblasts after 72 hours [69].

In vivo testing, commonly conducted in rat or rabbit models, also evaluates immune response, fibrous encapsulation, and tissue integration. For instance, collagen aerogels implanted subcutaneously in rats exhibited minimal inflammation after 4 weeks, reflecting good tissue compatibility. Such comprehensive tests, ranging from cellular to animal model levels, are critical to biomaterials' translation into safe and efficacious medical practices [70].

8.3.7 Water Vapor Transmission Rate

In wound dressing use, a perfect balance of moisture avoiding overdrying as well as maceration, is important for successful healing. Chitosan–gelatin composite films with a water vapor transmission rate (WVTR) of approximately 2000–2500 $g/m^2/day$ achieve this balance by providing a moist but aerated environment [71].

Keratin-derived aerogels also provide adjustable WVTR based on porosity and thickness of the film and can be customized to various types of wounds. These attributes are critically assessed according to standardized tests such as the ASTM E96/E96M test, used to determine moisture permeability for clinical effectiveness. By adjusting WVTR through material structure and composition, next-generation wound dressings can promote tissue regeneration while reducing complications [72].

8.3.8 Cross-Linking and Network Structure

Analytical methods such as FTIR spectroscopy (for monitoring changes in functional groups) and X-ray diffraction (XRD) (for determining crystallinity) are used to study these cross-linked networks. Such methods provide accurate control of material properties, balancing structural performance and biological safety for use in biomedical applications.

The extent and method of cross-linking (physical, chemical, and enzymatic) influences mechanical, degradation, and drug release profiles. The cross-linking procedure and network architecture of biomaterials play a critical role in governing their mechanical performance, degradation, and drug release patterns. Chemical cross-linking agents such as glutaraldehyde and EDC/NHS yield robust covalent networks with the potential to introduce cytotoxicity, confining their biocompatibility [73].

An alternative is offered by physical cross-linking techniques, e.g. ionic gelation (e.g. chitosan with tripolyphosphate) or freeze–thaw cycling (e.g. gelatin hydrogels) with nontoxic properties but possibly weaker mechanical strength. Cross-linking via enzymes, with agents such as transglutaminase or horseradish peroxidase (HRP), presents a biocompatible compromise, allowing the gelation kinetics to be modulated for specific application [74].

8.3.9 Summary Table of Key Properties and Methods

See Table 8.7.

Table 8.7 Summary table of key properties.

Property	Measurement technique	Relevance
Swelling ratio	Gravimetric in PBS	Drug delivery, wound healing
Porosity/pore size	SEM, mercury intrusion	Tissue engineering, cell infiltration
Mechanical strength	Tensile/compression rheology	Scaffold integrity
Thermal stability	TGA, DSC	Sterilization, material processing
Biodegradability	Enzymatic incubation, weight loss	Scaffold longevity
Surface morphology	SEM, AFM	Cell adhesion, proliferation
Cytotoxicity/ biocompatibility	MTT, live dead, animal studies	Safety, compatibility
WVTR	ASTM E96/E96 M	Moisture balance in wound dressing
Cross-linking efficiency	FTIR, XRD	Network strength, degradation rate

8.4 Biomedical Applications

Animal fiber-derived hydrogels and aerogels have emerged as popular biomedical models because of their biodegradability, mechanical stability, bioactivity, and biocompatibility. Their structural analogy to the ECM and the ease with which they can load bioactive agents have established them as promising models for wound healing, tissue engineering, drug delivery, biosensing, and other medical treatments, as shown in Figure 8.2.

8.4.1 Wound Healing and Skin Regeneration

Hydrogels are perfect for wound healing because of their water-holding capacity, oxygen permeability, and bioactivity. By contrast, aerogels are superior as wound dressings with drug delivery function because of their high porosity and absorptive ability. For example, diabetic wound healing is enhanced by chitosan hydrogels due to antimicrobial and hemostatic actions, whereas collagen hydrogels induce granulation tissue development in dermal substitutes such as Integra® [76].

Silk fibroin aerogels augment chronic wound healing through the delivery of antibacterial agents (i.e., silver nanoparticles) and growth factors. These technologies maximize hydration, create scaffolds for fibroblast migration, and function as antimicrobial barriers to *Staphylococcus aureus* and *Pseudomonas aeruginosa*. Research indicates that chitosan hydrogels decrease bacterial loads by >99% within 24 hours [57], and silk sponges enhance wound healing by approximately 35% versus gauze in rat models [77].

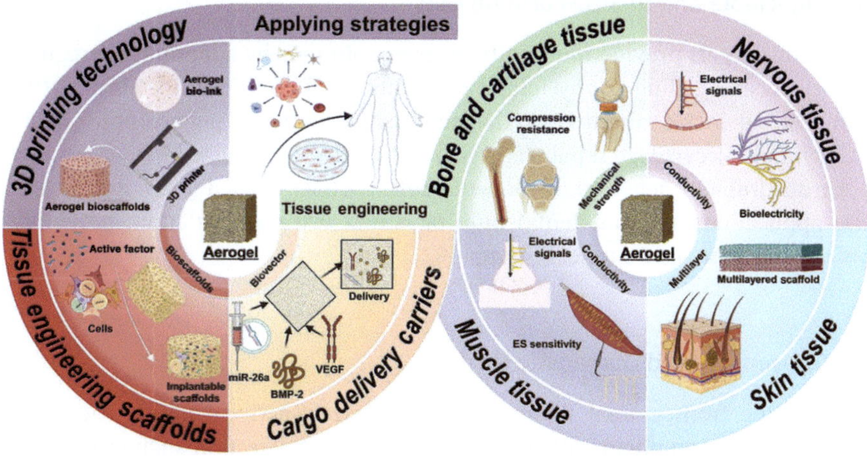

Figure 8.2 Applications of hydrogels and aerogels. *Source:* [75] / Elsevier / CC BY 4.0.

8.4.2 Tissue Engineering Scaffolds

Hydrogels and aerogels are biomimetic 3D matrices that mimic the natural ECM to support cell adhesion, migration, proliferation, and differentiation for bone regeneration, cartilage repair, nerve regeneration, and corneal stroma replacement. For example, collagen–hydroxyapatite aerogels with a compressive modulus of 0.2–1 MPa induce osteoinduction in bone scaffolds, and silk fibroin hydrogels facilitate chondrocyte proliferation for cartilage repair. Keratin hydrogels containing neurocompatible motifs such as LDV allow for neural cell attachment, supporting nerve regeneration. Recent developments involve gelatin–elastin hydrogels that replicate the elasticity of arteries for vascular grafts. These new materials fill the gap between synthetic and biological systems, allowing for customized tissue regeneration [58, 78, 79].

8.4.3 Drug Delivery Systems

Hydrogels and aerogels are commonly used in controlled drug delivery based on their tunable degradation, high water uptake, and stimuli-responsiveness (e.g. pH, temperature, and enzymes), making them ideal for oral, topical, injectable, and implantable formulations. For example, silk fibroin hydrogels permit sustained release of doxorubicin (70% over 5 days), efficiently suppressing tumor growth [80], whereas chitosan aerogels release tetracycline for extended wound therapy. Keratin hydrogels outperform in sequestering water-insoluble drugs due to their amphiphilic properties [81]. Clever formulations such as pH-sensitive chitosan–bioactive glass or temperature-responsive gelatin–PNIPAm hydrogels do even better by targeting acidic tumor microenvironments or temperature-triggered release. The therapies maximize therapeutic impact while reducing systemic side effects.

8.4.4 Injectable and In Situ-Forming Gels

Injectable hydrogels provide minimally invasive solutions by adapting to irregular tissue defects and gelling at physiological conditions, such as the silk fibroin-HRP system (setting in <5 minutes) or thermosensitive chitosan/β-glycerophosphate (37 °C gelling) for cartilage repair. The gels are applied in localized tumor therapy (e.g. for doxorubicin delivery), orthopedic defects (BMP-loaded gels), and regenerative applications such as dental pulp or ocular cell encapsulation, providing precision with decreased surgical trauma [82].

8.4.5 Biosensors and Diagnostic Devices

Aerogels made from animal fibers and having electroactive nanoparticles (e.g. carbon nanotubes, graphene) allow sensitive biomarker detection such as glucose, cortisol, and tumor markers. For example, keratin–graphene aerogels possess high conductivity (~0.1 S/cm) for wearable skin sensors, and silk protein films with fluorophores or chitosan-based biosensors with immobilized enzymes (e.g. glucose oxidase) improve optoelectronic and electrochemical sensing. These materials are flexible, biodegradable, and adaptive for implantable or skin-mountable devices, coupling sustainability with sophisticated diagnostics [83].

8.4.6 Hemostatic and Antibacterial Applications

Chitosan hydrogels are superior hemostatic agents because of their mucoadhesive, cationic properties, which make them perfect for emergency bleeding control in trauma or surgery. Chitosan sponges decreased rat femoral artery injury bleeding time by 50%, while silver nanoparticle-loaded collagen hydrogels had $>99.9\%$ bacterial killing. These polymers retain rapid clotting and antimicrobial activity, improving results in infected and acute wounds. Their adaptability facilitates both battlefield and therapeutic uses.

8.4.7 Ophthalmic and Dental Applications

Animal fiber-derived biomaterials are most influential in ocular and dental regeneration, with collagen hydrogels used to restore corneal transparency and silk membranes as tough corneal patches. In dentistry, keratin hydrogels stimulate pulp cell growth, and elastin–gelatin composites function as periodontal ligament scaffolds and dentin. These biomaterials bridge the gaps between biocompatibility and mechanical properties in a tailored fashion, filling essential needs in ophthalmology and dentistry [84].

8.4.8 Summary Table of Biomedical Applications

Application	Material used	Key benefits
Wound healing	Chitosan, collagen	Moisture, antimicrobial, hemostatic
Bone tissue engineering	Collagen-HA aerogel	Porosity, osteoconduction, mineral deposition
Drug delivery	Silk, chitosan hydrogels	Sustained, targeted release
Injectable gels	Silk-HRP, Chitosan-GP	Minimally invasive, in situ solidification
Biosensing	Keratin–graphene aerogel	Conductivity, flexibility, and biocompatibility
Hemostasis	Chitosan aerogel	Rapid bleeding control
Ophthalmic applications	Collagen silk membrane	Transparency, ECM mimicry

8.5 Conclusion

Animal fiber-derived hydrogels and aerogels are now considered to be next-generation biomaterials that fill the functional gap between biopolymeric biomaterials and engineered systems. Being derived from proteins like collagen, keratin, silk fibroin, gelatin, and chitosan, they possess natural benefits such as biocompatibility, biodegradability, cell recognition motifs, and modifiable mechanical properties. These materials have performed optimally in a broad range of biomedical applications, such as wound healing, drug delivery, biosensing, tissue engineering, ophthalmology, and dentistry.

What distinguishes these biomaterials is that they can closely replicate the ECM of nature, supporting cellular functions like adhesion, proliferation, and differentiation. Hydrogels provide hydration, conformability, and sustained release, and aerogels add porosity, mechanical strength, and structural integrity. With tunable architectures and multifunctional properties, they present clinicians and scientists with a very flexible toolkit to meet increasingly demanding medical challenges.

Even with their vast potential, many hurdles still need to be overcome. Some of these are scalability of production, batch-to-batch reproducibility, expense of high-purity biopolymers, mechanical robustness against physiological stress, and regulatory issues. Overcoming them needs a synergistic effort between material science, biotechnology, and regulatory science.

8.6 Future Trends

Emerging innovations in animal fiber-based hydrogels and aerogels will revolutionize the biomedical field by binding the power of smart materials, personalized medicine, and eco-friendly technology. New generations of biomaterials will have the capability to sense physiologically relevant cues, including pH, enzymes, or temperature, to deliver controlled drug release and in situ biosensing. The convergence of silk and keratin-based bioinks with 3D printing will enable patient-specific implants and vascularized tissue constructs. Multifunctional composites of animal fibers mixed with synthetic polymers, ceramics, or nanoparticles will provide superior antimicrobial, mechanical, and regenerative properties. At the same time, environmentally friendly extraction technologies—such as enzymatic and plasma-based processes—and human hair- or feather-based ethical sourcing will comply with ecological aims. Regulatory conditions are also changing to facilitate clinical translation, focusing on sterility, standardization, and scaling. In addition, aerogels from animal fibers are being targeted for use in digital health platforms, including keratin–graphene biosensors for wearables and implantable diagnostics, which represents a major move toward personalized, networked medicine.

References

1 Xu, Y., Yan, C., du, C. et al. (2023). High-strength, thermal-insulating, fire-safe bio-based organic lightweight aerogel based on 3D network construction of natural tubular fibers. *Composites Part B: Engineering* 261: 110809.

2 Salleh, K.M., Armir, N.A.Z., Mazlan, N.S.N. et al. (2022). Hydrogel- and aerogel-based composites: biodegradable hydrogel and aerogel polymer blend-based composites. In: *Biodegradable Polymers, Blends and Composites* (ed. S.M. Rangappa, J. Parameswaranpillai, M. Ramesh, and S. Siengchin), 355–388. Amsterdam, Netherlands: Elsevier.

3 Singh, N., Agarwal, S., Jain, A., and Khan, S. (2021). 3-Dimensional cross linked hydrophilic polymeric network 'hydrogels': an agriculture boom. *Agricultural Water Management* 253: 106939.

4 Costa, F., Silva, R., and Boccaccini, A. (2018). Fibrous protein-based biomaterials (silk, keratin, elastin, and resilin proteins) for tissue regeneration and repair. In: *Peptides and Proteins as Biomaterials for Tissue Regeneration and Repair* (ed. M. Barbosa and M.C.L. Martins), 175–204. Cambridge, UK: Woodhead Publishing / Elsevier.

5 Bakhshandeh, B., Ranjbar, N., Abbasi, A. et al. (2023). Recent progress in the manipulation of biochemical and biophysical cues for engineering functional tissues. *Bioengineering & Translational Medicine* 8 (2): e10383.

6 Peppas, N.A., Bures, P., Leobandung, W., and Ichikawa, H. (2000). Hydrogels in pharmaceutical formulations. *European Journal of Pharmaceutics and Biopharmaceutics* 50 (1): 27–46.

7 Zheng, L., Zhang, S., Ying, Z. et al. (2020). Engineering of aerogel-based biomaterials for biomedical applications. *International Journal of Nanomedicine* 15: 2363–2378.

8 Karamikamkar, S., Yalcintas, E.P., Haghniaz, R. et al. (2023). Aerogel-based biomaterials for biomedical applications: from fabrication methods to disease-targeting applications. *Advanced Science* 10 (23): e2204681.
9 Alaghawani, N.A., Alkhatib, H., Elmancy, L., and Daou, A. (2024). Harmonizing innovations: an in-depth comparative review on the formulation, applications, and future perspectives of aerogels and hydrogels in pharmaceutical sciences. *Gels* 10 (10): 663.
10 Bayır, E. (2023). Comparative evaluation of alginate-gelatin hydrogel, cryogel, and aerogel beads as a tissue scaffold. *Sakarya University Journal of Science* 27 (2): 335–348.
11 De France, K.J., Hoare, T., and Cranston, E.D. (2017). Review of hydrogels and aerogels containing nanocellulose. *Chemistry of Materials* 29 (11): 4609–4631.
12 Wei, Z., Wu, C., Li, R. et al. (2021). Nanocellulose based hydrogel or aerogel scaffolds for tissue engineering. *Cellulose* 28 (12): 7497–7520.
13 Adekomaya, O. and Majozi, T. (2021). Industrial and biomedical applications of fiber reinforced composites. In: *Fiber Reinforced Composites*, 753–783. Elsevier.
14 Rezvani Ghomi, E., Nourbakhsh, N., Akbari Kenari, M. et al. (2021). Collagen-based biomaterials for biomedical applications. *Journal of Biomedical Materials Research Part B: Applied Biomaterials* 109 (12): 1986–1999.
15 Alipal, J., Mohd Pu'ad, N.A.S., Lee, T.C. et al. (2021). A review of gelatin: properties, sources, process, applications, and commercialisation. *Materials Today: Proceedings* 42: 240–250.
16 Kumar, A. and Kumar, A. (2017). Chitosan as a biomedical material: properties and applications. In: *Biopolymers: Structure, Performance and Applications* (ed. M.A. Kumar and H.C. Mustansar), 139–153. Nova Science Publishers.
17 Grabska-Zielińska, S. and Sionkowska, A. (2021). How to improve physico-chemical properties of silk fibroin materials for biomedical applications? – blending and cross-linking of silk fibroin – a review. *Materials* 14 (6): 1510.
18 Sarma, A. (2022). Biological importance and pharmaceutical significance of keratin: a review. *International Journal of Biological Macromolecules* 219: 395–413.
19 Feroz, S., Muhammad, N., Ratnayake, J., and Dias, G. (2020). Keratin-based materials for biomedical applications. *Bioactive Materials* 5 (3): 496–509.
20 Altman, G.H., Diaz, F., Jakuba, C. et al. (2003). Silk-based biomaterials. *Biomaterials* 24 (3): 401–416.
21 Ravi, S. and Chaikof, E.L. (2010). Biomaterials for vascular tissue engineering. *Regenerative Medicine* 5 (1): 107–120.
22 Lazaris, A., Arcidiacono, S., Huang, Y. et al. (2002). Spider silk fibers spun from soluble recombinant silk produced in mammalian cells. *Science* 295 (5554): 472–476.
23 Murphy, S. and Atala, A. (2014). 3D bioprinting of tissues and organs. *Nature Biotechnology* 32 (8): 773–785.
24 Bhattacharjee, P., Kundu, B., Naskar, D. et al. (2017). Silk scaffolds in bone tissue engineering: an overview. *Acta Biomaterialia* 63: 1–17.

25 Shariatinia, Z. and Ziba, M. (2022). Smart pH-responsive drug release systems based on functionalized chitosan nanocomposite hydrogels. *Surfaces and Interfaces* 29: 101739.

26 Bugarčić, M., Jovanović, A., Petrović, J. et al. (2024). Advances in biopolymer production and applications: a comprehensive review of key biomaterials. *Metallurgical and Materials Data* 2 (3): 81–98.

27 F. B. Insights. Biomaterials Market to Exhibit 12.2% CAGR and Hit USD 245.6 Billion till 2027; Increasing Number of Novel Product Launches to Skyrocket Demand. https://www.globenewswire.com/news-release/2021/03/15/2192567/0/en/Biomaterials-Market-to-Exhibit-12-2-CAGR-and-Hit-USD-245-6-Billion-still-2027-Increasing-Number-of-Novel-Product-Launches-to-Skyrocket-Demand-Fortune-Business-Insights.html (accessed 21 May 2025).

28 F. B. Insights. Advanced Wound Dressings Market Size, Share & Industry Analysis, By Product Type (Alginate Dressings, Hydrogel Dressings, Film Dressings, Foam Dressings, Hydrocolloid Dressings, Antimicrobial Dressings, and Others), By Indication (Diabetic Foot Ulcers, Pressure Ulcers, Surgical Wounds, and Others), By End-user (Hospitals, Clinics, Homecare Settings, and Others), and Regional Forecast, 2024–2032. https://www.fortunebusinessinsights.com/industry-reports/advanced-wound-dressings-market-100384 (accessed 21 May 2025).

29 MarketsandMarkets. Orthobiologics Market by Product (Viscosupplementation, Synthetic Orthobiologics, DBM, BMP, PRP, BMAC, Allograft), Application (Fracture Recovery, Osteoarthritis, Spinal Fusion, Soft Tissue), End User (Hospitals, ASCs, Academia) - Global Forecast to 2022. https://www.marketsandmarkets.com/Market-Reports/orthobiologics-market-162747970.html (accessed 21 May 2025).

30 Xiao, M., Sun, L., Wu, K. et al. (2025). A sequential drug delivery system based on silk fibroin scaffold for effective cartilage repair. *Bioactive Materials* 49: 255–270.

31 Hong, S., Choi, D.W., Kim, H.N. et al. (2020). Protein-based nanoparticles as drug delivery systems. *Pharmaceutics* 12 (7): 604.

32 Liu, J., Pei, Y., Dong, Y. et al. (2021). Conversion of protein and polysaccharide wastes into value-added composite products. In: *Conversion of Renewable Biomass into Bioproducts* (ed. M.I. Sarker, L. Liu, M.P. Yadav, et al.), 219–260. Washington, DC: ACS Publications, American Chemical Society.

33 Wattie, B. (2016). Synthesis of keratin-based hydrogels and cryogels destined for environmental applications. Thesis. McGill University (Canada).

34 Ha, T.L.B., Quan, T.M., Vu, D.N., and Si, D.M. (2013). Naturally derived biomaterials: preparation and application. In: *Regenerative Medicine and Tissue Engineering* (ed. J.A. Andrades). London, UK: IntechOpen.

35 Rockwood, D.N., Preda, R.C., Yücel, T. et al. (2011). Materials fabrication from *Bombyx mori* silk fibroin. *Nature Protocols* 6 (10): 1612–1631.

36 Popescu, C. and Stanescu, M.D. (2024). Eco-friendly processing of wool and sustainable valorization of this natural bioresource. *Sustainability* 16 (11): 4661.

37 Friess, W. (1998). Collagen–biomaterial for drug delivery. *European Journal of Pharmaceutics and Biopharmaceutics* 45 (2): 113–136.

38 Rinaudo, M. (2006). Chitin and chitosan: properties and applications. *Progress in Polymer Science* 31 (7): 603–632.

39 Liu, X., Huang, Q., Pan, P. et al. (2023). Comparative study of the preparation of high-molecular-weight fibroin by degumming silk with several neutral proteases. *Polymers* 15 (16): 3383.

40 Viñas, M.Á., Sanz, V., Seoane, P.R. et al. (2023). Microwave-assisted extraction (MAE). In: *Green Extraction Techniques in Food Analysis* (ed. M. Plaza and M.L. Marin), 396–430. Bentham Science Publishers.

41 Aigner, T.B., DeSimone, E., and Scheibel, T. (2018). Biomedical applications of recombinant silk-based materials. *Advanced Materials* 30 (19): e1704636.

42 Shavandi, A., Silva, T.H., Bekhit, A.A., and Bekhit, A.E.-D.A. (2017). Keratin: dissolution, extraction and biomedical application. *Biomaterials Science* 5 (9): 1699–1735.

43 Heydari, M., Carbone, K., Gervasi, F. et al. (2023). Cold plasma-assisted extraction of phytochemicals: a review. *Food* 12 (17): 3181.

44 Shoulders, M.D. and Raines, R.T. (2009). Collagen structure and stability. *Annual Review of Biochemistry* 78 (1): 929–958.

45 Bhattacharya, J., Dutta, S., das, M. et al. (2023). Biopolymer nanocomposites and membranes in tissue engineering. In: *Nanostructured Materials for Tissue Engineering*, 337–372. Elsevier.

46 Xu, J., Liu, X., Ren, X., and Gao, G. (2018). The role of chemical and physical crosslinking in different deformation stages of hybrid hydrogels. *European Polymer Journal* 100: 86–95.

47 Yang, J., Chen, Y., Zhao, L. et al. (2023). Constructions and properties of physically cross-linked hydrogels based on natural polymers. *Polymer Reviews* 63 (3): 574–612.

48 Liu, S., Oderinde, O., Hussain, I. et al. (2018). Dual ionic cross-linked double network hydrogel with self-healing, conductive, and force sensitive properties. *Polymer* 144: 111–120.

49 Fichman, G. and Gazit, E. (2014). Self-assembly of short peptides to form hydrogels: design of building blocks, physical properties and technological applications. *Acta Biomaterialia* 10 (4): 1671–1682.

50 Yu, Y., Xu, S., Li, S., and Pan, H. (2021). Genipin-cross-linked hydrogels based on biomaterials for drug delivery: a review. *Biomaterials Science* 9 (5): 1583–1597.

51 Ifkovits, J.L. and Burdick, J.A. (2007). Photopolymerizable and degradable biomaterials for tissue engineering applications. *Tissue Engineering* 13 (10): 2369–2385.

52 Hu, T., Cai, Z., Yin, R. et al. (2023). 3D embedded printing of complex biological structures with supporting bath of pluronic F-127. *Polymers* 15 (17): 3493.

53 Harrison, H.B. (2021). Covalent functionalization of BNNTs for low density aerogels. Dissertatio. The University of North Carolina at Greensboro.

54 Thomas, N.G., Joy, S., Binci, P.K. et al. (2024). Biomaterials from sustainable natural sources for tissue engineering. In: *Application of Engineering Principles and Practices In Biotechnology* (ed. A.R. Abraham, L.S. Torre, T. Tatrishvili, and A.K. Haghi), 43–86. New York: Apple Academic Press.

55 Gurav, J.L., Jung, I.-K., Park, H.-H. et al. (2010). Silica aerogel: synthesis and applications. *Journal of Nanomaterials* 2010 (1): 409310.

56 Shi, F., Wang, L., and Liu, J. (2006). Synthesis and characterization of silica aerogels by a novel fast ambient pressure drying process. *Materials Letters* 60 (29–30): 3718–3722.

57 Jayakumar, R., Prabaharan, M., Kumar, P.S. et al. (2011). Biomaterials based on chitin and chitosan in wound dressing applications. *Biotechnology Advances* 29 (3): 322–337.

58 Rouse, J.G. and Van Dyke, M.E. (2010). A review of keratin-based biomaterials for biomedical applications. *Materials* 3 (2): 999–1014.

59 Katrilaka, C., Karipidou, N., Petrou, N. et al. (2023). Freeze-drying process for the fabrication of collagen-based sponges as medical devices in biomedical engineering. *Materials* 16 (12): 4425.

60 Çetin, A. (2025). Use of sandwich structures in biomedical applications: an innovative design for external ring fixators. *Biomedical Materials* 20 (2): 025040.

61 Pok, S., Myers, J.D., Madihally, S.V., and Jacot, J.G. (2013). A multilayered scaffold of a chitosan and gelatin hydrogel supported by a PCL core for cardiac tissue engineering. *Acta Biomaterialia* 9 (3): 5630–5642.

62 Dong, X., Wan, B., Zheng, M.S. et al. (2023). Versatile spider-web-like cross-linked polyimide aerogel with tunable dielectric permittivity for highly sensitive flexible sensors. *Chemical Engineering Journal* 465: 143034.

63 Yan, Y., Jin, B., Zhou, Q. et al. (2022). Preparation of a chitosan-lead composite carbon aerogel and its catalytic thermal decomposition performance on ammonium perchlorate. *Langmuir* 38 (28): 8623–8632.

64 Ng, H.W., Zhang, Y., Naffa, R., and Prabakar, S. (2020). Monitoring the degradation of collagen hydrogels by collagenase *clostridium histolyticum*. *Gels* 6 (4): 46.

65 Ashna, M., Senthilkumar, N., and Sanpui, P. (2023). Human hair Keratin-based hydrogels in regenerative medicine: current status and future directions. *ACS Biomaterials Science & Engineering* 9 (10): 5527–5547.

66 Grenier, J., Duval, H., Barou, F. et al. (2019). Mechanisms of pore formation in hydrogel scaffolds textured by freeze-drying. *Acta Biomaterialia* 94: 195–203.

67 Reizabal, A., Costa, C., Pérez-Álvarez, L. et al. (2023). The new silk road: silk fibroin blends and composites for next generation functional and multifunctional materials design. *Polymer Reviews* 63 (4): 1014–1077.

68 Chartier, C. (2023). Chitosan-based aerogels and cryogels for wound healing applications. Université Paris sciences et lettres.

69 Singh, P., Andrabi, S.M., Tariq, U. et al. (2023). Functionally multifaceted scaffolds delivering bioactive compounds for treatment of infectious chronic and ischemic wounds. *Chemical Engineering Journal* 457: 141359.

70 Zhao, Y., Cheng, C., Wang, X. et al. (2024). Aspirin-loaded anti-inflammatory ZnO-SiO_2 aerogel scaffolds for bone regeneration. *ACS Applied Materials & Interfaces* 16 (14): 17092–17108.

71 Esmaeilzadeh, J., Shabani, F., and Zak, A.K. (2024). Electrospun poly (ε-caprolactone)/gelatin nanofibrous mats with local delivery of vitamin C for wound healing applications. *Colloids and Surfaces A: Physicochemical and Engineering Aspects* 687: 133546.

72 Sanchez Ramirez, D.O., Carletto, R.A., and Truffa Giachet, F. (2019). Keratin processing. In: *Keratin as a Protein Biopolymer: Extraction from Waste Biomass and Applications* (ed. S. Sharma and A. Kumar), 77–121.

73 Grabska-Zielińska, S. (2024). Cross-linking agents in three-component materials dedicated to biomedical applications: a review. *Polymers* 16 (18): 2679.

74 Khanmohammadi, M., Dastjerdi, M.B., Ai, A. et al. (2018). Horseradish peroxidase-catalyzed hydrogelation for biomedical applications. *Biomaterials Science* 6 (6): 1286–1298.

75 Liu, H., Xing, F., Yu, P. et al. (2024). Multifunctional aerogel: a unique and advanced biomaterial for tissue regeneration and repair. *Materials & Design* 243: 113091.

76 Joshi, R. and Joshi, R. (2021). Collagen biografts for chronic wound healing. In: *Collagen Biografts for Tunable Drug Delivery*, 53–65.

77 Roh, D.-H., Kang, S.Y., Kim, J.Y. et al. (2006). Wound healing effect of silk fibroin/alginate-blended sponge in full thickness skin defect of rat. *Journal of Materials Science: Materials in Medicine* 17: 547–552.

78 Sawaragi, E., Yamanaka, H., Li, Y. et al. (2023). Comparisons of the effects of silk elastin and collagen sponges on wound healing in murine models. *Regenerative Therapy* 24: 385–397.

79 Kumar, V.A., Caves, J.M., Haller, C.A. et al. (2013). Acellular vascular grafts generated from collagen and elastin analogs. *Acta Biomaterialia* 9 (9): 8067–8074.

80 Wu, W., Chen, H., Shan, F. et al. (2014). A novel doxorubicin-loaded in situ forming gel based high concentration of phospholipid for intratumoral drug delivery. *Molecular Pharmaceutics* 11 (10): 3378–3385.

81 Han, S., Ham, T.R., Haque, S. et al. (2015). Alkylation of human hair keratin for tunable hydrogel erosion and drug delivery in tissue engineering applications. *Acta Biomaterialia* 23: 201–213.

82 Pandya, A.K., Vora, L.K., Umeyor, C. et al. (2023). Polymeric in situ forming depots for long-acting drug delivery systems. *Advanced Drug Delivery Reviews* 200: 115003.

83 Akter, M., Anik, H.R., Tushar, S.I. et al. (2024). Advances in functionalized applications of graphene-based wearable sensors in healthcare. *Advanced Sensor Research* 3 (2): 2300120.

84 Palchesko, R.N., Carrasquilla, S.D., and Feinberg, A.W. (2018). Natural biomaterials for corneal tissue engineering, repair, and regeneration. *Advanced Healthcare Materials* 7 (16): 1701434.

9

Advancements in Hydrogel and Aerogel Textile Technology

Syeda Rubab Batool[1], Rahima Hussain[1], and Muhammad Anwaar Nazeer[1,2]

[1]National Textile University, School of Engineering and Technology, Department of Clothing, Sheikhupura Road, 37610 Faisalabad, Pakistan
[2]National Textile University, Biomaterials and Tissue Engineering Research Laboratory, Sheikhupura Road, Faisalabad 37610, Pakistan

9.1 Introduction

Textiles are an essential part of everyday life, offering comfort, functionality, and protection across a wide range of applications. From garments and home furnishings to highly specialized industrial and medical products, textile materials are found in nearly every aspect of modern living. Whether used in bandages, bedding, filters, or insulation, textiles serve purposes that go well beyond basic clothing. They support personal well-being, public health, and environmental safety. From birth to old age, humans remain closely connected to textiles, whether through absorbent infant clothing, protective sports gear, or medical sutures that preserve life [1].

Over time, the demand for textiles with advanced features has grown significantly. Technical textiles, in particular, have seen increasing interest due to their specialized properties such as flame retardance, antibacterial function, moisture control, and thermal insulation [2]. These properties are either inherent to the fibers or added through physical or chemical modifications. Porosity is one of the most crucial characteristics in such fabrics, especially in applications involving insulation, filtration, and moisture management. While traditional fabrics, woven, knitted, or nonwoven, can provide some degree of porosity, their limitations become evident when extremely fine control over pore size and structure is required [3].

Conventionally, fibers offer pore sizes in the micrometer range, with woven and knitted fabrics shaped by factors such as yarn thickness and fabric structure. Techniques like electrospinning have been explored to create nanoporous fibers with diameters down to tens of nanometers. However, while electrospun fibers offer improved porosity, their mechanical weaknesses often hinder broader application. To address this challenge, researchers have turned to aerogels, highly porous, ultralightweight materials with outstanding thermal and acoustic insulation properties. When integrated with textile substrates, aerogels provide improved structural performance while retaining nanoscale porosity, making them ideal for protective and insulating fabrics. The development of textile-reinforced aerogel composites has therefore opened new possibilities in high-performance applications where both low weight and high insulation are required [4].

Simultaneously, hydrogels have attracted growing interest within the textile field. These are three-dimensional, water-swollen polymer networks known for their softness, biocompatibility, and responsiveness to environmental stimuli [5]. Unlike aerogels, which are predominantly valued for their insulation and structural characteristics, hydrogels are particularly useful for their moisture-retaining ability, flexibility, and interactive behavior, making them well-suited for wearable healthcare systems, soft robotics, drug delivery patches, and moisture management layers in clothing. Hydrogels can adapt to temperature, pH, or mechanical changes, enabling dynamic responses in textiles that support smart and bio-integrated functions [6].

Although hydrogels have been widely studied for their applications in comfort and healthcare, and aerogels for their insulation capabilities. There has been little focus on how these materials specifically influence fabric porosity and adaptability. Much of the literature tends to treat them separately, without fully exploring how their distinct properties might work together within textile systems. Hydrogels offer responsiveness and moisture management, while aerogels contribute strength and thermal control, making both valuable in designing multifunctional textiles [4].

This chapter will dive into how hydrogels and aerogels are reshaping the textile landscape. The discussion begins with their core characteristics and progresses through fabrication techniques and integration methods. We also highlight practical applications in wearable technology and sustainable textiles and conclude with current limitations and possible future developments in the field.

9.2 Fundamentals of Hydrogel and Aerogel Materials

In this chapter, we explore two textile materials that bring innovation in textiles: hydrogels and aerogels. Hydrogels are developed from water-soluble polymers that create flexible and gentle structures that soak and preserve a huge quantity of moisture. Hydrogels are utilized for eco-friendly purposes because of their soft nature and stretchability, which include smart textiles and digital wearables. On the other hand, aerogels are extremely lightweight and soft materials. These properties make it suitable for high-insulation fabrics. These aerogels have many advantages, such as they have heat and sound insulation properties, and they rely on the base material, which is silica, polymer, or carbon. Both materials, hydrogels and aerogels, play a significant role in smart and multifunctional textiles [7].

9.2.1 Hydrogels

Hydrogels have a long and evolving history, beginning in the late 1800s with the development of the first hydrogel, Ivalon, which was initially created for use in

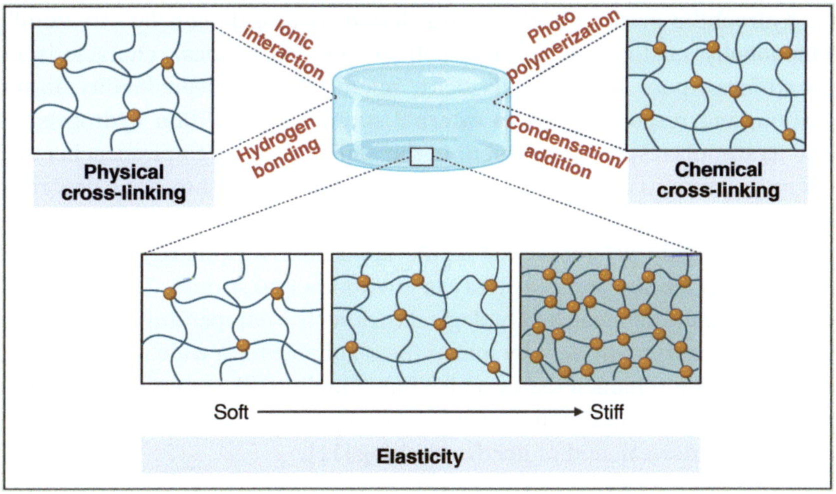

Figure 9.1 Schematic illustration of hydrogel network formation through physical and chemical cross-linking mechanisms. The lower panel demonstrates the influence of cross-link density on hydrogel elasticity, ranging from soft (low cross-link density) to stiff (high cross-link density) structures.

biomedical implants. A hydrogel is a three-dimensional (3D) polymeric network primarily composed of hydrophilic groups that enable it to absorb and retain substantial amounts of water while maintaining its structural integrity. Typically, hydrogels consist of polymer chains that are cross-linked either chemically or physically (Figure 9.1), with the cross-linking driven by covalent bonds, van der Waals interactions, or physical entanglements. Key factors such as the degree of cross-linking, ionization, and the hydrogel's hydrophilic or hydrophobic nature play crucial roles in regulating its equilibrium, swelling behavior, and phase transitions. Physically cross-linked hydrogels rely on non-covalent forces such as hydrogen bonds, ionic attractions, or hydrophobic interactions. These networks are often reversible and suitable for biological uses, especially when no synthetic cross-linkers are desired. Depending on the strength of these physical links, the gels may form either stable or transient structures. On the other hand, chemically cross-linked hydrogels form permanent covalent bonds between polymer chains. These are more robust and durable under stress. Functional groups such as –OH, –COOH, or $–NH_2$ on the polymer backbone allow such bonds to form, sometimes initiated by heat, light, or radiation (e.g. UV, γ-rays, and electron beams).

Hydrogels can also be classified based on their origin, natural or synthetic [8]. Natural hydrogels are derived from alginate, gelatin, or cellulose. These are known for their biocompatibility and environmental friendliness. In contrast, synthetic hydrogels, such as those based on polyacrylamide or polyethylene glycol, offer greater design flexibility, enabling researchers to fine-tune properties like mechanical strength, responsiveness, and stability. In some cases, blending natural and synthetic components results in hybrid hydrogels that aim to combine the advantages of both types [9]. All materials used to prepare hydrogels contain hydrophilic groups. With careful design, hydrogels can be structurally and functionally customized to adjust their physicochemical characteristics, such as stiffness, pore size, viscoelasticity, microstructure, degradability, ligand presentation, and responsiveness to external stimuli [10]. Certain hydrogels are classified as stimuli-responsive, exhibiting the ability to undergo physicochemical changes in response to external stimuli. These stimuli can be broadly categorized as physical, chemical, or biological. Hydrogels have been extensively studied and are considered highly promising for textile applications, primarily due to their intrinsic biocompatibility and sensitivity to environmental stimuli. Their ability to respond to physical, chemical, and biological cues such as temperature changes, pH fluctuations, and enzymatic activity makes them suitable for a wide range of smart textile systems. These characteristics enable functionalities like moisture regulation, controlled drug release, and environmental sensing, which are particularly valuable in healthcare, sportswear, and protective clothing [11].

9.2.2 Aerogels

Aerogels are highly porous materials, typically containing over 95% air, first developed by Kistler in 1931. Aerogels possess a three-dimensional nanoporous structure that resembles that of hydrogels; however, their fabrication involves distinct processing techniques. While hydrogels retain their water content, aerogels are produced via a sol–gel process followed by supercritical drying, which eliminates the liquid phase without causing pore collapse due to surface tension. This method results in an ultralightweight solid network characterized by extremely low density, high porosity, and excellent structural integrity. These properties make aerogels highly attractive for advanced textile applications, including thermal insulation, filtration, and wearable electronics. These materials are composed of a vast network of interconnected pores, typically in the mesoporous (2–50 nm) and macroporous (greater than 50 nm) ranges [11].

Aerogels are broadly classified based on their appearance, chemical composition, and microstructure. Microstructurally, they are categorized as microporous, mesoporous, or macroporous. Appearance-based classifications include monoliths, powders, films, and fibers. Chemically, aerogels are divided into three main types: inorganic, organic, and hybrid (Figure 9.2). Among these, inorganic aerogels, such as silica, alumina, titania, and zirconia-based materials, are the earliest and most widely studied. They are valued for their thermal insulation, low density, and structural properties. Silica aerogels are particularly prominent due to their mesoporous structure and extremely high porosity, making them ideal for applications in insulation, aerospace, electronics, and filtration. Efforts are ongoing to improve their brittleness through hybridization with organosilanes. Metal oxide aerogels like alumina and titania offer enhanced thermal stability, making them suitable for extreme environments such as furnaces and reactors. In the organic category, polysaccharide-based aerogels, particularly those derived from cellulose, are gaining traction for drug delivery due to their biocompatibility and adaptable pore

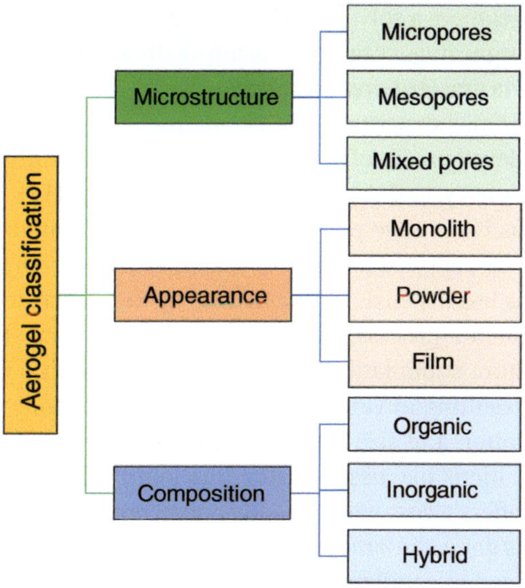

Figure 9.2 Schematic illustration of hydrogel network formation through physical and chemical cross-linking mechanisms.

structures. Carbon aerogels, known for their mechanical strength and conductivity, are used in energy storage, catalysis, and environmental remediation. Composite aerogels integrate polymers or biopolymers with inorganic matrices like silica to enhance mechanical strength, thermal insulation, and functional versatility. Silica-polymer composites use strategies like one-pot synthesis to reinforce fragile silica networks for uses like ballistic protection. Silica–biopolymer composites, including chitosan and silk fibroin hybrids, are being explored for sustainable and high-performance insulation [12]. Carbon-reinforced silica aerogels incorporate graphene or graphene oxide to improve surface area, hydrophobicity, mechanical strength, and thermal properties, suitable for supercapacitors, sensors, and oil absorption [13].

Aerogels are typically synthesized through various methods, with sol–gel polymerization being the most widely used, especially for inorganic aerogels such as silica-based types. This process involves several key steps: formation of a sol, gelation through hydrolysis and condensation of precursors (catalyzed to control reaction rates), and subsequent aging to enhance mechanical strength (Figure 9.3). After

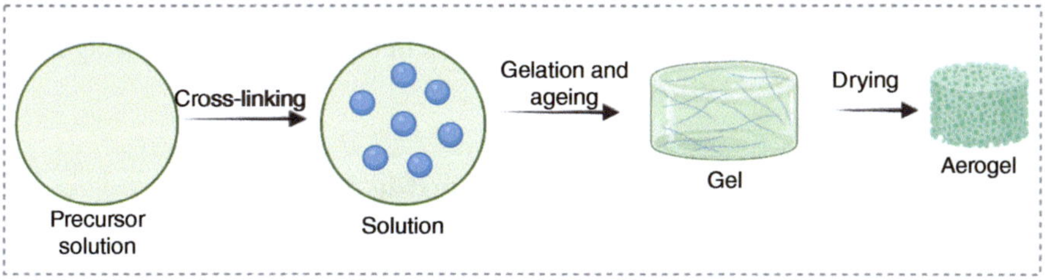

Figure 9.3 Schematic of aerogel synthesis from sol formation to porous aerogel via gelation and drying.

gelation, a solvent exchange step is carried out to replace pore liquids and eliminate impurities, commonly using alcohol. The final and critical stage is drying, performed via ambient pressure, freeze drying, or supercritical drying, to preserve the aerogel's porous structure without collapse [14]. These drying processes impact the structural development of aerogels and also notably influence their properties such as porosity, density, and thermal insulation capacity [15]. Supercritical drying is more desirable as it can save porous structures, but it entails high cost and safety issues. It is the most applied method for the production of low-density aerogels [16]. Freeze-drying is less expensive but can destroy the structure because of solvent crystallization. In such a process, the water is removed from the gel by rapidly freezing it and vacuum sublimation of ice. A process like this prevents shrinkage and stress within, resulting in very porous structures. Depending on freezing conditions, pore orientation and size can be controlled, and freeze-dried aerogels can be used for applications like tissue scaffolding and flexible insulation [4]. Ambient pressure drying offers a cost-effective alternative but requires longer processing times. This process is done at room temperature or in a controlled oven. While solvent evaporation may lead to capillary tension, resulting in partial pore shrinkage, recent developments have greatly enhanced the outcome. For instance, cellulose acetate/polybenzoxazine hybrid aerogels developed via this process have exhibited low shrinkage, mechanical stability at cryogenic temperatures, and superior flame resistance [17].

For textile applications, aerogels are particularly valued for their thermal and acoustic insulation, achieved through nanoscale pores that trap air and disrupt heat and sound transmission. Their insulating power comes from three mechanisms: reduced solid conduction, suppressed gas-phase transfer (via the Knudsen effect), and minimal radiative transfer. Silica aerogels, for instance, can have thermal conductivities lower than still air (under 0.015 W/m·K), making them excellent for cold-weather or protective clothing. However, aerogels in monolithic form are often brittle [18]. To make them compatible with flexible materials, researchers have developed aerogel fibers and flexible composites. These materials can be woven, bent, or stretched while retaining their insulation properties, making them ideal for integration into smart fabrics, outdoor gear, and wearable devices [19]. Beyond thermal performance, aerogels are also emerging as potential materials for acoustic insulation. Their ability to absorb and attenuate sound waves stems from their internal pore geometry and density. As sound passes through the aerogel matrix, energy is lost due to changes in medium and internal friction, leading to a measurable reduction in sound intensity. Though this application is still relatively new, it points to further opportunities for aerogels in environments requiring both thermal and noise control [20].

9.2.3 Comparison and Complementarity in Textile Applications

Hydrogels and aerogels, with their unique nanostructured properties, are indeed bringing about significant variations and advancements in the textile industry by offering enhanced functionalities like thermal insulation, mechanical reinforcement, and specialized absorption capabilities. Aerogels are better inductors contrary to noise, electrical energy, and heat because they are known for their porous structure. Aerogels are perfect options for protective garments and in medical textiles. If they are prepared with silica or aluminum oxide, they have heat-resistant properties and protect against strong chemicals. They are used in various applications, such as medical textiles, including use in breathable dressings, drug delivery fabrics, surgical implants, and wound healing materials. When combined with high-performance fibers such as aramid, carbon, or nanocomposites, aerogels exhibit enhanced strength and functionality [21].

On the other hand, hydrogels have a soft, flexible structure and are composed of interconnected hydrophilic polymers capable of holding a large amount of water. Their excellent moisture retention makes them ideal for wearable fabrics and medical applications involving direct contact with the skin or body [22]. Synthetic polymer-based hydrogels offer structural control, mechanical stability, and biocompatibility, making them suitable for applications such as nerve regeneration and smart sensors. Although they are nondegradable, these hydrogels are highly versatile and available in responsive forms, including pH-sensitive, temperature-sensitive, light-sensitive, and electro-responsive types. Recent studies have shown that nanocellulose-based hydrogels, produced through eco-friendly methods, outperform conventional hydrogels in terms of mechanical strength, responsiveness, and stability. These materials hold significant potential for future textile innovations, especially if production costs can be reduced [7]. Meanwhile, aerogels continue to stand out as leading candidates for next-generation thermal insulation in textiles and industrial applications. Their extremely low density, exceptionally high surface area (up to 1600 m^2/g), and air content ranging from 80% to 99.8% give them a distinct advantage in energy storage, purification, and catalytic systems [23]. With their unique porous network, aerogels offer superior insulation while maintaining a lightweight profile, an advantage that could eventually replace traditional insulating materials.

9.3 Fabrication and Integration Methods

This section focuses on the primary methods for fabricating and integrating hydrogels and aerogels into textile structures.

9.3.1 Hydrogel–Textile Integration Techniques

Hydrogels are gaining increasing attention in the textile sector due to their exceptional properties such as high-water absorption, softness, flexibility, biocompatibility, and responsiveness to environmental stimuli like pH, temperature, and humidity. These characteristics make them more suitable candidates for application in smart textiles, biomedical textiles, wearable technology, personal care, and wound dressings. Hydrogels can be incorporated into a fabric in many ways, depending on the application. Some of the commonly employed methods in incorporating hydrogel in textiles involve dip coating, in situ polymerization, electrospinning, grafting, freeze-drying, and 3D printing (Figure 9.4).

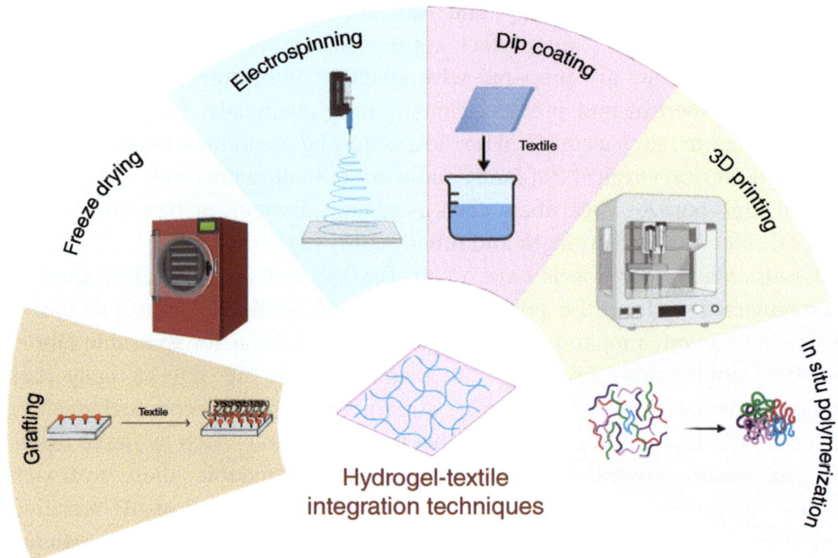

Figure 9.4 Schematic illustration of hydrogel integration methods into textiles

Hydrogel coating of textile substrates is a generic approach toward imparting the fabrics with sophisticated functionalities like moisture management, drug delivery, or sensing properties. In the coating process, hydrogel solutions are applied directly to the fabric surface employing different techniques such as dip-coating, spray coating, or brushing. This is helpful in achieving controlled layering, particularly in the context of moisture-control or sensor-integrated textiles. The coating process starts by preparing a hydrogel precursor solution, typically consisting of monomers, polymers (e.g. polyethylene glycol, polyacrylamide, or natural polymers such as alginate). The fabric substrate (e.g. cotton, polyester, or nylon) is submerged in the solution, and the precursor permeates the porous structure of the fabric. The immersion duration and solution concentration are key factors determining the uniformity and thickness of the hydrogel film. Gelation is subsequently triggered by chemical or physical cross-linking. Surface coating is one of the simplest and most commonly used methods [24]. Pal et al. developed a simple two-step dip-coating method to make cotton fabric hydrophobic using silver nanoparticles and a fluorine-free silane monomer. The modified fabric showed excellent water repellency, oleophilic behavior, and strong durability against mechanical stress, pH variations, and washing. It also demonstrated self-cleaning and oil–water separation properties, highlighting its potential for both household and industrial use [25].

Grafting is another technique that involves forming covalent bonds between functional molecules or polymers and the surface of textile fibers or fabrics. Grafting enables the incorporation of specific properties such as water repellency, flame retardancy, antimicrobial properties, and self-cleaning characteristics in textiles. The commonly used methods of surface grafting modification of materials mainly include chemical grafting, ultraviolet light-induced grafting, plasma-induced grafting, and so on.

Chemical grafting refers to the use of chemical modification to functional groups onto fibers or fabrics. It is categorized into direct and pretreatment methods. Direct chemical grafting, also referred to as surface grafting or chemical surface modification, establishes covalent bonds between functional groups and textile surfaces. In contrast, pretreatment grafting modifies the fabric prior to finishing or dyeing to enhance its functional properties. Han et al. prepared SCG nanogels through single-step polymerization of styrene (St), polycaprolactone-hydroxyethyl methacrylate (PCL-HEMA), and polyhexamethylene guanidine hydrochloride (M-PHGC) monomers through free radical polymerization. The NCO-SCG nanogel was isocyanate-modified and grafted onto cotton fabric using chemical grafting. The treated antibacterial fabric exhibited satisfactory antibacterial and heat stability [26]. Ultraviolet (UV) grafting process is primarily employed using UV irradiation to generate free radicals on the polymer surfaces to induce free radical graft copolymerization [26].

Plasma grafting creates free radicals on the fabric surface to facilitate attachment of polymer or functional groups. There are two categories of plasma grafting: plasma polymerization, where thin polymer films are deposited by monomer polymerization on the surface of the fabric, and plasma activation, which forms the reactive sites for grafting without the creation of a polymer layer. Tourrette et al. used three types of low-temperature plasma gases (oxygen, nitrogen, and argon) to treat the surface of cotton fibers. This treatment increased the number of functional groups on the fiber surface and enhanced surface roughness through etching. As a result, the contact between the fibers and the microgel particles improved. This led to better attachment of the PNIPAAm/chitosan (PN/CS) hydrogel on the fabric surface [27].

3D printing is a promising and versatile technique for the fabrication of hydrogel-based structures in the textile industry. Due to its precision and ease of customizability, it is particularly interesting for integrating smart and functional materials into textiles. 3D printing with specially engineered bioinks can create complex hydrogel geometries that can be tailored for flexibility, breathability, and responsiveness, which are key features for biomedical and wearable textile products [28, 29]. New methods such as fused deposition modeling (FDM), direct ink writing (DIW), inkjet printing, and stereolithography (SLA) have all gained success in generating intricate hydrogel structures [30]. A good example of such a strategy is presented by Li et al., who created a high-performance strain sensor by means of 3D printing a new hydrogel ink made of polyvinyl alcohol (PVA), Poly(3,4-ethylenedioxythiophene):poly(styrenesulfonate) (PEDOT:PSS), hydroxypropyl methylcellulose (HPMC), and dopamine hydrochloride (DA-HCl) [31]. Electrospinning is a widely applied method in the fabrication of fibers, particularly in the production of nanofibrous mats. It employs an electric field to stretch a polymer solution into very thin fibers that accumulate on a surface in a flexible, porous mesh [32]. It is particularly valuable for use in wound care fabrics and biosensing layers [33]. Wet spinning and microfluidic spinning are efficient methods for the continuous fabrication of hydrogel fibers. In wet spinning, a polymer solution is pumped into a coagulation bath, where it becomes solidified in the form of fibers. This process is especially useful for fabricating hydrogel fibers from biopolymers like gelatin and alginate. Integration strategies involve printing onto an electrospun membrane, nanofiber coating of printed scaffolds, fiber infusion into printed matrices, and sequential layer-by-layer assembly of alternating layers of both architectures [34]. Freeze-drying is also widely followed in creating porous hydrogel structures by evaporation of water via sublimation. This can be done prior to or following the cross-linking step, depending on the desired structural nature. The freezing conditions have a great effect on the pore structure and size. Quick freezing with liquid nitrogen produces fine and isotropic pores, while slow or directional freezing gives more ordered and layered structures. Such designed porous hydrogels are particularly beneficial for use in tissue engineering because of their huge surface area and tailor-made structure [35]. Nonwoven materials are often used in most instances to incorporate hydrogel because they offer high surface area, porosity, flexibility, and processability. Hydrogel-coated fabrics are capable of responding dynamically to environmental stimuli, for example, swelling in humid environments or releasing active ingredients (e.g. drugs and moisturizers), and hence are useful in smart functional fabrics.

9.3.2 Aerogel Textile Integration Techniques

Aerogels are investigated in textiles for their superior thermal insulation properties, though with obstacles such as brittleness. Their incorporation, particularly with nonwovens, enhances the strength of the fabric and provides an inexpensive, scalable option. There are various methods for developing aerogel-incorporated textiles, including padding, thermal bonding, lamination, and electrospraying. Padding is a traditional, scalable method used to apply functional materials to textiles by immersing the fabric in a liquid solution and passing it through rollers for even distribution. This method has been adapted to integrate aerogels, known for their low density, high porosity, and excellent thermal insulation, into textiles for applications such as protective and outdoor clothing. Despite challenges like aerogel brittleness and dust generation, studies have shown padding to be effective in reducing thermal conductivity and enhancing comfort. Silica aerogel was applied to polyester fabrics using a low-energy, single-step impregnation process. The treatment provided superhydrophobicity (contact angle ~170°), improved soil release, and had no negative effect on fabric strength or appearance, making it an efficient multifunctional finishing method [36]. Jin et al. used the padding method to apply D50 nanogel to nonwoven fabrics, which improved thermal protection but reduced water vapor permeability. Although simple, the method lacks data on the flexibility of the final product [37]. Lamination is another technique for integrating aerogel in textiles, mostly nonwoven fabrics. In this method, aerogel particles are sprayed upon the nonwoven fabric surface, and then a nonwoven layer or mesh of nanofibers is hot-pressed on the surface of the nonwoven fabric. Researchers developed advanced heat-resistant fabrics by combining silica aerogel with aramid nonwovens using sol–gel and dip-coating methods. Laminated fabrics reinforced with materials like polyimide and zeolitic imidazolate framework-L (ZIF-L) showed significantly better thermal insulation and durability than standard firefighter suits. The best-performing fabric had low thermal conductivity, high heat protection, and strong wear resistance, indicating great potential for use in protective clothing [38]. Thermal bonding is another method, where aerogel particles are embedded within the fiber matrix of nonwoven fabrics through the application of heat. This heat softens the adhesive component present in the nonwoven fibers, enabling it to secure the aerogel particles within the fabric structure. A notable example is the work by Xiong et al. designed a laminated aerogel nonwoven composite using thermal bonding. They placed aerogel particles and low-melting powder between a needle-punched nonwoven layer and an electrospun nanofiber layer. After thermal bonding, the composite showed significantly improved thermal insulation. Interestingly, the aerogel addition had minimal impact on the air and water vapor permeability, preserving breathability, a crucial property for wearable textiles. [39]. These studies illustrate that thermal bonding offers a scalable, binder-free, and structurally stable approach for integrating aerogels into textile materials, particularly for thermal insulation applications such as in firefighting suits, cold-weather apparel, or aerospace textiles. Another approach for integrating aerogels into textiles involves blending aerogel granules with a polymer solution, or in its molten form, and applying the mixture through the electrospinning technique. For example, Bhuiyan et al. deposited silica aerogel particles onto viscose fabric using electrospun polyacrylonitrile (PAN) fibers. Initially, the silica aerogel granules were ground to reduce their size to the micron scale. These fine particles were then combined with a PAN solution, which was fed into an electrospinning device to produce nanofibers embedded with aerogel. These nanofibers were subsequently layered between two viscose nonwoven fabrics to create a multilayer composite [40]. In another study, Xiong et al. utilized needleless electrospinning to incorporate aerogel particles into a polymer solution, which was then sprayed onto fabric surfaces. They added silica aerogel powder to a polytetrafluoroethylene (PTFE) solution and sprayed it onto a polypropylene sheet, resulting in a microporous membrane structure [41].

9.4 Conductive Hydrogel and Aerogel

Conductive aerogels and hydrogels hold great promise for future textile innovations based on their special properties, such as flexibility, conductivity, biocompatibility, and heat insulation. Conductive hydrogels, with their high-water content and electrical conductivity, have great potential in wearable electronics, sensors, and smart textiles, whereas aerogels, due to their remarkable heat insulation and lightness, can improve textile attributes such as water repellence and fire resistance.

9.4.1 Material Design and Composites

Conductive hydrogels, made of 3D polymers having electrical or ionic conductivity, facilitate rapid ion/electron transport and are dynamically tunable for the adjustment of desirable specific conductivity and mechanics. These hydrogels are formed by incorporating ions or conductive particles into the hydrogel matrix. Electron-conductive materials are suited for biomedical applications owing to their biocompatibility, while ion-conductive hydrogels are most suitable in energy devices and displays for their ion gradients and transparency. These hydrogels have emerged as a groundbreaking material in smart textile technology, offering a new fusion of flexibility, water absorption, and electric conductivity.

Electron-conductive hydrogels can be classified into four general groups according to their conductive components: metal nanoparticles, carbon materials, conductive polymers, and hybrids. Metal-containing hydrogels employ nanoparticles of metals such as silver, gold, and copper because of their high electrical conductivity and multifunctionality. Silver nanoparticles, for instance, have been incorporated into networks of hydrogels via in situ reduction processes and thus yield highly conductive, pH- and thermoresponsive materials. Gold and copper nanoparticles have also been added to synthesize conductive hydrogels for thermal sensing and vapor sensing applications. Hydrogels based on carbon make use of materials like carbon nanotubes (CNTs), graphene, and graphene oxide (GO) to form electron pathways in hydrogel matrices. These materials not only facilitate electrical conduction but also enhance mechanical strength. Conducting polymer hydrogels have properties such as flexibility, biocompatibility, and tunable conductivity.

These hydrogels are prepared by techniques such as in situ polymerization and freeze–thaw processes. Some of the applications are glucose sensors, transparent conductors, and electrodes of supercapacitors. For example, polyaniline with PVA or polydopamine–polypyrrole (PPy) systems has resulted in high-performance electroactive hydrogels. PPy is a promising conductive polymer due to its easy synthesis route, excellent environmental stability, superior electrical conductivity, biocompatibility, and good adhesion to various substrate surfaces. Loste et al. developed a simple and effective in situ polymerization method to create metal-free conductive textile electrodes using fabrics such as cotton, wool, silk, and polyester. The PPy-coated textiles exhibited low sheet resistance, high conductivity, enhanced mechanical strength, and maintained breathability and comfort. Knitted cotton electrodes showed the highest capacitance and flexibility, demonstrating strong potential for wearable electronic applications [42]. Hybrid conductive hydrogels combine various conductive materials to promote both electrical and mechanical properties. A few of these include polyaniline hydrogels with Pt nanoparticles for biosensing, or SWCNTs embedded in PVA-hydrogels that catalyze the polymerization of aniline, leading to enhanced capacitive behavior. Furthermore, GO hydrogels with copper sulfide and carbon dots have been utilized in asymmetric supercapacitors with enhanced energy density and cyclic stability.

Ion-conductive hydrogels are water-dominant polymer networks that facilitate ion transport, and they find applications in flexible electronics, solid-state batteries, and actuators. They are based on free ions of acids, salts, and ionic liquids. They have distinctive benefits over electron-conductive hydrogels in conductivity and flexibility. Having both ion- and electron-conductive components in a single hydrogel can lead to better performance, particularly for devices such as supercapacitors and biomedical devices. Some examples include adhesive cardiac patches and polyaniline-based supercapacitor electrodes. New designs employing materials such as copper phthalocyanine-3,4′,4″,4‴-tetrasulfonic acid tetrasodium salt (CuPcTs) with PPy have demonstrated dramatically greater conductivity than conventional conductive hydrogels.

Various methods are employed to embed conductive hydrogels into textiles. Coating is an easy and cost-effective method involving dip-coating, spray-coating, or screen-printing hydrogel solutions onto fabric surfaces. For example, a multilayer coating with L-cysteine, silver nanoparticles, and a hydrogel blend onto cotton fabric resulted in textiles with high electrical stability and mechanical flexibility. Embedding is done by adding hydrogels into the process of textile production, providing strong integration but with more sophisticated methods. Spinning techniques such as wet spinning or electrospinning can also be employed to create conductive hydrogel fibers that are thereafter woven or knitted into fabrics. These fibers have good breathability, flexibility, and mechanical toughness, as indicated in a test with sodium alginate/polyacrylamide fibers having strain sensing over a 0–1000% range, perfect for high-performance stretchable fabrics.

Conductive aerogels are ultralightweight, high-surface-area, electrically conductive, and thermally insulating materials with a great potential to be used in advanced textiles. The most common materials for constructing conductive aerogels are carbon nanomaterials (e.g. graphene, CNTs, and carbon nanofibers), conductive polymers (e.g. PPy and PEDOT:PSS), and metallic nanostructures (e.g. silver or gold nanoparticles). Pure silica aerogels are extremely fragile and nearly transparent in the 3–8 μm infrared range. They are best known for their exceptional thermal insulation, boasting thermal conductivities even lower than that of still air, making them some of the most insulating solid materials available [43]. On the other hand, carbon aerogels were the first electrically conductive type of aerogel and have been extensively studied. These aerogels stand out from traditional porous materials because they combine charge-carrying performance with flexible mechanical properties and can perform reliably even in harsh conditions. These strengths make them an excellent fit for cutting-edge energy systems and wearable tech applications [44]. Whereas, graphene aerogels embedded into textiles have exhibited significant promise for strain sensing based on their mechanical deformation sensitivity and durability on repeated mechanical deformation. Such aerogels can be simply coated onto textiles or embedded into fiber matrices through techniques like freeze-drying, vacuum filtration, or 3D printing [30]. MXene-based aerogel addresses the conductivity limitations of traditional materials like biomass and silica aerogels. MXenes enhance energy storage, photothermal efficiency, and gas-sensing properties. Their piezoresistive response makes them ideal for pressure and strain sensors. Moreover, recent developments have included doping MXene-based aerogels with metal oxides and sulfides to further boost electrochemical performance, such as in nitrogen electroreduction. Innovative approaches like combining noble metals with biological molecules have also led to the creation of aerogel fibers with tailored functionalities [33]. To improve the effectiveness of hydrogels and aerogels in smart textiles, conductive nanomaterials such as graphene, MXenes, CNTs, and metallic nanoparticles are being increasingly integrated. These additives enhance conductivity and also enable the fabrics to respond to external stimuli like heat, light, and mechanical stress.

9.4.2 Electrical and Electrochemical Properties

The conductive hydrogels' electrical conductivity is typically between 0.1 and 1 S/m, depending on the type and concentration of conductive filler, and hydrogel network cross-link density. For instance, CNT-filled hydrogels can achieve up to 0.69 S/m conductivity. The embedded conductive network facilitates electron transfer, while the hydrogel matrix offers mechanical strength and elasticity [45]. Conductive hydrogels exhibit excellent electrochemical behavior, which is important for their use in bioelectronics and soft energy storage. With high water content and porosity, hydrogels exhibit paths of low resistance for ion diffusion and hence superior capacitive characteristics. These properties are advantageous for the construction of flexible supercapacitors and stretchable electrodes [46]. For example, hydrogels of PPy have shown high specific capacitance and satisfactory cycling stability and would be a suitable candidate for energy harvesting or storage in the smart textile system [47].

Conductive aerogels, though structurally different from hydrogels, show excellent electrical and electrochemical properties due to their highly porous, low-density, and interconnected network structure. Aerogels are synthesized by drying gel precursors without structural collapse, resulting in ultralow-density materials with high surface area and enhanced electron transport properties. Conductivity of such aerogels can be more than 1 S/m, especially when filled with conducting nanomaterials such as CNTs, graphene, or metal nanowires. These fillers form percolation networks that ensure high electron mobility throughout the aerogel structure [48]. In a recent work, a multifunctional aerogel film was prepared from a hybrid material consisting of aramid nanofibers, conductive CNTs, and hydrophobic fluorocarbon resin. The obtained film is light, durable, and thin, with a high specific surface area (232.8 m^2/g), good electrical conductivity (230 S/m), and excellent hydrophobicity (contact angle as high as 137°). It exhibits superior self-cleaning capability, effective low-voltage Joule heating, and outstanding electromagnetic interference (EMI) shielding performance, with an EMI shielding effectiveness of 54.4 dB at only 568 μm thickness. The specific EMI SE is as high as 33,528.3 $dB{\cdot}cm^2/g$, surpassing most traditional metal- and carbon-based composites. The above characteristics render the aerogel film extremely suitable for smart clothing, thermal management, and EMI shielding in wearable electronics applications [49]. Aerogel-based electrochemical sensors possess high conductivity and an open porous structure. They show rapid response and high sensitivity toward gas, biomolecule, and environmental pollutants. Their incorporation into fabrics may result in smart fabrics for environmental monitoring or even wearable diagnostics. Aerogel-based electrodes are also shown to have good cycling stability and mechanical integrity under repeated bending or stretching, which are necessary for wearable technology in practice [50]. Hybrid aerogels that blend conductive polymers with carbon nanostructures demonstrate synergistic features, including increased capacitance, improved conductivity, and mechanical strength.

One advantage of aerogels in clothing electronics is that they are thermally stable. Hydrogels, on the other hand, would dehydrate or degrade under high temperatures. Aerogels are structurally stable and can be applied for extreme or high-temperature uses. This property comes in handy when designing intelligent uniforms for firefighters under extreme conditions. In addition, incorporation of conductive aerogels in textiles can go beyond electrical conduction to multifunctionality, for example, fire resistance, insulation from heat, and strain sensing. For example, cellulose nanofibril (CNF)–graphene aerogel composites have demonstrated exceptional conductivity and thermal resistance, paving the way to multifunctional smart fabrics.

9.4.3 Applications for E-Textiles

Electronic textiles (e-textiles) embed electronic elements such as sensors, actuators, and energy harvesters into fabrics to facilitate applications such as health monitoring, environmental adjustment, and wearable electronics. Hydrogels and aerogels, due to their special properties, are being researched more for the development of e-textile functionality. This section consolidates data from published articles, emphasizing their applications, properties, and prospects in e-textiles.

9.4.3.1 Hydrogel in E-Textiles

Hydrogels are 3D, cross-linked polymer networks with high water content, softness, and sensitivity to environmental stimuli like temperature, pH, or humidity. The presence of these properties renders hydrogels particularly well-suited for incorporation into electronic textiles (e-textiles), especially in those applications that demand biocompatibility, flexibility, and active interaction with the human body. A prominent application is moisture management, where hydrogels control sweat and provide comfort to the wearer. Wang et al. synthesized a thermoresponsive hydrogel-coated fabric based on poly(ethylene glycol) (PEG) and poly(ε-caprolactone) (PCL) cross-linked with hexamethylene diisocyanate (HMDI). The intelligent fabric showed a sol–gel transition temperature near 34 °C, allowing one-way liquid transport from the hydrophobic-coated side to the uncoated side. It recorded a high-water vapor transmission rate (WVTR) of $2209 \pm 54\,g/m^2/day$ and a one-way transport index of $612.8 \pm 7.18\%$ (grade 5), indicating it is very breathable and suitable for wearable devices [51]. Hydrogels are developed not only for comfort but also for drug delivery in e-textiles. Mostafalu et al. developed an advanced wound dressing that uses composite fibers connected to a heating element, enclosed within a hydrogel layer containing thermoresponsive drugs for controlled drug release [52]. Likewise, Cuturicu et al. explored chitosan-based hydrogel treatment on textiles for long-term wound therapy, specifically for the treatment of tuberculosis. These examples point to the capabilities of hydrogels for sustained, controlled therapeutic release in wearable medical devices. In addition, their therapeutic implications include treatment of wounds and skin. Their ability to retain large amounts of moisture facilitates wound healing as it creates an environment that is conducive to tissue regeneration. Chitosan hydrogels, for instance, have been proven to facilitate reepithelialization, showing compatibility with wearable healthcare systems intended to contact the skin [53]. Overall, the multifunctional nature of hydrogels renders them useful elements in next-generation smart textile development for health, hygiene, and wellness monitoring. Shuai et al. fabricated stretchable conductive fibers composed of acrylamide (AAm) and *N*-acryloylglycinamide (NAGA) and the physically cross-linked poly (NAGA-co-AAm) hydrogel using a dry-wet spinning technique, aiming at applications in strain sensing and triboelectric energy-harvesting electronic textiles [54]. Similarly, Ding et al. integrated p-type and n-type hydrogel fibers in an alternating pattern through direct weaving to produce a soft, flexible thermogalvanic fabric capable of efficiently capturing body heat [55]. PVA-based gels have emerged as superior and stretchable alternatives to traditional electrolytes. They offer excellent leak resistance, long-term stability, and compatibility with fabric structures. When further enhanced with conductive additives such as MXene and PEDOT:PSS, these PVA-based gels demonstrate significantly improved electrochemical performance, making them ideal for use in flexible and robust e-textile applications [56].

9.4.3.2 Aerogel in E-Textiles

Aerogels are nanostructured materials characterized by ultra-low density (down to ~0.003 g/cm^3), high porosity, and superior thermal insulation capacities. Such properties render them especially beneficial for use in electronic textiles (e-textiles), particularly where light thermal regulation, sustainability, and multifunctionality are necessary. Thermal insulation is one of the major uses of aerogels in e-textiles. A review by McNeil et al. identified the diverse applications of aerogels in textiles, highlighting not only their well-established use in thermal insulation but also their potential to improve mechanical strength, acoustic performance, water repellency, and protection from radiation, chemicals, and fire. They also outlined the use of aerogels in wastewater treatment and noted that future growth, tied to advances in energy, aerospace, and biomedical applications, may encourage the application of biopolymers and hybrids, supporting new textile technology innovation [21]. Aerogel films on the substrates of textiles, like silica aerogels on nonwovens, greatly enhance the mechanical strength of the fabric and make it more resistant to everyday use and mechanical strain. For example, MXene-based aerogel composites have exhibited top-notch mechanical resilience with compressive stresses of up to 16 kPa and sustaining more than 24,000 cycles of compression, appropriate for usage like flexible piezoresistive sensors integrated into clothing [57]. Aerogels are also revolutionizing energy storage in e-textiles. Nanocellulose-based aerogels, and especially those involving cellulose nanocrystals (CNCs) and conductive fillers such as polypyrrole and CNTs, have shown promising performance for wearable supercapacitors. In a study, a flexible all-solid-state supercapacitor was developed using CNF, reduced graphene oxide (RGO), and CNT hybrid aerogels as electrodes, along with an H_2SO_4/PVA gel electrolyte. The device was fabricated without binders, current collectors, or electroactive additives. Thanks to the porous structure and excellent electrolyte absorption capacity of the aerogel electrodes, the supercapacitor exhibited a high specific capacitance of 252 F/g, retained over 99.5% of its capacitance after 1000 cycles, and showed excellent areal capacitance, power density, and energy density. These results highlight the potential of such aerogel-based systems for low-cost, scalable, and eco-friendly flexible energy storage applications [58]. Overall, the incorporation of aerogels in fabrics imparts them with improved functionality in areas of thermal control, mechanical robustness, and energy storage, making them pivotal materials in the development of future e-textiles.

9.5 Responsive and Adaptive Textile Systems

Responsive and adaptive textile systems are a revolutionary breakthrough in material science and textile engineering, where fabrics can respond to environmental stimuli and user-defined conditions in real time. Such systems integrate smart materials, such as hydrogels, aerogels, and shape-memory polymers, with embedded electronics and sensors to offer dynamic functions like thermal regulation, pressure responsiveness, or structural adaptability.

9.5.1 Stimuli-Responsive Hydrogels and Aerogels

Stimuli-sensitive hydrogels and aerogels are sophisticated materials that experience dynamic alterations in their function or structure upon response to slight differences in external stimuli. These external cues are divided into three categories: physical, chemical, and biological.

9.5.1.1 Physical Stimuli

Physical stimuli involve alterations in physical parameters of the external environment, including temperature, light, pressure, or electromagnetic fields. These are among the most widely studied and practically implemented in smart textile systems. These hydrogels undergo a reversible sol–gel transition in response to changes in external temperature. These hydrogels are classified into two main types based on their thermal response: LCST (Lower Critical Solution Temperature) and UCST (Upper Critical Solution Temperature) hydrogels. LCST-type hydrogels shrink as the temperature increases because hydrophilic interactions and hydrogen bonding with water dominate at lower temperatures, causing the hydrogel to remain swollen. As the temperature rises, these hydrogen bonds weaken and hydrophobic interactions take over, causing the hydrogel to collapse. In contrast, UCST-type hydrogels swell with increasing temperature, as hydrogen bonding becomes stronger at higher temperatures, enhancing their water solubility [8]. Poly(*N*-isopropylacrylamide) (PNIPAM) is a commonly used thermoresponsive hydrogel, particularly in biomedical fields, due to its volume phase transition temperature (VPTT) of approximately 32 °C, which is close to human body temperature. This transition point can be fine-tuned by copolymerization of PNIPAM with various co-monomers. Zhu et al. developed dynamic, structurally colored hydrogel fibers using soft, thermoresponsive microgels. These fibers, fabricated from a scalable exclusion process, become color-changing with temperature due to PNIPAM microgel shrinkage. The microgels also exhibit sensitivity to humidity and ionic strength, making real-time monitoring of the environmental conditions possible. When deposited on textiles, they confer thermochromic as well as sensing properties, demonstrating great promise for intelligent textiles and wearable sensors [59]. In another study, dual-stimuli responsive cotton fabric was prepared by functionalizing the fabric with a spiropyran–NIPAAm hydrogel, making it possible for the fabric to change dimensions in response to visible light and thermal stimuli. The volume changes occur due to the capacity of the fabric to absorb or release water, either when the temperature increases above the LCST in the dark or when exposed to sunlight or artificial white light. The responsive behavior was attained by grafting photo- and thermoresponsive monomers onto the cotton fibers directly through a straightforward ARGET ATRP-controlled polymerization method [60].

Current research, including the creation of wearables aerogel for personal thermal management (PTM), demonstrates an increasing application of aerogels in textiles operating under harsh conditions. These aerogels are used in thermal control through insulating properties and their integration with temperature feedback systems, demonstrating their importance in intelligent textile design [61]. Apart from thermal control, hydrogel and aerogels are also being investigated for wearable sensing technology applications. Wang et al. fabricated a flexible and sustainable pressure sensor based on a hierarchical carbon aerogel obtained from cellulose fibers (CFs). The aerogel was prepared by sequential microfibrillation, periodate oxidation, and tannic acid (TA) cross-linking, and then freeze-drying and high-temperature pyrolysis. This method yielded a lightweight, compressible carbon aerogel with superior mechanical strength and electrical conductivity. When employed as a sensing medium, the aerogel-based sensor exhibited high sensitivity (4 kPa^{-1}), rapid response and recovery times (32/30 ms), a low detection limit (0.8 Pa), and exceptional durability over 3000 cycles. Aside from human motion and spatial pressure distribution detection, the sensor also displayed linear sensitivity to temperature and humidity, and thus has the potential to be used as a multifunctional wearable sensor for health monitoring and intelligent textiles [62]. Similarly, Li et al. developed conductive ionic polyimide hydrogel fibers with excellent mechanical strength and chemical stability, addressing key limitations in wearable and flexible electronics. Unlike conventional hydrogel fibers that degrade in harsh environments, such as alkaline solutions or organic solvents, these fibers exhibited remarkable chemical resilience. These fibers are produced rapidly through a facile wet spinning process using an organosoluble polyimide salt. The fibers achieved a conductivity of ~21 mS/cm, a tensile strength of 2.5 MPa, and 215% elongation. The robust and ion-rich structure allowed the fibers to be woven into textiles suitable for wearable use. Assembled into strain sensors, the fibers demonstrated high sensitivity, linear response, and strong cycling stability, making them highly promising for smart textiles and strain-sensing applications in complex or extreme environments [63]. Complementing these advances in sensing capabilities, researchers have also explored new frontiers in electromagnetic protection within wearable systems. Jafri et al. highlighted the development of smart stimuli-responsive materials for EMI shielding, which offer dynamic and tunable protection unlike traditional static materials. These smart materials adjust their shielding performance in response to mechanical, chemical, or environmental stimuli, making them ideal for modern applications in wearable electronics, telecommunications, aerospace, and defense. The review also discusses recent progress, challenges, and future directions for advancing this adaptive technology [64]. In another study, Shui et al. engineered aerogels by incorporating highly conductive carbon nanoparticles into natural wood layers, creating a material capable of toggling between electromagnetic wave (EMW) transmission and shielding states. This reversible functionality is enabled by cyclic compression and decompression, allowing real-time tuning of the material's electromagnetic interference shielding effectiveness (EMI-SE). Such adaptive behavior makes these aerogels especially suitable for applications that demand flexible and responsive EMI protection, including wearable electronics, aerospace technologies, and smart devices [65]. In the context of hydrogel, a novel thermochromic hydrogel-based smart window was developed with adjustable light transmission, shape flexibility, and tunable EMI shielding. Using a simple layer-by-layer polymerization and solvent exchange method, the hydrogel achieved high transparency (80.87%), fast response time (16 seconds), and adjustable EMI shielding from 0 to 27.64 dB. This multifunctional material addresses the challenge of balancing optical clarity with EMI protection, showing strong potential for use in next-generation smart windows, especially in military and energy-efficient buildings [66].

9.5.1.2 Chemical Stimuli

Chemical stimuli include alterations in the local chemical environment, for example, pH, ionic strength, solvent composition, or exposure to certain chemical agents (e.g. gases and oxidants). pH-sensitive hydrogels bear functional groups (e.g. $-COOH$ and $-NH_2$) that ionize or become protonated under varying pH. This ionic alteration alters their swelling ratio. Such hydrogels are useful in wound dressings that respond to pH shifts, indicating infection. Tamayol et al. developed pH-responsive hydrogel fibers designed for continuous monitoring of epidermal wound conditions. These fibers incorporate pH-sensitive dyes loaded into mesoporous microparticles, which are embedded into the hydrogel matrix using a microfluidic spinning technique. The resulting microfibers are flexible, skin-conformal, and capable of providing real-time pH mapping of wounds, which is critical for tracking healing progress. The device's pH response can be visualized and recorded using a smartphone camera, enabling easy on-site readouts. The scalable fabrication method supports the production of clinically relevant wound dressings, making it a promising point-of-care solution for wound management [67].

Similarly, Jocic et al. developed temperature and pH-responsive microhydrogels for functional finishing of cotton fabric. In this research, a new approach to cotton functionalization using dual-responsive microparticulate hydrogels derived from temperature-sensitive poly(NiPAAm) and pH-sensitive chitosan was proposed. Cotton was chemically treated by carboxymethylation and aminization to make it easier for hydrogels to be attached. The modified fabrics were found to have evident sensitivity toward pH, temperature, and humidity, as attested by microscopic and spectroscopic analysis, with potential application in smart textiles [68]. An epichlorohydrin-cross-linked red radish pigment-based pH-sensitive aerogel film (R-CEF) was synthesized from red radish pigment (R) and epichlorohydrin as a natural pH indicator and cross-linking agent, respectively. The obtained film has a highly porous network structure that allows water uptake and visible color response to pH change. Structural characterizations (XPS, FTIR) established firm hydrogen bonding and molecular interactions between the cellulose matrix and the pigment, whereas SEM images indicated that water uptake converts the material from sponge-likc to layered morphology. The film displayed significant color changes, from pink to red under acidic conditions (pH 3–5) and yellow to pink under alkaline conditions (pH 10–13), showing evident pH sensitivity. Furthermore, the material provided antibacterial functionality, qualifying it as a potential green smart composite for pH and water-sensing applications [69]. Another interesting development in moisture sensing is the application of hybrid aerogels coupling PNIPAM and gold nanoparticles (AuNPs). These hygroscopic aerogels have strong water affinity, which allows them to rapidly adsorb and desorb water. Their porous structure also renders them highly sensitive to slight vibrations, including those created through respiration or sound waves, pointing toward their applications in next-generation wearable and environmental sensing technologies [70]. In 2021, Li et al. created a pH-sensitive injectable hydrogel of *N*-carboxyethyl chitosan, aldehyde hyaluronic acid, and adipic acid dihydrazide for the therapy of diabetic foot ulcers. Delivery of insulin, which was encapsulated in the hydrogel dressing, reduced glucose levels, facilitating the healing of the wound. The extended pH-mediated release of insulin is permitted due to the acylhydrazone and imine bonds, which are subject to breakage in acidic pH conditions and greatly influence the hydrogel networks, promoting a drug release [71].

The classification of stimuli into physical, chemical, and biological categories provides a structured understanding of how responsive materials function in textile systems. Hydrogels, with their flexibility and water-holding capacity, are ideal for biological and chemical stimuli, especially in medical and wearable health applications. Aerogels, known for their insulating properties and surface customizability, excel in physical stimuli applications such as thermal regulation and mechanical sensing.

9.5.2 Adaptive Clothing Systems

The incorporation of stimuli-responsive hydrogels and aerogels in textile systems has provided new possibilities for adaptive clothing, especially in thermal regulation, comfort, and wearability. Hydrogels like PNIPAM/SA/CNC and PAA/SA/CNC, when organized in bilayer composite form, allow clothing to respond dynamically to body motion, controlling fit, and shape in real time. These intelligent materials respond to environmental stimuli like humidity and temperature, enabling fabrics to shrink, swell, or curve depending on their internal cross-linking structure. Such moisture-sensitive systems surpass conventional fabrics by actively regulating body heat and moisture content, enhancing user comfort under diverse conditions [70, 72]. Concurrently, 3D fiber-based aerogels, designed from nanoscale pieces, imbue clothing with increased durability, flexibility, and high surface permeability. These aerogels, typically produced by advanced processes to obtain highly porous structures, are increasingly being integrated into wearables for energy efficiency, particularly in applications such as soft robotics and artificial skin [73]. Recent breakthroughs have also investigated aerogel composites with phase change materials (PCMs) that store and release heat, allowing clothing to transition with changing weather conditions. Adaptive solutions also include integrating shape memory alloys (SMAs) into thermoplastic nanocomposites (TPCs), which are heat-responsive and provide active thermal comfort [74]. Although aerogels provide very good thermal insulation, one typical problem is their poor moisture transport ability, reducing breathability in clothing. Thus, current research also centers on achieving the balance between insulation and permeability to moisture for maximum performance for personal thermal regulation. Eventually, adaptive clothing systems based on responsive hydrogels and aerogels seek to provide stretchy, climate-sensitive wearables that adapt smoothly to changes in environment and physiology.

9.6 Sustainability and Biocompatibility

The increasing need for environmentally friendly and sustainable materials has greatly developed the hydrogel and aerogel technologies in the textile industry. Besides solving ecological problems, these materials also possess outstanding properties like biocompatibility, softness, and functionality diversity, making them perfect candidates for advanced smart and biomedical textiles. The natural polymers of cellulose, chitosan, sodium alginate, gelatin, and starch form the building blocks of green hydrogels and aerogels. These are biodegradable, nontoxic, renewable, and lightweight. These are the most relevant characteristics to minimize the environmental impact of traditional textiles. Of these, hydrogels and aerogels based on cellulose, such as foams and sponges, are extensively researched because they are abundant, have structural strength, and possess physicochemical advantages. Nanocellulose forms, such as CNFs, CNCs, and bacterial cellulose (BC), play a central role in developing next-generation hydrogel and aerogel systems with enhanced mechanical strength, porosity, and functional tunability. These nanostructured forms are now being used to produce composites for thermal regulation, EMI shielding, and wearable sensors [75, 76].

Aerogels derived from renewable resources such as plant oils, proteins, starch, and lignin have demonstrated great potential in energy-efficient and biomedical textile applications. For instance, starch- and silk fibroin-based aerogels produced via green synthesis routes exhibit high porosity, cytocompatibility, and biodegradability, making them suitable for therapeutic textiles. The environmental urgency surrounding fossil-derived polymers has spurred a shift toward these bio-based alternatives, which offer safer degradation profiles and eliminate the risks associated with non-recyclable synthetic residues. [77, 78].

In wearable textile systems, particularly in smart clothing and biosensing garments, material selection must prioritize skin safety, flexibility, and long-term comfort. Chitosan-based hydrogels (CS-Gels) are increasingly used in such applications for their stretchability, breathability, and biocompatibility [79]. These hydrogels are also biodegradable and, in some cases, edible, qualities that enhance their environmental profile and make them safe for direct skin contact. Similarly, sodium alginate, derived from brown seaweed, forms biocompatible matrices ideal for textile-integrated wound dressings. Researchers have also explored the use of industrial cotton waste nonwovens as reinforcing substrates for hydrogel systems, leveraging cotton's softness, breathability, and moisture-absorbing properties to enhance the performance of biomedical textiles.

Hydrogels and aerogels incorporated into textiles are being designed more and more to satisfy performance requirements like mechanical stability, washability, and durability. Pressure and electrical activity sensing in cellulose-based conductive hydrogels enables them to be used as multifunctional wearable electronics and health monitoring clothing. Their lightweight and flexible structure make them easily integrated into fabric systems without compromising wearability. To enhance their processing and textile compatibility, new cross-linking techniques and composite formulations are being developed, providing increased fiber adhesion and long-term durability. Processing and environmental concerns have also spurred development. Ionic liquid (IL) use provides a green pathway for the preparation of aerogels, cryogels, and xerogels from polysaccharide materials. The solvents can be recovered and recycled, with the ability to facilitate closed-loop, environmentally friendly processes of manufacture [77]. Replacing synthetic, harmful cross-linkers such as resorcinol with natural additives like lignin not only makes the resulting materials more biodegradable but also their structure, conductivity, and mechanical performance better—of relevance for e-textile usage. Biocompatibility is particularly essential for textiles that will have prolonged skin contact or are therapeutic in nature. Hydrogels based on chitosan, alginate, gelatin, and cellulose are a great platform for the design of textile-based wound care devices, biosensors, and soft actuators. These systems replicate the extracellular matrix (ECM) and facilitate biological processes such as cell adhesion, proliferation, and tissue regeneration, degrading into nontoxic byproducts [78]. Aerogels have been recently used for a broader spectrum of biomedical textiles, such as drug delivery devices, tissue scaffolds, and wound dressings. For instance, silica-silk fibroin aerogels have been reported to possess good porosity, cytocompatibility, and mechanical stability, facilitating bone regeneration through increased osteoblast activity. Likewise, silica aerogels provide high surface area and tunable pore architectures, making them extremely efficient drug delivery systems. Based on whether they are hydrophilic or hydrophobic, these aerogels can be designed for slow or fast drug release, facilitating controlled therapeutic effects. Overall, the incorporation of sustainable, biocompatible hydrogel and aerogel systems in textile systems is consistent with the increasing demand for environmentally friendly and health-centric technologies. They enable a new generation of smart, responsive, and green textile products that can solve problems in healthcare, personal comfort, and environmental sustainability [77].

9.7 Challenges

Hydrogel and aerogel integration into textile systems presents a wide range of challenges that limit their widespread adoption, despite their excellent properties, such as lightweight structure, flexibility, thermal insulation, and responsiveness. Aerogels have densities ranging to as low as ~0.003 g/cm^3 and are typically brittle, with the properties causing dusting or fragmentation upon handling or abrasion. Upon application onto fabric, oxide-based aerogels tend to be non-durable since their brittleness leads to particle attrition by mechanical forces. Adhesives or binders added to improve adhesion can undermine insulation properties by triggering pore wetting [80]. Another major issue lies in the absence of a universal sol–gel synthesis strategy for nanoscale building blocks, which makes standardization of aerogel fiber production difficult. Industrial-scale fabrication is still limited due to challenges in continuous processing and the need for precise ambient pressure drying techniques [81]. Thermal and optical characterization of single aerogel fibers also requires sophisticated microfabrication and microfluidic-spectrometer systems. Additionally, aerogels tend to have poor mechanical strength, which restricts their use in wearable or deformable textile systems. Their integration into smart textiles is further hindered by structural fragility, lack of macroporosity (critical for biomedical applications such as tissue engineering), and difficulties in heat dissipation especially in protective garments like firefighting suits, where they can trap body heat. High production costs, limited commercial availability (primarily restricted to a few companies like Cabot Corporation and Aspen Aerogels), complex synthesis methods, lack of automation, and insufficient use of AI-guided design tools all contribute to slow commercialization. Moreover, there is a growing need for smart, eco-friendly, and multifunctional aerogels that can perform under extreme or variable conditions [78].

Hydrogels, on the other hand, also face several integration challenges in textile and wearable systems. Many hydrogel formulations are non-biodegradable, requiring surgical removal that can lead to infections [78]. In wearable electronics, hydrogels tend to lose their flexibility and conductivity over time as water evaporates, impacting long-term durability and performance. They also struggle with the controlled release of drugs, especially those with poor water solubility [77, 82]. The complex synthesis processes, lack of standardized and automated fabrication techniques (such as robotic freeze-drying), and high production costs further hinder their scalability [83]. Moreover, hydrogels exhibit sensitivity to environmental factors such as temperature and humidity, limiting their reliability in real-world textile applications. In spite of their benefits like softness, stretchability, and biocompatibility, these barriers need to be overcome in order to realize their full potential in wearable electronics, smart textiles, biomedical devices, and human–machine interface systems. Breakthroughs in artificial intelligence, materials engineering, and automation of the processes are crucial to overcoming these technical and business barriers for hydrogel as well as aerogel-based textile technologies.

9.8 Conclusions

Hydrogels and aerogels have transformed textile science by making smart, responsive, and high-performance materials available for a broad spectrum of applications such as wearable electronics, biomedical devices, and energy storage systems. Some of the most significant developments are the creation of conductive, stimuli-responsivc, and eco-friendly composites with both flexibility and improved functionality. These advanced materials are increasingly taking center stage in the development of adaptive clothing, stretchable sensors, artificial skin, and biocompatible systems for real-time health monitoring and therapy. In spite of current limitations like limited mechanical strength, degradation due to drying, and intricate processing methods, advancements in nanotechnology, artificial intelligence (AI), and green chemistry keep advancing their role. Smart fabrication methods, bio-based polymers, and integrated electronics are driving the advancement of hydrogel and aerogel textiles toward greater efficiency, comfort, and personalization. These textiles are rapidly becoming core components in next-generation smart wearables that not only react to environmental stimuli but also interact with users and machines. Looking ahead, future textile technologies envision multifunctional, AI-guided, and self-powered systems capable of real-time sensing, data processing, and sustainable performance. Hydrogel- and aerogel-based textiles are expected to shape a new era of intelligent clothing one that merges sustainability, functionality, and human-centered design.

References

1. Sajovic, I., Kert, M., and Boh Podgornik, B. (2023). Smart textiles: a review and bibliometric mapping. *Applied Sciences* 13 (18): https://doi.org/10.3390/app131810489.
2. Hossain, M.T., Shahid, M.A., Limon, M.G.M. et al. (2024). Techniques, applications, and challenges in textiles for a sustainable future. *Journal of Open Innovation: Technology, Market, and Complexity* 10 (1): 100230. https://doi.org/10.1016/j.joitmc.2024.100230.
3. Peng, Y. and Cui, Y. (2024). Thermal management with innovative fibers and textiles: manipulating heat transport, storage and conversion. *National Science Review* 11 (10): nwae295. https://doi.org/10.1093/nsr/nwae295.
4. Azam, F., Ahmad, F., Ulker, Z. et al. (2022). The role and applications of aerogels in textiles. *Advances in Materials Science and Engineering* 2022 (1): 2407769.

5 Batool, S.R., Nazeer, M.A., Ekinci, D. et al. (2020). Multifunctional alginate-based hydrogel with reversible crosslinking for controlled therapeutics delivery. *International Journal of Biological Macromolecules* 150: 315–325. https://doi.org/10.1016/j.ijbiomac.2020.02.042.

6 Du, J., Ma, Q., Wang, B. et al. (2023). Hydrogel fibers for wearable sensors and soft actuators. *iScience* 26 (6): 106796. https://doi.org/10.1016/j.isci.2023.106796.

7 Yang, L., Yuan, Q.Y., Lou, C.W. et al. (2025). Modified nanocellulose hydrogels and applications in sensing fields. *Gels* 11 (2): 1–25. https://doi.org/10.3390/gels11020140.

8 Nazeer, M.A., Batool, S.R., and Kizilel, S. (2021). Stimuli-responsive drug delivery hydrogels. In: *Soft Matter for Biomedical Applications* (ed. H.S. Azevedo, J.F. Mano, and J. Borges). The Royal Society of Chemistry https://doi.org/10.1039/9781839161124-00542.

9 Batool, S.R., Nazeer, M.A., Yildiz, E. et al. (2021). Chitosan-anthracene hydrogels as controlled stiffening networks. *International Journal of Biological Macromolecules* 185: 165–175. https://doi.org/10.1016/j.ijbiomac.2021.06.023.

10 Bashir, S. et al. (2020). Fundamental concepts of hydrogels: synthesis, properties, and their applications. *Polymers (Basel)* 12 (11): 1–60. https://doi.org/10.3390/polym12112702.

11 Fricke, J. and Emmerling, A. (1992). Aerogels. *Journal of the American Ceramic Society* 75 (8): 2027–2035.

12 Khan, N.R., Sharmin, T., and Bin Rashid, A. (2024). Exploring the versatility of aerogels: broad applications in biomedical engineering, astronautics, energy storage, biosensing, and current progress. *Heliyon* 10 (1): e23102.

13 Aegerter, M.A., Leventis, N., Koebel, M., and Steiner, S.A. III, (2023). *Springer Handbook of Aerogels*. Cham: Springer International Publishing.

14 Liu, Q., Yan, K., Chen, J. et al. (2021). Recent advances in novel aerogels through the hybrid aggregation of inorganic nanomaterials and polymeric fibers for thermal insulation. *Aggregate* 2 (2): e30.

15 Yin, M., Fu, Z., Yu, X. et al. (2025). The effect of drying methods on the pore structure of Balsa wood aerogels. *Polymers* 17: https://doi.org/10.3390/polym17121686.

16 Cuce, E., Cuce, P.M., Wood, C.J., and Riffat, S.B. (2014). Toward aerogel based thermal superinsulation in buildings: a comprehensive review. *Renewable and Sustainable Energy Reviews* 34: 273–299. https://doi.org/10.1016/j.rser.2014.03.017.

17 Hou, Z., Zhao, C., Cheng, J. et al. (2024). Preparation and properties of fiber-reinforced polybenzoxazine composite aerogels based on freeze drying and ambient pressure drying methods. *Colloids and Surfaces A: Physicochemical and Engineering Aspects* 687: 133510. https://doi.org/10.1016/j.colsurfa.2024.133510.

18 Chen, Z., Li, D., Yao, S. et al. (2025). Effects of silica aerogel content on the properties of waterborne organic thermal insulation coatings. *Gels* 11 (7): 547.

19 Hou, X., Chen, J., Chen, Z. et al. (2024). Flexible aerogel materials: a review on revolutionary flexibility strategies and the multifunctional applications. *ACS Nano* 18 (18): 11525–11559.

20 Duong, H.M., Ling, N.R.B., Thai, Q.B. et al. (2021). A novel aerogel from thermal power plant waste for thermal and acoustic insulation applications. *Waste Management* 124: 1–7.

21 McNeil, S.J. and Gupta, H. (2022). Emerging applications of aerogels in textiles. *Polymer Testing* 106: 107426. https://doi.org/10.1016/j.polymertesting.2021.107426.

22 Sayed, U. and Deshmukh, I. (2021). Hydrogels for textile applications - review. *International Journal of Advanced Science and Engineering* 7 (4): 1968–1974. https://doi.org/10.29294/ijase.7.4.2021.1968-1974.

23 Lee, K.H., Arshad, Z., Dahshan, A. et al. (2023). Porous aerogel structures as promising materials for photocatalysis, thermal insulation textiles, and technical applications: a review (Catalysts, (2023), 13, 9, (1286), 10.3390/catal13091286). *Catalysts* 13 (12): https://doi.org/10.3390/catal13121498.

24 Liu, J., Qu, S., Suo, Z., and Yang, W. (2021). Functional hydrogel coatings. *National Science Review* 8 (2): https://doi.org/10.1093/nsr/nwaa254.

25 Pal, S., Mondal, S., Pal, P. et al. (2022). Fabrication of AgNPs/Silane coated mechanical and washing durable hydrophobic cotton textile for self-cleaning and oil–water separation application. *Journal of the Indian Chemical Society* 99 (1): 100283. https://doi.org/10.1016/j.jics.2021.100283.

26 Han, H., Zhu, J., Wu, D.Q. et al. (2019). Inherent guanidine nanogels with durable antibacterial and bacterially antiadhesive properties. *Advanced Functional Materials* 29 (12): 1806594.

27 Tourrette, A., De Geyter, N., Jocic, D. et al. (2009). Incorporation of poly(N-isopropylacrylamide)/chitosan microgel onto plasma functionalized cotton fibre surface. *Colloids and Surfaces A: Physicochemical and Engineering Aspects* 352 (1): 126–135. https://doi.org/10.1016/j.colsurfa.2009.10.014.

28 Erkoc, P., Uvak, I., Nazeer, M.A. et al. (2020). 3D printing of cytocompatible gelatin-cellulose-alginate blend hydrogels. *Macromolecular Bioscience* 20 (10): 2000106.

29 Nazeer, M.A., Onder, O.C., Sevgili, I. et al. (2020). 3D printed poly(lactic acid) scaffolds modified with chitosan and hydroxyapatite for bone repair applications. *Mater Today Commun* 25: 101515. https://doi.org/10.1016/j.mtcomm.2020.101515.

30 Chan, Y.L., Widodo, R.T., Ming, L.C. et al. (2025). Review on 3D printing filaments used in fused deposition modeling method for dermatological preparations. *Molecules* 30 (11): 2411.

31 Li, G., Cheng, J., and Lu, B. (2025). 3D printing of highly robust and bioadhesive strain sensors for human motion monitoring. *Macromolecular Chemistry and Physics* 226: e00536.

32 Nazeer, M.A., Yilgor, E., and Yilgor, I. (2019). Electrospun polycaprolactone/silk fibroin nanofibrous bioactive scaffolds for tissue engineering applications. *Polymer (Guildf)* 168: 86–94. https://doi.org/10.1016/J.POLYMER.2019.02.023.

33 Zainab, I., Naseem, Z., Batool, S.R. et al. (2025). Polyurethane/silk fibroin-based electrospun membranes for wound healing and skin substitute applications. *Beilstein Journal of Nanotechnology* 16: 591–612. https://doi.org/10.3762/bjnano.16.46.

34 Athukorala, S.S., Tran, T.S., Balu, R. et al. (2021). 3D printable electrically conductive hydrogel scaffolds for biomedical applications: a review. *Polymers* 13 (3): 1–21. https://doi.org/10.3390/polym13030474.

35 Grenier, J., Duval, H., Barou, F. et al. (2019). Acta Biomaterialia Mechanisms of pore formation in hydrogel scaffolds textured by freeze-drying. *Acta Biomaterialia* 94: 195–203. https://doi.org/10.1016/j.actbio.2019.05.070.

36 Altay, P., Atakan, R., and Özcan, G. (2021). Silica aerogel application to polyester fabric for outdoor clothing. *Fibers and Polymers* 22 (4): 1025–1032. https://doi.org/10.1007/s12221-021-0420-4.

37 Jin, L., Hong, K., and Yoon, K. (2024). Effect of aerogel on thermal protective performance of firefighter clothing. *Journal of Fiber Bioengineering and Informatics* 6: 315–324. https://doi.org/10.3993/jfbi09201309.

38 Xue, R., Chang, Y., and Liu, F. (2024). Fabrication and performance evaluation of aramid laminated fabrics incorporated with silica composite aerogel for protective clothing. *Ceramics International* 50 (7): 10348–10354. https://doi.org/10.1016/j.ceramint.2023.12.346.

39 Xiong, X., Yang, T., Mishra, R., and Militky, J. (2016). Transport properties of aerogel-based nanofibrous nonwoven fabrics. *Fibers and Polymers* 17 (10): 1709–1714.

40 Bhuiyan, M.A.R., Wang, L., Shanks, R.A. et al. (2020). Electrospun polyacrylonitrile–silica aerogel coating on viscose nonwoven fabric for versatile protection and thermal comfort. *Cellulose* 27 (17): 10501–10517.

41 Xiong, X., Venkataraman, M., Yang, T. et al. (2020). Transport properties of electro-sprayed polytetrafluoroethylene fibrous layer filled with aerogels/phase change materials. *Nanomaterials* 10 (10): 2042.

42 Loste, J., Lopez-Cuesta, J.-M., Billon, L. et al. (2019). Transparent polymer nanocomposites: an overview on their synthesis and advanced properties. *Progress in Polymer Science* 89: 133–158. https://doi.org/10.1016/j.progpolymsci.2018.10.003.

43 Zhang, H., Zhang, C., Ji, W. et al. (2018). Experimental characterization of the thermal conductivity and microstructure of Opacifier-Fiber-Aerogel Composite. *Molecules* 23: 2198. https://doi.org/10.3390/molecules23092198.

44 Araby, S., Qiu, A., Wang, R. et al. (2016). Aerogels based on carbon nanomaterials. *Journal of Materials Science* 51 (20): 9157–9189. https://doi.org/10.1007/s10853-016-0141-z.

45 Wang, J., Yang, B., Jiang, Z. et al. (2024). Recent advances of conductive hydrogels for flexible electronics. *Electronic materials* 5: 101–131. https://doi.org/10.3390/electronicmat5030008.

46 Zhang, X., Chen, X., Ye, Z. et al. (2023). Conductive hydrogels for bioelectronics: molecular structures, design principles, and operation mechanisms. *Journal of Materials Chemistry C: Materials for Optical and Electronic Devices* 11 (32): 10785–10808. https://doi.org/10.1039/D3TC01821K.

47 Meenakshy, S., Jesslyn, J., and Anas, S. (2025). Development and applications of polypyrrole-based conductive inks: an overview. *Advanced Materials Technologies* 10 (10): 2401216.

48 Worsley, M.A., Pauzauskie, P.J., Olson, T.Y. et al. (2010). Synthesis of graphene aerogel with high electrical conductivity. *Journal of the American Chemical Society* 132 (40): 14067. https://doi.org/10.1021/ja1072299.

49 Wei, H., Lei, T., Ma, L., and Li, W. (2024). Aramid nanofiber-reinforced carbon nanotubes@ cobalt ferrite nanoparticles aerogel films achieve excellent electromagnetic interference shielding, photothermal and joule heating performance. *Ceramics International* 50 (23): 50388–50396.

50 Dube, A., Malode, S.J., Alodhayb, A.N. et al. (2025). Conducting polymer-based electrochemical sensors: progress, challenges, and future perspectives. *Talanta Open* 11: 100395. https://doi.org/10.1016/j.talo.2024.100395.

51 Wang, X., Hu, H., Yang, Z. et al. (2014). Smart hydrogel-functionalized textile system with moisture management property for skin application. *Smart Materials and Structures* 23 (12): 125027.

52 Mostafalu, P., Kiaee, G., Giatsidis, G. et al. (2017). A textile dressing for temporal and dosage controlled drug delivery. *Advanced Functional Materials* 27 (41): 1702399.

53 Cuturicu, L.L. and Radu, C.D. (2023). A hydrogel from a textile support used in therapy of wounds. In: *The 18th Romanian Textiles and Leather Conference*, 42–52. Sciendo https://doi.org/10.2478/9788367405133-007.

54 Shuai, L., Guo, Z.H., Zhang, P. et al. (2020). Stretchable, self-healing, conductive hydrogel fibers for strain sensing and triboelectric energy-harvesting smart textiles. *Nano Energy* 78: 105389.

55 Ding, T., Zhou, Y., Wang, X.Q. et al. (2021). All-soft and stretchable thermogalvanic gel fabric for antideformity body heat harvesting wearable. *Advanced Energy Materials* 11 (44): 2102219.

56 Ribeiro, A.I., Alves, C., Fernandes, M., and Zille, A. (2025). *Gel Electrolytes in the Development of Textile-Based Power Sources*, vol. 11.

57 Niu, F., Qin, Z., Min, L. et al. (2021). Ultralight and hyperelastic nanofiber-reinforced MXene–graphene aerogel for high-performance piezoresistive sensor. *Advanced Materials Technologies* 6 (11): 2100394.

58 Zheng, Q., Cai, Z., Ma, Z., and Gong, S. (2015). Cellulose nanofibril/reduced graphene oxide/carbon nanotube hybrid aerogels for highly flexible and all-solid-state supercapacitors. *ACS Applied Materials & Interfaces* 7 (5): 3263–3271.

59 Zhu, L., Meng, Y., Zhou, J. et al. (2024). Thermoresponsive structural coloration of hydrogel fibers. *ACS Applied Polymer Materials* 6 (3): 1720–1729.

60 ter Schiphorst, J., van den Broek, M., de Koning, T. et al. (2016). Dual light and temperature responsive cotton fabric functionalized with a surface-grafted spiropyran–NIPAAm-hydrogel. *Journal of Materials Chemistry A* 4 (22): 8676–8681. https://doi.org/10.1039/C6TA00161K.

61 Wu, B., Qi, Q., Liu, L. et al. (2024). Wearable aerogels for personal thermal management and smart devices. *ACS Nano* 18 (14): 9798–9822. https://doi.org/10.1021/acsnano.4c00967.

62 Wang, G., Liu, X., Song, Z. et al. (2023). Multifunctional flexible pressure sensor based on a cellulose fiber-derived hierarchical carbon aerogel. *ACS Applied Electronic Materials* 5 (3): 1581–1591. https://doi.org/10.1021/acsaelm.2c01628.

63 Li, M., Chen, X., Li, X. et al. (2021). Wearable and robust polyimide hydrogel fiber textiles for strain sensors. *ACS Applied Materials & Interfaces* 13 (36): 43323–43332. https://doi.org/10.1021/acsami.1c10055.

64 Jafari, A., Al-Ostaz, A., and Nouranian, S. (2025). Recent advances in stimuli-responsive materials for electromagnetic interference shielding. *Polymers for Advanced Technologies* 36 (2).

65 Liu, X., Li, Y., Sun, X. et al. (2021). Off/on switchable smart electromagnetic interference shielding aerogel. *Matter* 4 (5): 1735–1747. https://doi.org/10.1016/j.matt.2021.02.022.

66 Hu, X., Cheng, Y., Wei, Z. et al. (2024). Smart windows with adjustable electromagnetic interference shielding using hydrogel technology. *Chemical Engineering Journal* 493: 152533. https://doi.org/10.1016/j.cej.2024.152533.

67 Tamayol, A., Akbari, M., Zilberman, Y. et al. (2016). Flexible pH-sensing hydrogel fibers for epidermal applications. *Advanced Healthcare Materials* 5 (6): 711–719.

68 Jocic, D., Tourrette, A., Glampedaki, P., and Warmoeskerken, M.M.C.G. (2009). Application of temperature and pH responsive microhydrogels for functional finishing of cotton fabric. *Materials Technology* 24 (1): 14–23. https://doi.org/10.1179/175355509X417963.

69 Liu, B., Zhong, H., and Hu, D. (2023). Construction of network-like cross-linked cellulose aerogel films with water-responsive properties for visualization of pH changes. *Colloids and Surfaces A: Physicochemical and Engineering Aspects* 656: 130420. https://doi.org/10.1016/j.colsurfa.2022.130420.

70 Ali, I., Chen, L., Huang, Y. et al. (2018). Humidity-responsive gold aerogel for real-time monitoring of human breath. *Langmuir* 34 (16): 4908–4913. https://doi.org/10.1021/acs.langmuir.8b00472.

71 Serpico, L., Dello Iacono, S., Cammarano, A., and De Stefano, L. (2023). Recent advances in stimuli-responsive hydrogel-based wound dressing. *Gels* 9: https://doi.org/10.3390/gels9060451.

72 Arsuffi, B., Siqueira, G., Nyström, G. et al. (2024). Programmable multi-responsive nanocellulose-based hydrogels with embodied logic. *Advanced Functional Materials* 34 (51): 2409864.

73 Zhao, G., Shi, L., Yang, G. et al. (2023). 3D fibrous aerogels from 1D polymer nanofibers for energy and environmental applications. *Journal of Materials Chemistry A* 11 (2): 512–547. https://doi.org/10.1039/d2ta05984c.

74 Zhang, X., Wang, Z., Huang, G. et al. (2024). Soft robotic textiles for adaptive personal thermal management. *Advanced Science* 11 (21): e2309605-9. https://doi.org/10.1002/advs.202309605.

75 Ahmed, B., Hossain, M.J., al Parvez, A. et al. (2024). recent advancements of mxene/nanocellulose-based hydrogel and aerogel: a review. *Advanced Energy and Sustainability Research* 5 (3): https://doi.org/10.1002/aesr.202300231.

76 Salman, S.M., Azam, F., Ahmad, S. et al. (2025). Development of environment friendly biodegradable biopolymer-based hydrogel composites incorporated with waste cotton for wound dressings. *Journal of Industrial Textiles* 55: 1–17. https://doi.org/10.1177/15280837251324895.

77 Mandić, Vilko, Arijeta Bafti, Ivana Panžić, and Floren Radovanović-Perić. "Bio-based aerogels in energy storage systems." *Gels*, vol. 10, no. 7, pp. 438, 2024, https://doi.org/10.3390/gels10070438.

78 Garcia-Garcia, A., Muñana-González, S., Lanceros-Mendez, S., Ruiz-Rubio, L., Alvarez, L. P., and Vilas-Vilela, J.L. "Biodegradable natural hydrogels for tissue engineering," *Polymers*, vol. 16, *no*. 18, pp. 2599, 2024, https://doi.org/10.3390/polym16182599

79 Azeem, M., Shahid, M., Masin, I., and Petru, M. (2025). Design and development of textile-based wearable sensors for real-time biomedical monitoring; a review. *Journal of the Textile Institute* 116 (1): 80–95. https://doi.org/10.1080/00405000.2024.2318500.

80 Mekonnen, B.T., Ding, W., Liu, H. et al. (2021). Preparation of aerogel and its application progress in coatings: a mini overview. *Journal of Leather Science and Engineering* 3 (1): 25. https://doi.org/10.1186/s42825-021-00067-y.

81 Tadesse, M.G. and Lübben, J.F. (2023). Review on hydrogel-based flexible supercapacitors for wearable applications. *Gels* 9 (2): 1–20. https://doi.org/10.3390/gels9020106.

82 Ghosh, A., Kumar, S., Singh, P.P. et al. (2024, https://doi.org/10.1021/acsomega.4c04851). Dynamic metal-coordinated adhesive and self-healable antifreezing hydrogels for strain sensing, flexible supercapacitors, and EMI shielding applications. *ACS Omega* 9: 33204–33223.

83 Shrestha, S., Barvenik, K.J., Chen, T. et al. (2024). Machine intelligence accelerated design of conductive MXene aerogels with programmable properties. *Nature Communications* 15 (1): 1–14. https://doi.org/10.1038/s41467-024-49011-8.

10

Applications of Aerogels in Protective Clothing

Abdul Salam[1], Farooq Azam[2], Yasir Nawab[2,3], and Jibran Khaliq[4]

[1]North Dakota State University, Department of Mechanical Engineering, Fargo, ND, 58102, USA
[2]National Textile University, School of Engineering and Technology, Department of Textile Technology, Sheikhupura Road, 37610 Faisalabad, Pakistan
[3]University of Kamalia, QAED Academy, Rajana Rd, Kamalia, Pakistan
[4]Northumbria University, Faculty of Engineering and Environment, Department of Mechanical and Construction Engineering, Newcastle upon Tyne, NE1 8ST, UK

10.1 Introduction

Protective textiles are a specialized class of clothing designed to ensure functional safety in environments where individuals face hazardous conditions. Their applications range widely across laboratories, hospitals, industrial sites, rescue, and military operations. Unlike conventional apparel that prioritizes aesthetics, protective clothing is engineered with a focus on performance, safety, and durability. Recent advancements have aimed at developing lightweight, high-strength, and multifunctional garments to meet the evolving demands of modern workplaces and extreme environments [1].

The increasing complexity of hazards (ranging from extreme temperatures to chemical exposure) has driven the development of innovative textile systems capable of providing both protection and wearer comfort. Modern protective clothing is expected to go beyond basic shielding; it must regulate body temperature, resist microbial growth, prevent chemical infiltration, and, often, integrate with electronic or smart technologies [2]. To achieve these multifaceted demands, researchers and manufacturers are turning to novel materials that can deliver high performance without adding significant bulk. Among these, aerogels have emerged as one of the most promising candidates.

Aerogels are ultralightweight materials renowned for their exceptional thermal insulation, high porosity, and large surface area. Initially developed for aerospace and cryogenic applications due to their outstanding ability to inhibit heat transfer, aerogels are now being adapted for wearable technologies. Their unique porous network structure makes them highly effective as thermal barriers [3, 4]. Today, aerogel-based textile composites are being increasingly explored for use in advanced protective clothing applications. Their combined properties, that is, thermal resistance, impact absorption, chemical repellency, and minimal weight, make them ideal for next-generation garments used in firefighting, military operations, cold-climate environments, and hazardous material handling. This chapter explores the diverse applications of aerogels in protective clothing, focusing on their functional roles, fabrication methods, structural design, and performance outcomes. It also highlights recent innovations driving their integration into cutting-edge textile engineering.

10.2 Mechanism of Thermal Insulation

Aerogels are revolutionizing thermal insulation in textiles through their unique nanoporous architecture and ultralow thermal conductivity, enabling the development of lightweight, high-performance protective gear for extreme environments.

Characterized by extremely high porosity (typically >90%) and nanoscale pore sizes (ranging from a few to hundreds of nanometers), aerogels possess a structural advantage that significantly inhibits heat transfer. In a typical solid material, heat is primarily transferred through the vibration of atoms (phonons). However, the sparse and discontinuous solid network of aerogel limits phonon transport, thereby minimizing conductive heat transfer through the solid phase. In porous materials, convection can occur if pore sizes are large enough to allow gas movement. However, in aerogels, the pore size is smaller than the mean free path of air molecules (~70 nm), which means gas molecules cannot effectively collide or move freely. This suppresses convective heat transfer within the pores. Additionally, some aerogels, especially those doped or composited with opacifying agents such as carbon black or titanium dioxide, can scatter or absorb infrared radiation, thereby further reducing radiative heat transfer [5]. These combined mechanisms make aerogels exceptionally efficient thermal insulators and highly suitable for integration into protective textiles. The schematic in Figure 10.1 illustrates the pathways of heat transfer in aerogel material.

The overall thermal conductivity (k) of aerogels is remarkably low, typically in the range of 0.012 to 0.020 W/m·K, which is even lower than air (~0.025 W/m·K). In

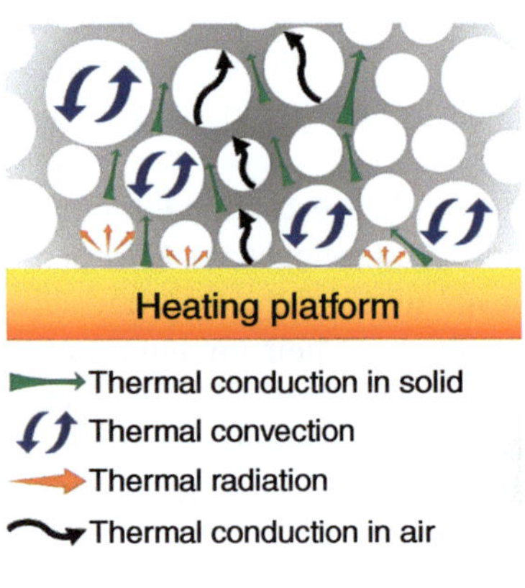

Figure 10.1 The schematic diagram of heat transfer in porous aerogel. *Source:* [6] / with permission of Elsevier.

porous materials such as aerogel, the total thermal conductivity (k_{total}) is the sum of three components:

- k_{solid}—heat transfer from the solid framework (very low in aerogels)
- k_{gas}—heat transfer from the air inside pores (minimized due to nanoscale pores)
- $k_{radiation}$—heat from infrared radiation (suppressed by optical additives or scattering agents)

In aerogels, k_{solid} remains minimal due to the high porosity and discontinuous skeleton. k_{gas} is reduced by the Knudsen effect, as the pores are smaller than the mean free path of gas molecules. Meanwhile, $k_{radiation}$ becomes negligible in doped aerogels due to high scattering and absorption of IR. Thanks to this combination, aerogels act as super-insulators, outperforming conventional insulating materials like fiberglass or polystyrene foam. As a result, aerogels are widely considered as an ideal candidate for applications requiring superior heat insulation, including cryogenic insulation, building materials, and aerospace applications [5].

10.3 Types of Aerogels

Based on chemical origin, aerogels can be broadly classified into inorganic and organic types, each exhibiting distinct characteristics derived from their molecular building blocks. This classification not only facilitates a systematic understanding of aerogel diversity but also serves as a guide for selecting suitable materials for specific industrial and scientific applications.

10.3.1 Inorganic

Oxide-based aerogels, such as silica (SiO_2), alumina (Al_2O_3), titania (TiO_2), zirconia (ZrO_2), and magnesia (MgO), are gaining significant attention for protective applications due to their unique thermal, structural, and chemical properties. Silica aerogels are widely recognized for their ultralightweight and highly porous structure, exceptional thermal insulation (~13 mW/m·K), and high surface area. These properties make them ideal for integration into nonwoven textiles for thermal and chemical protection [7, 8]. Alumina aerogels, featuring nanoporous structures and superior thermal stability above 800 °C, are well-suited for aerospace and high-temperature industrial insulation [9, 10]. Titania aerogels leverage their photocatalytic activity for environmental uses, such as water purification and chemical decontamination, offering potential for multifunctional protective clothing [11–13]. Zirconia aerogels offer excellent thermal resistance and mechanical strength at temperatures exceeding 1000 °C, making them suitable for extreme conditions in aerospace and industrial applications. They also exhibit photocatalytic properties useful for environmental remediation [14–16]. Magnesia aerogels provide chemical inertness, structural stability at high temperatures, and low thermal conductivity, serving as effective thermal insulators in harsh environments. Their nontoxic and eco-friendly nature enhances their appeal for sustainable protective clothing and industrial use [17, 18].

Carbide-based aerogels, such as silicon carbide (SiC) aerogels, are advanced non-oxide ceramic materials known for their exceptional thermal stability, chemical resistance, and mechanical strength, making them ideal for ultrahigh-temperature protective applications. They can endure temperatures above 1300 °C in oxidative environments and even higher in inert atmospheres, outperforming traditional oxide aerogels such as silica. Their low thermal conductivity, high porosity, and excellent thermal shock resistance render them suitable as thermal barrier layers in aerospace, firefighting, and industrial safety gear. Similarly, molybdenum carbide (MoC) aerogels combine the porosity and high surface area of traditional aerogels with the electrical conductivity and catalytic activity of MoC. Typically synthesized via sol–gel and carbothermal reduction, these lightweight and thermally stable materials are ideal for energy conversion, electrocatalysis, and protective coatings in harsh environments [19–21].

Nitride-based aerogels like boron nitride (BN) aerogels, represent a class of ultra-light, thermally stable, and chemically inert materials with exceptional potential for protective clothing and thermal shielding applications. With porosity exceeding 96%, specific surface areas up to 982 m^2/g, and ultralow thermal conductivity (~26 mW/m · K), BN aerogels provide outstanding thermal insulation. Their high oxidation resistance and ability to bear temperatures up to 1200 °C make them ideal for use in aerospace, defense, and wearable electronics. Moreover, their high reflectivity in the visible and near-infrared range allows for laser protection in high-energy environments. These multifunctional properties positioning BN aerogels as next-generation materials for smart, thermally adaptive, and laser-resistant protective clothing [22–24].

Metal aerogels such as gold (Au) aerogels are porous metallic nanostructures that combine high surface area and low density with the intrinsic electrical, optical, and catalytic properties of gold. Typically formed through sol–gel or dopamine-assisted methods, their 3D nanowire networks offer high conductivity and tunable porosity, making them suitable for electrocatalysis, biosensing, and wearable health monitoring system, especially where biocompatibility and rapid response are required [25–28]. Silver (Ag) aerogels, formed from interconnected Ag nanowires, offer ultralight structure, excellent thermal and electrical conductivity, and outstanding electromagnetic interference (EMI) shielding performance. Their interconnected 3D networks provide thermal interface efficiency, shielding effectiveness above 100 dB, and mechanical strength, making them ideal for electronic and communication-based protective wearables [29–31]. Iron (Fe) aerogels offer magnetic and catalytic functionality with high surface area and porosity, often derived from Fe(III) salt sol–gel processes. With mesoporous structures, superparamagnetic behavior, and potential hybridization with carbon networks, they are promising for magnetic sensing, smart textiles, environmental remediation, and energy-efficient protective applications [32–34].

10.3.2 Organic Aerogels

Cellulose aerogels are sustainable, lightweight, and highly porous materials ideal for applications in thermal insulation, adsorption, biomedical, and sensing. Compared to brittle inorganic aerogels, they offer superior elasticity, biocompatibility, and lower thermal conductivity, making them suitable for wearable and eco-friendly protective systems. Their mechanical limitations can be addressed through reinforcement with fibers such as cotton or glass, or by using crosslinking agents like polyvinyl alcohol (PVA). These aerogels are being developed for advanced textiles, including space suits, thermal wear, and smart fabrics with integrated sensing and energy management capabilities [35–37]. Chitosan (CS)-based aerogels, derived from marine biomass, are biodegradable, biocompatible, and inherently flame-retardant due to their nitrogen content. To meet the high-performance demands of protective applications, they are often modified with inorganic fillers, surface coatings, or phosphorus-based treatments that promote char formation and enhance flame resistance. These enhancements make CS-based aerogels viable for insulation in construction, electronics, and protective clothing [38, 39]. Similarly, sodium alginate (SA)-based aerogels are biodegradable, biocompatible with 3D porous structure. They form mechanically strong networks through ionic crosslinking with multivalent ions. Despite challenges such as moisture sensitivity and shrinkage, recent innovations, including the incorporation of flame retardants and thermosensitive materials like graphene oxide, have extended applications to intelligent fire-resistant clothing, building insulation, and aerospace applications, supporting sustainable and high-performance solutions [40, 41].

Resorcinol–formaldehyde (RF) aerogels are organic aerogels synthesized through base- or acid-catalyzed polycondensation of resorcinol and formaldehyde in aqueous media, typically using sodium carbonate as a catalyst. Due to resorcinol's high reactivity, RF gels form readily at lower temperatures, producing a crosslinked network that typically dries via supercritical CO_2 to yield lightweight, mesoporous aerogels. RF aerogel exhibits high surface area, low thermal conductivity, and tunable nanostructures, with morphologies influenced by the resorcinol-to-catalyst (R/C) ratio. Notably, RF aerogels serve as precursors to electrically conductive carbon aerogels, making them highly attractive for applications in thermal insulation, energy storage, and electronic devices. Their structural resemblance to base-catalyzed silica aerogels, combined with the versatility of organic system, has positioned RF aerogels as a foundational material for the development of functional porous networks [42].

Polyurea aerogels are formed by the reaction of triisocyanates with water, resulting in a robust, crosslinked polymer network. Initially used as a strengthening agent in polymer-crosslinked silica aerogels, they have since evolved into a standalone material with exceptional mechanical strength, low density, and high porosity. Depending on synthesis conditions like solvent and catalyst concentration, these aerogels can exhibit diverse nanostructures with enhanced elastic modulus, yield strength, and energy absorption, making them ideal for impact-resistant materials, thermal insulation, and structural applications. Polyurethane (PU) aerogels, particularly those derived from thermoplastic polyurethane (TPU), are gaining attention for passive daytime radiative cooling (PDRC) and personal thermal management (PTM). TPU offers transparency, flexibility, and durability. Recent innovations using non-solvent phase separation have led to white films with high solar reflectance and infrared emissivity, capable of reducing temperature by up to 10 °C without external energy. When combined with superhydrophobic silica aerogels, these TPU films also acquire water resistance, making them suitable for wearable cooling textiles. Their scalable fabrication and multifunctional performance make aerogel-functionalized TPU films a sustainable solution for energy-efficient cooling in medical wearables, protective clothing, and hot climate applications [43–45].

Graphene aerogels, composed of lightweight and porous graphene sheet networks, offer exceptional mechanical strength, thermal and electrical conductivity, and antimicrobial activity. Their integration into textiles enhances enhancing personal protective equipment (PPE) by offering flame resistance, UV protection, abrasion resistance, and barriers against heat and gases, while retaining flexibility and breathability. These attributes enable the creation of lightweight, comfortable, and highly protective clothing for emergency responders, healthcare workers, and military personnel [46]. Similarly, carbon nanotube (CNT)-based aerogels, feature a 3D-interconnected structure that delivers properties such as high tensile strength, electrical conductivity, thermal stability, and chemical resistance. These properties make CNT-aerogel excellent candidates for PPE requiring EMI shielding, thermal insulation, impact resistance, and chemical sensing. Their integration into smart textiles supports the development of wearable electronics and responsive gear, offering reliable performance under extreme conditions for military, aerospace, and emergency operations [47, 48].

10.4 Impact Resistance and Chemical Protection of Aerogels

Mechanical reinforcement of aerogels is a pivotal strategy for enhancing their structural integrity, flexibility, and suitability in wearable and protective applications. Early research focused on integrating aerogels with various fiber types to improve mechanical robustness while maintaining their inherently low thermal conductivity. Tang et al. [49] pioneered the use of five fiber-reinforced silica aerogel composites as insulation layers for space suits, evaluating their thermal efficiency and structural stability under repeated mechanical flexing. Similarly, Paul and Diller [50] investigated the performance of insulating fibers embedded with aerogel in various configurations, highlighting the effect of fiber denier and orientation on thermal performance.

To advance from experimental models to commercial viability, recent efforts have incorporated commercially available aerogel fiber composites with additional functional materials. For instance, Zhang et al. [51] combined these composites with microencapsulated phase change materials (PCMs) to enhance thermal protection in firefighting suits. However, a persistent issue has been the detachment and airborne dispersion of aerogel particles, leading to functional degradation and potential health risks. To mitigate this, lamination and encapsulation techniques like nylon coverings or dual-sided films have been employed to contain particles effectively, though often at the expense of breathability and ease of processing.

To address these limitations, researchers introduced aerogel foam composites that embed aerogel particles within a flexible matrix. The integration of aerogels in syntactic foams or polyurethane (PU) foam enhances compressive properties while maintaining flexibility, which is essential for applications like footwear and wetsuit. Reinforcement by fabric coating has also shown promise. Aramid-blended fabrics coated with silica aerogels and hybrid aerogel/PCM layers significantly improved thermal protection in firefighter gear. These coatings not only enhance resistance to heat flux and delayed second-degree burns but also improve flame resistance of otherwise flammable PCM materials [52]. Further, the structural enhancement of aerogel composites through stitching or pattern-based mechanical integration has also been validated. Stitched aerogel nonwoven composites demonstrated improved tensile strength, although at the cost of higher thermal conductivity. Nonetheless, studies confirmed that properly reinforced aerogel composites retain their insulation properties even after extensive mechanical cycling, making them a strong candidate for commercial application in footwear, PPE [53, 54].

10.4.1 Use of Aerogels in Ballistic-Resistant Textiles

Ballistic-resistant textiles are designed to absorb and dissipate the kinetic energy of projectiles, thereby protecting the wearer from penetration and blunt trauma. Conventional materials like Kevlar® (para-aramid) and Twaron® (aramid) fibers are widely used due to their high tensile strength, lightweight, and excellent energy-absorption capabilities. However, recent advancements have focused on the integration of aerogels – ultralight, nanoporous materials with exceptional thermal and mechanical properties to enhance ballistic performance. With porosity level reaching 88% and bulk densities as low as $0.17\,g/cm^3$, aerogels are significantly lighter than traditional ballistic materials like steel or ceramic plates.

Aerogels have demonstrated significant potential for enhancing the mechanical properties of protective textiles when strategically integrated into fabric assemblies. Recent studies show that incorporating silica aerogels (1–7% by mass) within aramid fabric layers can substantially improve ballistic resistance by increasing inter-fiber friction and potentially creating inter-fiber bonds. Figure 10.2 demonstrates the bullet-stopping capability of aerogel-reinforced Kevlar protective clothing. Innovative approaches using laser-cut voids in nonwoven polyester and polyurethane foam, subsequently filled with hydrophobic silica aerogel, have revealed important structure–property relationships. Although some configurations may experience a reduction in compression resistance, the integration of aerogels consistently improves thermal insulation and can enhance impact resistance—particularly in technical textiles where increased stiffness is acceptable [56].

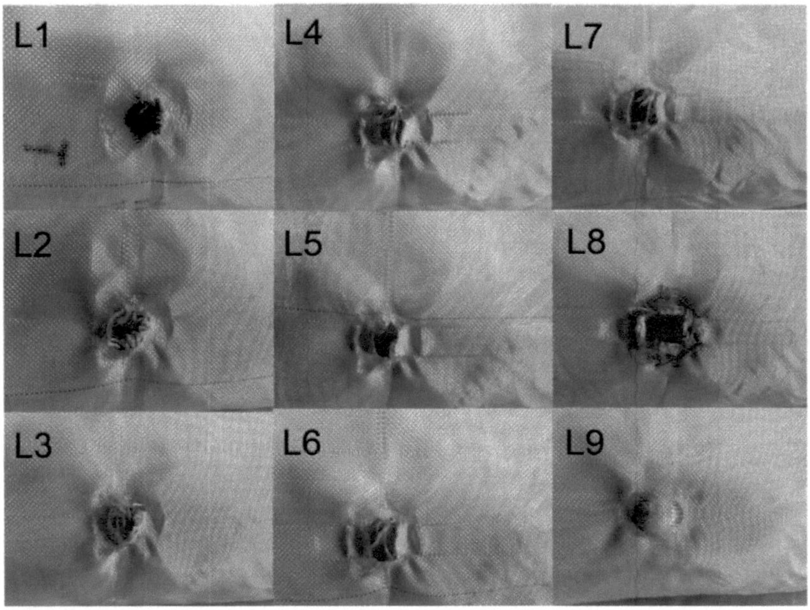

Figure 10.2 Fabric failure until the bullet stopped. *Source:* [55] / with permission of Elsevier.

The high porosity and lightweight nature of aerogels also enable integration into weight-sensitive applications such as spacecraft shielding (e.g., stuffed Whipple shields for orbital debris protection) without compromising structural integrity [57]. Orbital debris, traveling at hypervelocities (7–10 km/s), poses critical risks to spacecraft, necessitating advanced shielding solutions. While traditional Whipple shields use Kevlar/Nextel layers, their bulkiness limits efficiency. This work explores lightweight polyimide aerogels—with superior mechanical properties over silica aerogels—as next-generation stuffing materials, demonstrating their potential to enhance debris containment while reducing shield mass and volume [58]. Additionally, the combination of aerogel foams with shear thickening fluid (STF) has yielded promising results. A low-cost composite of baked starch-based sSTF and aerogel foam demonstrated superior impact resistance, reducing deflection by 69% compared to pristine aerogel in ballistic tests, while successfully preventing projectile penetration in hygroelectric devices [59]. These findings highlight the versatility of aerogel–textile composites, which can be engineered to optimize specific protective functions such as impact resistance and ballistic performance while maintaining the fabric functionality.

10.4.2 Laboratory Test Methods

To evaluate the performance of aerogel-enhanced protective clothing, a range of standardized testing methods is employed to assess impact resistance, ballistic performance, and fluid barrier capabilities under various conditions. In the context of ballistic protection, the NIJ Standard-0115.00, established by the US National Institute of Justice, provides the test procedures for ballistic and stab resistance of personal body armor. This standard evaluates a material's ability to resist penetration from various threats, such as knives and spikes. While traditionally applied to aramid or ceramic-based armors, recent studies suggest that laminated and hybrid aerogel structures hold promise for multi-threat protection. Due to their unique nanoporous morphology, aerogels can effectively absorb impact energy and redistribute force across a wider area, making them candidates for inclusion in soft, lightweight body armor systems that prioritize both flexibility and safety [55].

For evaluating high-velocity impact resistance in textile materials, ASTM F2912 outlines procedures for gas-gun testing, where projectiles are launched at controlled speeds to simulate ballistic and fragment impacts. This approach is particularly useful for investigating the dynamic behavior of aerogel-reinforced fabrics under rapid loading conditions. The standard provides crucial insight into how aerogels perform when subjected to sudden kinetic forces, as in explosions or high-speed collisions, enabling developers to fine-tune the composition and thickness of aerogel layers in protective gear [59].

Another critical parameter for protection clothing, especially in chemical, biological, or blood-borne, is fluid penetration resistance under pressure. The ISO 13994 standard evaluates a material's ability to withstand liquid penetration when subjected to hydrostatic pressure, thereby simulating exposure to pressurized hazardous fluids. Aerogel-based composites, particularly those integrated with breathable yet hydrophobic matrices, are evaluated under this protocol to ensure adequate barrier performance without compromising comfort and air permeability. Properly engineered aerogel fabrics can repel hazardous fluids while allowing moisture vapor transmission, which is vital for extended wear in hazardous environments [60].

10.4.3 Chemical Protection

The unique structural and surface properties of aerogels make them promising candidates for chemical protective clothing, particularly in hazardous material (hazmat) suits and military gear. Traditional chemical protective materials often rely on activated carbon textiles for adsorption of toxic vapors; however, aerogels, particularly silica-based and functionalized organic–inorganic hybrids, offer superior barrier properties, chemical tunability, and reduced weight, positioning them as next-generation materials for advanced chemical defense.

Aerogels exhibit an exceptionally high surface area (typically 600–1000 m^2/g), interconnected nanoporous structure, and the ability to be functionalized chemically, which enables them to act as efficient barriers against a wide range of chemical agents. Silica aerogels, functionalized with amine, thiol, or fluorinated groups, can be engineered to absorb or repel specific toxic compounds, such as acidic vapors, volatile organic solvents, and even chemical warfare agents. The tortuous path provided by the porous network of aerogels effectively restricts the diffusion of hazardous vapors, thereby extending breakthrough time and offering prolonged protection. Moreover, functionalization enhances selectivity and stability, hydrophobicity, and humidity/thermal stability, ensuring consistent performance in harsh operational environments. Studies have demonstrated that aerogel composites significantly delay the permeation of sulfur mustard (HD), a blister agent, compared to uncoated textiles. Fluorinated silica aerogels have shown exceptional repellency against both polar and nonpolar solvents, including dimethyl sulfoxide (DMSO), hydrochloric acid vapors, and organophosphate compounds. Additionally, aerogel coatings have demonstrated resistance to corrosion and degradation even after exposure to aggressive chemicals, a critical requirement for field-deployable protective garments [61].

Due to their low density and flexibility when incorporated as thin films, coated fabrics, or foam composites, aerogels can be seamlessly integrated into multilayer hazmat suits. Typically, such suits consist of an outer layer for mechanical and flame resistance, a chemical barrier layer, and a comfort or moisture transport layer. Aerogels, due to their multifunctionality, can simultaneously serve as thermal insulators and chemical barriers within the same layer, thus simplifying suit architecture and reducing weight. Hybrid laminates using aerogel-infused nonwovens or coated membranes have been experimentally tested in composite protective assemblies. Beyond chemical protection, aerogels also contribute to thermophysiological comfort by permitting moisture transport and heat dissipation, addressing limitations associated with traditional impermeable membranes [62].

Hydrophobicity is a key factor in the chemical resistance of aerogels. While pristine silica aerogels are inherently hydrophilic due to surface silanol (–SiOH) groups, post-synthesis modification with alkylsilanes or perfluorinated silanes imparts hydrophobicity and oleophobic properties. These surface treatments reduce surface energy and enabling the material to repel aqueous and organic chemicals. Superhydrophobic aerogels not only resist wetting by toxic liquids but also prevent the capillary uptake of vapors, thereby preserving dry thermal and chemical barrier properties. Moreover, surface-engineered aerogels can be designed to selectively adsorb chemical agents while repelling water and oils. This dual function is particularly important in environments where chemical agents are present alongside water vapor, which might otherwise saturate conventional adsorbents like activated carbon. Some advanced formulations have shown self-cleaning behavior, where droplets of hazardous liquid roll off the surface without wetting or absorption, greatly enhancing the reusability and safety of the protective garment [63, 64].

Although activated carbon-based textiles have long been the industry standard for chemical protective clothing due to their high adsorption capacity and low cost, they present several drawbacks such as poor thermal insulation, bulkiness, limited selectivity toward specific chemical threats, and performance degradation in humid environments. In contrast, aerogels, especially when chemically functionalized, offer tailored selectivity, lower density, and superior multifunctionality. While activated carbon passively adsorbs a broad range of compounds, aerogels can be engineered for reactive adsorption, and enabling neutralization of specific chemical agents. Additionally, aerogels provide excellent thermal insulation, an added advantage in extreme temperature environments. The combination of lightweight, tunable chemical functionality, and form factor flexibility makes aerogels superior alternatives to activated carbon in high-performance chemical protective clothing. Although current commercial implementation remains limited, ongoing research aims to develop scalable processing techniques and hybrid aerogel–carbon composites, merging the strengths of both technologies for future protective systems [65, 66].

10.5 Aerogels as Lightweight Protective Gear

Aerogels represent one of the most promising classes of materials for next-generation protective clothing owing to their exceptional combination of ultralow density, high thermal insulation, and mechanical resilience. Structurally, aerogels are composed of up to 99.8% air and feature a highly porous network of nanostructured solids, which dramatically reduces their weight while maintaining sufficient structural integrity. The ultralightweight nature of aerogels typically ranging from 0.003 to 0.5 g/cm^3 offers significant ergonomic and mobility benefits, particularly in applications requiring prolonged wear and physical exertion, such as military operations, mountaineering expeditions, firefighting, and emergency rescue missions. This lightweight property enhances mobility and comfort. For instance, military personnel require gear that minimizes fatigue while providing thermal protection in extreme environments; aerogel-integrated apparel has been tested in such scenarios, demonstrating superior heat retention without the bulk of conventional insulation. Similarly, mountaineers and rescue workers benefit from aerogel-based suits that offer warmth at sub-zero temperatures while allowing freedom of movement, critical in life-threatening situations [67, 68]. A lightweight thermal-insulating shear stiffening gel (LT-SSG) was developed by incorporating hollow glass microspheres (BR20), demonstrating both impact resistance and ultra-lightweight properties as shown in Figure 10.3. This innovation enabled the creation of a multifunctional Kevlar safeguarding suit (MKSS) with exceptional flame retardancy and temperature-regulating capabilities, effectively isolating heat in high-temperature environments while preventing heat loss in cold conditions.

In protective clothing, mobility and comfort are critical parameters, especially in high-risk or physically demanding environments. Conventional insulation materials

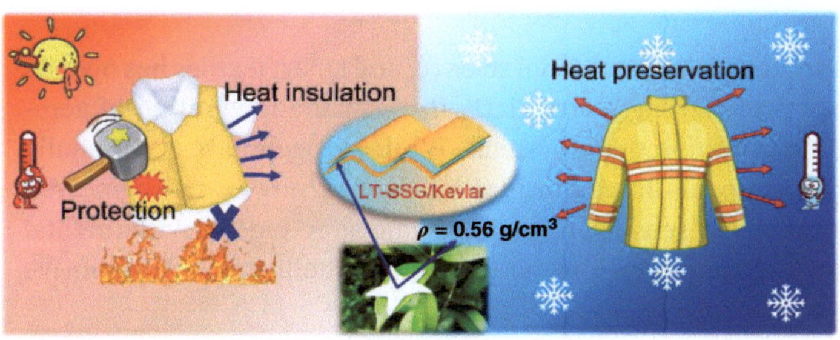

Figure 10.3 Schematics for multifunctional Kevlar safeguarding suit with flame-retardant and temperature-regulating capabilities. *Source:* [69] / with permission of Elsevier.

like down, foam, or heavy multilayered fabrics often add bulk and weight, leading to wearer fatigue and restricted motion. Aerogels, by contrast, provide equivalent or superior thermal protection at a fraction of the weight. For instance, silica aerogels have a thermal conductivity as low as 0.013 W/m K, even lower than air, making them ideal insulators even in thin, flexible forms. This property allows designers to produce slimline garments that maintain a high level of protection without burdening the wearer. The reduction in weight translates directly into improved endurance, agility, and reduced heat stress, which are essential for professionals operating in extreme conditions [70].

Several case studies and commercial applications have highlighted the practical integration of aerogels into protective gear. NASA has employed silica aerogels for spacesuit insulation in extraterrestrial missions, where minimal weight and extreme thermal protection are both imperative. In terrestrial applications, mountaineering and arctic expeditions have adopted aerogel-based liners in boots and jackets, such as the *Klymit Ulaar* and *Norrona Trollveggen* series, for improved insulation without sacrificing maneuverability. Likewise, military gear also benefits from aerogel technology, especially for cold-weather warfare, where body temperature regulation is crucial, and reduced load can enhance mission performance. Aerogel panels have also been incorporated into sleeping bags and shelter systems used by armed forces in extreme environments [3, 71].

The comparison between aerogels and conventional materials underscores their advantages in protective apparel. For example, polyfurethane foams and wool-based insulators typically have densities between 20 and 50 kg/m^3, while aerogel composites can be engineered with densities as low as 3–10 kg/m^3 without losing insulation effectiveness. Moreover, traditional flame-resistant garments often require heavy fabric layers (e.g. aramid-based textiles), which can be complemented or replaced with aerogel-infused layers. This approach can reduce total garment weight by up to 40% while maintaining or enhancing thermal protection. This is particularly beneficial in composite fabric systems where aerogels are laminated or embedded between breathable outer shells and comfort linings [72].

10.6 Hybrid Applications and Smart Integration of Aerogels in Protective Textiles

The convergence of nanotechnology, material science, and smart textiles has opened new frontiers in protective clothing, particularly through the integration of aerogels with electronics and multifunctional systems. While aerogels have traditionally been valued for their exceptional thermal insulation, they are now being engineered into hybrid platforms capable of sensing, responding, and adapting to environmental stimuli, making them ideal candidates for next-generation intelligent and responsive garments.

Recent innovations have demonstrated the successful incorporation of aerogels into electronic textile systems for thermal sensing and self-regulating functions, as shown in Figure 10.4. For example, Yang et al. [74] developed a flexible aerogel film

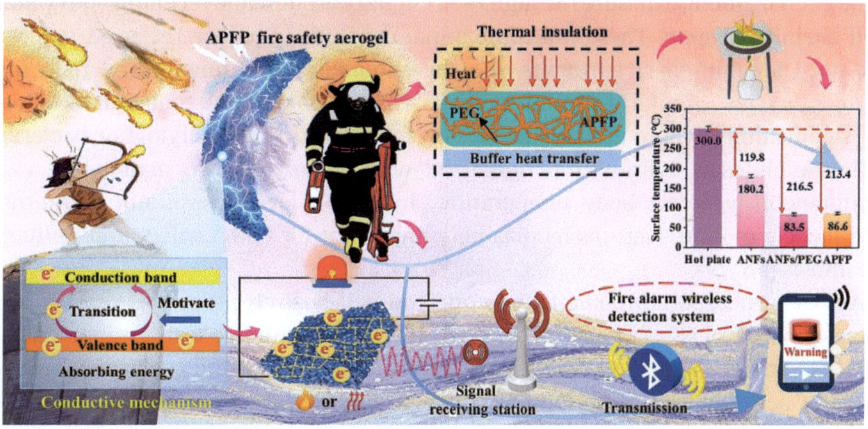

Figure 10.4 Schematic diagram for the development of smart thermal protective composite aerogel via integration of aramid nanofibers, polyethylene glycol, Fe_3O_4 nanowires, and polyaniline via freeze-drying and in situ polymerization method for ultrasensitive fire warning response (1.3 seconds), and flame retardancy (self-extinguishing in 1.5 seconds). *Source:* [73] / with permission of Elsevier.

using aramid nanofibers (ANF) and MXene, which not only provided structural integrity and thermal insulation but also functioned as a high-performance pressure sensor. The device exhibited rapid response (~100 ms) and high sensitivity (~37.4 kPa^{-1}). The film's conductive properties enable integration into thermal shielding wearables capable of electronic heating and pressure detection, key features for smart firefighting suits or military gear exposed to variable environments.

Similarly, carbon-based aerogels derived from cellulose nanofibers (CNFs) and graphene oxide (GO) have shown promise as lightweight thermal sensors with integrated Wi-Fi communication, rapid pressure response (~130 ms), and strain sensitivity. These systems allow not only body motion detection but also real-time monitoring and communication in hostile environments, such as firefighting or rescue missions. Of particular interest are self-powered aerogel fiber textiles, a novel class of materials combining Fe_3O_4, Ag nanowires, and calcium alginate. These aerogels can withstand temperatures from 100 to 400 °C and serve as temperature alarms or emergency location beacons, transmitting distress signals help to locate trapped firefighters in emergencies. This demonstrates how aerogels can evolve from passive barriers into active life-saving components within protective systems [70].

In addition to sensing function, aerogels can be functionalized to offer simultaneous resistance to fire, chemicals, and microbial agents, thereby surpassing the performance of conventional single-function protective materials. For instance, a phytic acid-coated bacterial cellulose aerogel embedded with silver nanoparticles exhibited not only high flame retardancy (self-extinguishing behavior) but also antibacterial activity and sensitivity to pressure, making it a multifunctional sensing material ideal for chemical spill response suits [75]. Additionally, superhydrophobic aerogels based on collagen fiber, MXene, and Ag nanoparticles have been designed to detect hazardous exposure through rapid signal output while repelling both water and chemical agents. These multifunctional capabilities reduce the need reduces the need for multilayer garments, thereby enhancing performance and reducing garment weight. Similarly, multiresponsive aerogels also extend to health monitoring applications, where aerogels fabricated from gelatin methacryloyl or bacterial cellulose/chitosan hybrids, functionalized with MXene and Ag nanowires, can simultaneously detect body temperature, hydration levels, respiratory patterns, and muscle motion, features increasingly important for both civilian and military wearables [76, 77].

Aerogels incorporating shape-memory and self-healing properties represent a transformative class of smart materials with vast potential in textiles, biomedical devices, environmental remediation, and wearable electronics. Shape-memory polymer aerogels (SMPAs) uniquely combine the ultralow density, high porosity, and large surface area of aerogels with the ability of polymers to retain a deformed (temporary) shape and recover their original configuration upon exposure to external stimuli such as temperature, moisture, or light. Despite criticism in earlier literature regarding whether aerogels can exhibit true shape-memory behavior, emerging research demonstrates that porous structural changes in aerogels under variable conditions substantiate their capacity for programmable deformation and recovery [78].

Furthermore, self-healing capabilities are enhanced in aerogels by utilizing polysaccharides such as chitosan, cellulose, hyaluronic acid, and alginate, which possess abundant functional amine and aldehyde groups. These groups enable dynamic covalent bonding that allows materials to autonomously repair micro-damages. In literature, studies have developed chitosan-itaconic acid-based aerogels exhibit ultrafast self-healing at room temperature through strong electrostatic interactions between oppositely charged components, repairing damage within 30 seconds upon contact with water. These aerogels also respond to acidic and alkaline gases, further enhancing their multifunctionality. Together, these findings confirm that integrating stimuli-responsive shape-memory behavior with autonomous self-repair mechanisms in aerogels paves the way for durable, multifunctional textile, and environmental materials. Continued investigation is essential to better understand the shape-memory mechanisms specific to aerogels and to refine their definitions in the context of emerging multifunctional smart materials [79].

10.7 Manufacturing Techniques and Textile Integration

The integration of aerogels into textile substrates has opened new avenues for the development of advanced functional materials, particularly in thermal insulation, chemical protection, and energy-related applications. The development of aerogel–textile composites generally involves two core stages: (1) synthesis of the aerogel and (2) incorporation into or onto textile materials. This section details the most widely used synthesis and drying methods of aerogels, followed by an in-depth discussion of textile integration strategies, including lamination, embedding, and electrospraying, while highlighting key challenges and performance characteristics.

10.7.1 Synthesis of Aerogels

Among several synthesis techniques, sol–gel polymerization remains the most widely adopted method for preparing inorganic aerogels such as silica aerogels. The general process for synthesis of aerogels involve several key steps.

Precursors, often silicon alkoxides, undergo hydrolysis, catalyzed by acids or bases, followed by condensation reactions that form a three-dimensional interconnected gel. The wet gel is aged to enhance its mechanical stability by strengthening the polymeric network. Pore-filling liquids are exchanged (typically with alcohol) to eliminate impurities and facilitate subsequent drying. Finally, wet gels, also called alcogels, are dried using one of the following approaches to form porous, lightweight aerogels.

Supercritical drying: First introduced by Kistler in 1931, supercritical drying involves raising the solvent above its critical temperature and pressure to avoid the formation of a vapor–liquid interface, which would collapse the gel's delicate structure [83]. Typically, ethanol or methanol is used, and drying occurs in an autoclave. Although it ensures excellent pore preservation, this technique involves fire hazards and high costs. To mitigate risks and reduce costs, liquid CO_2 is used as a safer alternative due to its lower critical point.

Freeze-drying: Freeze-drying involves freezing the wet gel at low temperatures (typically between −50 and −80 °C), followed by sublimation of the solvent under vacuum conditions [89–91]. This method has been employed for aerogels based on cellulose, graphene, clay, and alginate. However, the formation of solvent crystals can lead to microstructural cracks, limiting their suitability for industrial-scale production.

Ambient pressure drying: This cost-effective technique avoids high-pressure equipment and involves surface silylation (e.g. trimethylsilyl substitution) to prevent structural collapse. Although ambient-pressure-dried aerogels show promising porosity and thermal conductivity [95–97], the process typically takes 4–7 days, making it time-consuming for mass production. Among all techniques, supercritical drying is most preferred for commercial-scale applications due to its ability to preserve aerogel porosity and structural integrity without pore collapse [80].

10.7.2 Aerogel–Textile Integration Techniques

The incorporation of aerogels into textiles is generally classified into three main approaches: post-synthesis deposition, in situ synthesis, and hybrid electrospraying/coating. Each approach significantly affects the mechanical strength, flexibility, and thermal behavior of the final composite.

Padding technique: The simplest approach to integrate aerogels into textiles is through padding, where a dispersion of aerogel particles is applied to a fabric using a padder. Jin et al. utilized D50 silica nanogel (7–11 μm) on a nonwoven sheet, enhancing thermal protection at the expense of reduced water vapor permeability. This method is advantageous due to its simplicity but lacks data on product flexibility and durability.

Thermal bonding: In thermal bonding, low-melting fibers in a nonwoven matrix act as adhesives to trap aerogel particles upon heating. This approach with bi-component polyester fibers was used to produce compressible, bendable webs with low thermal conductivity (28 mW/mK).

Lamination: The lamination process involves spraying aerogel granules onto fabric, followed by overlaying with nanofibrous mats or fusible linings, and then hot-pressing to fix the structure. Sandwich-type nonwovens were developed by incorporating 100–120 μm silica aerogel granules using polyester/viscose webs and fusible linings. The chemical and thermal protection of the resulting structures was significantly enhanced.

Electrospraying and electrospinning: A more advanced method is electrospraying, wherein aerogel powders are incorporated into polymer solutions (e.g. PAN or PTFE), then electrospun or sprayed onto textile substrates. Layered structures are created by embedding silica aerogel in electrospun polyacrylonitrile fibers placed between viscose layers and achieved hydrophobicity by spraying a fluorocarbon–aerogel mix onto woven polyester fabrics. These techniques allow for fine control over particle distribution and composite porosity.

In situ incorporation during aerogel synthesis: A sophisticated strategy involves immersing textiles directly into the sol of aerogel precursors before gelation. During the sol–gel transition, fibers/textile substrates are embedded within the forming gel matrix, resulting in robust composite aerogels. Natural fibers such as silk, ramie, and jute are embedded into aerogel precursors, improving flexibility and maintaining low thermal conductivity (~0.026 W/mK). Similarly, alginate aerogels reinforced with polyester nonwovens were fabricated, yielding higher mechanical strength and thermal resistance. PVDF nanofiber-reinforced aerogels showed better hydrophobicity and flexibility. PAN, carbon, and PET woven fabrics were used as substrates to grow aerogels for applications such as wearable supercapacitors. Inorganic fibers such as glass and ceramic were also employed to produce aerogel composites with enhanced strength (up to 4.8×10^4 Pa) but slightly higher thermal conductivity (~0.026 W/mK) compared to pure aerogels. This in situ approach offers dual benefits i-e strengthening fragile aerogels and upgrading textile properties (e.g. thermal insulation, filtration efficiency, and hydrophobicity) [66].

10.7.3 Structural Challenges in Aerogel–Textile Composites

Despite their promising properties, integrating aerogels into textiles presents structural and processing challenges: Achieving strong adhesion between aerogel particles and fibers is crucial. Inadequate bonding can lead to delamination or particle shedding. Aerogels are inherently brittle. Embedding fibers improves toughness but may reduce thermal resistance. Supercritical drying ensures structural retention but requires costly, high-pressure equipment. Freeze-drying may cause cracking due to ice crystallization, while ambient pressure drying is time-consuming. Most aerogel-treated textiles show reduced drape and increased stiffness, limiting their use in wearable applications unless flexible aerogel matrices or nanofiber-reinforced aerogels are used.

The integration of aerogels with textiles is an interdisciplinary achievement, bridging chemistry, material science, and textile engineering. Each processing route, whether padding, thermal bonding, electrospraying, or in situ synthesis has its own trade-offs in terms of structural integrity, performance, and scalability. Drying methods critically influence the final microstructure and hence the thermal and mechanical properties of the composite. As the demand for multifunctional textiles increases, particularly for thermal insulation, energy devices, and personal protection, further innovation in aerogel–textile composite manufacturing is expected to enable commercial-scale production.

10.8 Conclusion

Aerogels have emerged as transformative materials in the field of revolutionized protective textiles, offering a unique combination of ultralow density, exceptional thermal insulation, and multifunctional adaptability. Their nanoporous structure enables superior thermal resistance, making them ideal for high-performance applications in firefighting, military, medical, and industrial protective gear. Beyond thermal regulation, aerogels contribute to impact absorption, chemical resistance, and ballistic protection, all while maintaining lightweight and flexible properties. Their integration into textiles has opened avenues for smart functionalities such as sensing, self-healing, and shape memory. However, challenges remain, including brittleness, high production costs, limited scalability, and reduced comfort due to low breathability. Addressing these issues requires innovations in flexible composites, cost-effective drying techniques, and designs that enhance moisture permeability. Future advancements should focus on hybrid aerogel systems, sustainable materials, and multifunctional performance. Crucially, interdisciplinary collaboration between materials scientists, textile engineers, and designers is essential to optimize aerogel–fiber integration, develop real-world testing protocols, and incorporate smart textile technologies. Such collaboration will drive the next generation of high-performance, intelligent protective clothing.

References

1 Islam, S.U. and Butola, B.S. (2019). *Advanced Functional Textiles and Polymers: Fabrication, Processing and Applications*. John Wiley & Sons.

2 Shaker, K. and Nawab, Y. (2020). *Fibers for Protective Textiles* (ed. S. Ahmad, A. Rasheed, and Y. Nawab), 65–91. Springer Nature.

3 Hrubesh, L.W. (1998). Aerogel applications. *Journal of Non-Crystalline Solids* 225: 335–342.

4 Shaid, A., Bhuiyan, M.A.R., and Wang, L. (2021). Aerogel incorporated flexible nonwoven fabric for thermal protective clothing. *Fire Safety Journal* 125: 103444.

5 Aegerter, M.A., Leventis, N., Koebel, M., and Steiner, S.A. III, (ed.) (2023). *Springer Handbook of Aerogels*. Cham: Springer International Publishing.

6 Qin, G., Liu, Y., Yan, Y. et al. (2025). Combined experiment and simulation on pore structure of graphene aerogel for microwave absorption and thermal insulation. *Composites. Part B, Engineering* 298: 112397.

7 Yu, Y., Wu, X., and Fang, J. (2015). Superhydrophobic and superoleophilic "sponge-like" aerogels for oil/water separation. *Journal of Materials Science* 50 (15): 5115–5124.

8 Amonette, J.E. and Matyáš, J. (2017). Functionalized silica aerogels for gas-phase purification, sensing, and catalysis: a review. *Microporous and Mesoporous Materials* 250: 100–119.

9 Poco, J., Satcher, J., and Hrubesh, L. (2001). Synthesis of high porosity, monolithic alumina aerogels. *Journal of Non-Crystalline Solids* 285 (1–3): 57–63.

10 Zu, G., Shen, J., Zou, L. et al. (2013). Nanoengineering super heat-resistant, strong alumina aerogels. *Chemistry of Materials* 25 (23): 4757–4764.

11 Schneider, M. and Baiker, A. (1997). Titania-based aerogels. *Catalysis Today* 35 (3): 339–365.

12 Ren, J., Zhao, Z., Kong, Y. et al. (2024). General approach to the synthesis of metal hybrid carbon/titania aerogel for the oxygen reduction reaction. *Energy & Fuels* 38 (9): 8262–8276.

13 Malinowska, B., Walendziewski, J., Robert, D. et al. (2003). Titania aerogels: preparation and photocatalytic tests. *International Journal of Photoenergy* 5 (3): 147–152.

14 Ben Hammouda, L., Mejri, I., Younes, M.K., and Ghorbel, A. (2023). Zirconia aerogels. In: *Springer Handbook of Aerogels* (ed. M.A. Aegerter, N. Leventis, M. Koebel, and S.A. Steiner III,), 459–476. Springer Nature.

15 Liu, B., Liu, X., Zhao, X. et al. (2019). High-strength, thermal-stable ZrO_2 aerogel from polyacetylacetonatozirconium. *Chemical Physics Letters* 715: 109–114.

16 Qin, Z., Jiang, Z., Zhou, L. et al. (2024). Construction and properties of ultralow thermal conductivity and high strength zirconia aerogel composites by freeze-drying. *ACS Applied Materials & Interfaces* 16 (44): 61026–61035.

17 Štengl, V., Bakardjieva, S., Maříková, M. et al. (2004). Aerogel nanoscale magnesium oxides as a destructive sorbent for toxic chemical agents. *Open Chemistry* 2 (1): 16–33.

18 Yam, B.J.Y., Le, D.K., Do, N.H. et al. (2020). Recycling of magnesium waste into magnesium hydroxide aerogels. *Journal of Environmental Chemical Engineering* 8 (5): 104101.

19 Yu, F., Wang, J., Yan, L. et al. (2022). Bio-inspired molybdenum carbide/carbon-based aerogel with advanced thermal management as a solar evaporator. *Solar Energy Materials & Solar Cells* 243: 111738.

20 Pérez-Cadenas, A.F., Maldonado-Hódar, F.J., and Moreno-Castilla, C. (2005). Molybdenum carbide formation in molybdenum-doped organic and carbon aerogels. *Langmuir* 21 (23): 10850–10855.

21 Lori, O., Zion, N., Honig, H.C., and Elbaz, L. (2021). 3D metal carbide aerogel network as a stable catalyst for the hydrogen evolution reaction. *ACS Catalysis* 11 (21): 13707–13713.

22 Chai, Y., Li, G., Ji, X., and Zhang, X. (2023). Super-white boron nitride aerogel-enabled high-energy laser irradiation protection. *Advanced Functional Materials* 33 (43): 2304355.

23 Liu, J., An, X., Yang, Y. et al. (2024). Flexible and hierarchical structured alumina boron nitride nanofibrous aerogel for thermal superinsulation in extreme conditions. *ACS Applied Nano Materials* 7 (22): 25954–25962.

24 Wang, B., Li, G., Xu, L. et al. (2020). Nanoporous boron nitride aerogel film and its smart composite with phase change materials. *ACS Nano* 14 (12): 16590–16599.

25 Anderson, M.L., Morris, C.A., Stroud, R.M. et al. (1999). Colloidal gold aerogels: preparation, properties, and characterization. *Langmuir* 15 (3): 674–681.

26 Guan, S., Xu, B., Yang, Y. et al. (2022). Gold nanowire aerogel-based biosensor for highly sensitive ethanol detection in simulated sweat. *ACS Applied Nano Materials* 5 (8): 11091–11099.

27 Wen, D., Liu, W., Haubold, D. et al. (2016). Gold aerogels: three-dimensional assembly of nanoparticles and their use as electrocatalytic interfaces. *ACS Nano* 10 (2): 2559–2567.

28 Ali, I., Chen, L., Huang, Y. et al. (2018). Humidity-responsive gold aerogel for real-time monitoring of human breath. *Langmuir* 34 (16): 4908–4913.

29 Munakata, S., Kobayashi, S., Sugime, H. et al. (2022). Ag nanoparticle-based aerogel-like films for interfacial thermal management. *ACS Applied Nano Materials* 5 (10): 15755–15761.

30 Cai, X., Wei, J., Bian, F. et al. (2024). Anisotropic silver nanowires/activated nanofiber aerogels for electromagnetic shielding. *ACS Applied Nano Materials* 7 (9): 10205–10215.

31 Peng, F., Zhu, W., Fang, Y. et al. (2023). Ultralight and highly conductive silver nanowire aerogels for high-performance electromagnetic interference shielding. *ACS Applied Materials & Interfaces* 15 (3): 4284–4293.

32 Jiang, X., Du, R., Hübner, R. et al. (2021). A roadmap for 3D metal aerogels: materials design and application attempts. *Matter* 4 (1): 54–94.

33 Leventis, N., Chandrasekaran, N., Sotiriou-Leventis, C., and Mumtaz, A. (2009). Smelting in the age of nano: iron aerogels. *Journal of Materials Chemistry* 19 (1): 63–65.

34 Long, J.W., Logan, M.S., Rhodes, C.P. et al. (2004). Nanocrystalline iron oxide aerogels as mesoporous magnetic architectures. *Journal of the American Chemical Society* 126 (51): 16879–16889.

35 Chen, Y., Zhang, C., Tao, S. et al. (2023). High-performance smart cellulose nanohybrid aerogel fibers as a platform toward multifunctional textiles. *Chemical Engineering Journal* 466: 143153.

36 Yin, S., Zhang, X., Hu, G. et al. (2022). In situ crosslinking of mechanically robust waterproof and moisture permeable cellulose diacetate nanofiber aerogels for warm clothing. *Chemical Engineering Journal* 444: 136528.

37 Su, G., Jiang, P., Guo, L. et al. (2024). Robust cellulose composite aerogels with enhanced thermal insulation and mechanical properties from cotton waste. *Industrial Crops and Products* 211: 118242.

38 Sun, Y., Chu, Y., Deng, C. et al. (2022). High-strength and superamphiphobic chitosan-based aerogels for thermal insulation and flame retardant applications. *Colloids and Surfaces A: Physicochemical and Engineering Aspects* 651: 129663.

39 Du, C., Xu, Y., Yan, C. et al. (2024). Facile construction strategy for intrinsically fire-safe and thermal-insulating bio-based chitosan aerogel. *Sustainable Materials and Technologies* 39: e00794.

40 Chen, X., Li, M., Yu, M., and Li, Y. (2024). Preparation and performance study of waterproof and breathable layer of alginate/aramid-based fabrics and flame-retardant multilayer combination. *Designed Monomers and Polymers* 27 (1): 1–14.

41 Shi, Y., Xu, Y., Xu, K. et al. (2024). Fire-resistant and thermal-insulating alginate aerogel with intelligent bionic armor for exceptional mechanical and fire early-warning performance. *Chemical Engineering Journal* 498: 155181.

42 Mulik, S. and Sotiriou-Leventis, C. (2011). Resorcinol–formaldehyde aerogels. In: *Aerogels Handbook* (ed. M.A. Aegerter, N. Leventis, and M.M. Koebel), 215–234. New York, NY: Springer New York.

43 Rahman Bhuiyan, M.A., Wang, L., Shaid, A. et al. (2019). Polyurethane-aerogel incorporated coating on cotton fabric for chemical protection. *Progress in Organic Coatings* 131: 100–110.

44 Xu, L., Liu, Y., Xin, B., and Zhou, Y. (2022). Preparation and properties of functional fabric coating based on SiO_2-aerogel/polyurethane. *Fibers and Polymers* 23 (7): 1870–1880.

45 Shan, X., Liu, L., Wu, Y. et al. (2022). Aerogel-functionalized thermoplastic polyurethane as waterproof, breathable freestanding films and coatings for passive daytime radiative cooling. *Advancement of Science* 9 (20): e2201190.

46 Bhattacharjee, S., Joshi, R., Chughtai, A.A., and Macintyre, C.R. (2019). Graphene modified multifunctional personal protective clothing. *Advanced Materials Interfaces* 6 (21): 1900622.

47 Bryning, M.B., Milkie, D.E., Islam, M.F. et al. (2007). Carbon nanotube aerogels. *Advanced Materials* 19 (5): 661–664.

48 Zou, J., Liu, J., Karakoti, A.S. et al. (2010). Ultralight multiwalled carbon nanotube aerogel. *ACS Nano* 4 (12): 7293–7302.

49 Tang, H.H., Orndoff, E.S., and Trevino, L.A. (2003). Mechanical properties and durability study of aerogel-base thermal insulation for advanced space suit. International Conference on Environmental Systems, July 2003.

50 Paul, H.L. and Diller, K.R. (2003). Comparison of thermal insulation performance of fibrous materials for the advanced space suit. *Journal of Biomechanical Engineering* 125 (5): 639–647.

51 Zhang, H., Song, G., Su, H. et al. (2017). An exploration of enhancing thermal protective clothing performance by incorporating aerogel and phase change materials. *Fire and Materials* 41 (8): 953–963.

52 Shaid, A., Furgusson, M., and Wang, L. (2014). Thermophysiological comfort analysis of aerogel nanoparticle incorporated fabric for fire fighter's protective clothing. *Chemical and Materials Engineering* 2 (2): 37–43.

53 Crowell, C., Reynolds, C., Stutts, A., and Taylor, H. (2015). Aerogel fabrics in advanced space suit applications. *Journal of Undergraduate Materials Research* 5: 1–7.

54 Yang, G., Liu, X., and Lipik, V. (2018). Evaluation of silica aerogel-reinforced polyurethane foams for footwear applications. *Journal of Materials Science* 53 (13): 9463–9472.

55 Ayten, A.İ., Taşdelen, M.A., and Ekici, B. (2020). An experimental investigation on ballistic efficiency of silica-based crosslinked aerogels in aramid fabric. *Ceramics International* 46 (17): 26724–26730.

56 McNeil, S.J. and Gupta, H. (2022). Emerging applications of aerogels in textiles. *Polymer Testing* 106: 107426.

57 Malakooti, S., Vivod, S.L., Ruggeri, C.R. et al. (2022). Ballistic performance of polyimide aerogels. 2022 SEM Annual Conference and Exposition on Experimental and Applied Mechanics, Pittsburgh, PA, USA.

58 Malakooti, S., Vivod, S.L., Pereira, M. et al. (2022). Polyimide aerogels for ballistic impact protection. *Scientific Reports* 12 (1): 13933.

59 LaRocco, J., Eom, T., Duggisani, T. et al. (2025). Ballistic testing of an aerogel/starch composite designed for use in wearable protective equipment. *Technologies* 13 (5): 199.

60 Islam, S.U., Majumdar, A., and Butola, B.S. (2023). *Advances in Healthcare and Protective Textiles*. Elsevier.

61 Zhao, Z., Li, P., Li, Y., and Wang, S. (2023). Durable thermal fluid super-repellency of elastic fluorine-modified SiO_2@sponge composite aerogel. *Chemical Engineering Journal* 454: 140247.

62 Bhuiyan, M.A.R., Wang, L., Shaid, A. et al. (2019). Advances and applications of chemical protective clothing system. *Journal of Industrial Textiles* 49 (1): 97–138.

63 Tabata, M., Adachi, I., Ishii, Y. et al. (2010). Development of transparent silica aerogel over a wide range of densities. *Nuclear Instruments and Methods in Physics Research Section A: Accelerators, Spectrometers, Detectors and Associated Equipment* 623 (1): 339–341.

64 Parale, V.G., Lee, K.-Y., Jung, H.-N.-R. et al. (2018). Facile synthesis of hydrophobic, thermally stable, and insulative organically modified silica aerogels using co-precursor method. *Ceramics International* 44 (4): 3966–3972.

65 Park, S., Hong, Y., and Choi, H. (2025) Enhanced performance of highly activated carbon and surface-treated porous polymers as physical adsorbents for chemical warfare agents.

66 Azam, F., Ahmad, F., Ulker, Z. et al. (2022). The role and applications of aerogels in textiles. *Advances in Materials Science and Engineering* 2022: 1–22.

67 Gurav, J.L., Jung, I.-K., Park, H.-H. et al. (2010). Silica aerogel: synthesis and applications. *Journal of Nanomaterials* 2010 (1): 409310.

68 Venkataraman, M., Mishra, R., Jasikova, D. et al. (2015). Thermodynamics of aerogel-treated nonwoven fabrics at subzero temperatures. *Journal of Industrial Textiles* 45 (3): 387–404.

69 Xuan, T., Wang, S., Sun, S. et al. (2023). Lightweight shear stiffening gel with enhanced mechanical and thermal-insulating performance for advanced fire suit. *Composites. Part A, Applied Science and Manufacturing* 164: 107312.

70 Wu, B., Qi, Q., Liu, L. et al. (2024). Wearable aerogels for personal thermal management and smart devices. *ACS Nano* 18 (14): 9798–9822.

71 Viggiano, R. (2017). Aerogels: overview and outlook for future space and terrestrial applications at NASA. Lecture on New Materials Technology Innovations at NASA, Erie, PA, USA (GRC-E-DAA-TN50197).

72 Koebel, M., Rigacci, A., and Achard, P. (2012). Aerogel-based thermal superinsulation: an overview. *Journal of Sol-Gel Science and Technology* 63 (3): 315–339.

73 Yu, Z., Wan, Y., Qin, Y. et al. (2023). High fire safety thermal protective composite aerogel with efficient thermal insulation and reversible fire warning performance for firefighting clothing. *Chemical Engineering Journal* 477: 147187.

74 Yang, B., Wang, L., Zhao, J. et al. (2022). A robust, flexible, hydrophobic, and multifunctional pressure sensor based on an MXene/aramid nanofiber (ANF) aerogel film. *ACS Applied Materials & Interfaces* 14 (41): 47075–47088.

75 Huang, Y., Zhou, P., and Zhang, X. (2022). Green synthesis of Ag-doped cellulose aerogel for highly sensitive, flame retardant strain sensors. *Cellulose* 29 (16): 8719–8731.

76 Bi, L., Yang, Z., Chen, L. et al. (2020). Compressible AgNWs/$Ti_3C_2T_x$ MXene aerogel-based highly sensitive piezoresistive pressure sensor as versatile electronic skins. *Journal of Materials Chemistry A* 8 (38): 20030–20036.

77 Zhu, Y., Haghniaz, R., Hartel, M.C. et al. (2023). A breathable, passive-cooling, non-inflammatory, and biodegradable aerogel electronic skin for wearable physical-electrophysiological-chemical analysis. *Advanced Materials* 35 (10): 2209300.

78 Xiao, X., Panahi-Sarmad, M., Xu, R. et al. (2022). Aerogels with shape memory ability: are they practical?—a mini-review. *European Polymer Journal* 179: 111531.

79 Wong, E.C., Lim, Y.H., Siew, M.P. et al. (2023). A review of self-healing composite films and its development in membrane for water filtration. *Journal of Water Process Engineering* 55: 104123.

80 Aegerter, M.A., Leventis, N., and Koebel, M.M. (2011). *Aerogels Handbook*. New York, NY: Springer New York.

11

Challenges and Solutions in the Use of Hydrogels and Aerogels for Textiles

Farooq Azam[1] *and Shahid Rasul*[2]

[1]*National Textile University, School of Engineering & Technology, Department of Textile Technology, Sheikhupura Road, 37610, Faisalabad, Pakistan*
[2]*Northumbria University, School of Engineering, Physics and Mathematics, Faculty of Engineering and Environment, Newcastle Upon Tyne, NE1 8ST, UK*

11.1 Introduction

Hydrogels and aerogels, due to their unique physical and chemical characteristics, have attracted much attention in the textile field. Because of the hydrophilic polymer-based network and large water-holding capacity of hydrogels, they are widely used in health-care therapies, wearable sensors, and moisture-control layers. Aerogels are ideal for tasks such as filtration, insulation, and smart fabrics due to their extreme porosity, low thermal conductivity, and lightness. These are advanced in textile engineering on both technology and function.

Trying to put into practice hydrogels and aerogels into industrial textiles is met with practical challenges, regardless of their promising performance. Such challenges encompass high production costs, environmental adherence, limited postuse disposal options, safety and biocompatibility concerns, resource-intensive energy expenditure, lack of industry familiarity, and customer awareness. Addressing these challenges will help enhance scalability and user adoption as well as long-term economic viability while ensuring ecological compliance and legal framework alignment. Fixing these challenges is important as multinational corporations transition toward circular economies and greener products.

The primary focus of this chapter is to analyze the application of hydrogel and aerogel in textiles and provide empirical solutions as well as strategies that are actionable in nature based on a thorough analysis of the challenges. The discussion is structured into six pivotal thematic parts: (1) environmental impact assessment, (2) recycling and upcycling strategies, (3) biocompatibility concerns, (4) consumer acceptance and education, and (5) economic feasibility and lifecycle analysis. Integrating the most recent hydrogels and aerogel innovations for sustainable development, each of these components together outline how such textiles can be produced, commercialized, and used throughout the textile industries.

11.2 Environmental Impact Assessment

The ecological impact of hydrogel- and aerogel-based textiles is becoming an increasingly important issue as their usage spreads into performing wear, hygiene textiles, and biomedical apparel. While hydrogels and aerogels do possess unique advantages such as high porosity, thermal performance, and light weight, the production methods associated with them do pose serious concerns of sustainability due to ecological risks such as hydrogel and aerogel textile production methods. The manufacturing steps taken regarding the energy and chemical resources constitute a problem of sustainability. The sol–gel and drying methods to produce silica precursor tetraethyl orthosilicate (TEOS)-based aerogels apply either freeze-drying or supercritical drying, both of which are costly in terms of energy and solvents. Freeze-drying, although efficient for pore structure retention, has a huge carbon footprint due to energy-intensive processes. The same applies for hydrogel synthesis, where customizing porosity requires a great deal of energy. Hydration drying might be less damaging, yet without some form of solvent manipulation, the structural integrity of the sample is compromised [1].

Aerogels and synthetic hydrogels often resist degradability and are potentially permanent resources in the environment. Aerogels, on the contrary, especially those produced from cellulose and other related biomaterials, are fully biodegradable. Typical bio-based hydrogels and aerogels are composed of renewable resources such as cellulose, chitosan (from crustacean shells), starch, pectin, and other biopolymers. Because these materials originate from plants or byproducts from other industries, these types of hydrogel and aerogels reduce reliance on fossil-fuel based feedstocks and promote sustainability [2].

The biodegradability and the end-of-life processes of hydrogels and synthetic aerogels pose yet another challenge. Many traditional formulations, due to their conventional hydrophobicity, lend themselves frugally toward environmental conservation and protection. Moreover, these materials often undergo chemical modification and thus require solvents, monomers, and cross-linkers, such as ethanol, methanol, acrylamide, acrylic acid, glutaraldehyde, and formaldehyde, which can remain and vaporize into the environment as waste. These molecules pose ecotoxicological risks such as carcinogenicity toward the aquatic and soil microorganisms and release volatile organic compounds (VOCs) that affect the atmospheric air quality. Surface modification using acrylamide and silane showed low biodegradability with considerable toxicity to marine ecosystems, with remaining cross-linkers potentially disrupting soil microbial ecosystems. Life cycle assessment (LCA), along with eco-toxicological modeling tools, evaluate the global warming potential (GWP), freshwater aquatic toxicity (FAETP), human toxicity potential (HTP), and terrestrial ecotoxicity (TETP) these materials possess and strive to minimize these environmental risks. There is neglect of information regarding the processes of degradation and the scalability of greener substitutes [3].

Many sustainable solutions have been established through research, and some have even hit the shots in light of the environmental challenges. Most notably, replacing fossil-derived precursors for biobased components such as cellulose, starch, wheat straw, and konjac glucomannan is one of the more promising routes. When analyzing biopolymer-based formulations with the conventional formulations, it is apparent that these sustainable feedstocks have considerable environmental benefits, shown to be less likely to contribute to global warming (76% decrease), nonrenewable energy necessity (92% decline), and terrestrial acidification (85% decline). These biopolymers are also naturally biodegradable and are well suited for textile applications using design-for-degradation principles.

The elimination or substitution of hazardous cross-linker and solvent usage is another important aspect of minimizing environmental impacts. There are greener cross-linkers such as citric acid, tannic acid, and enzymatic cross-linking systems that are safer alternatives to traditional cross-linking agents without compromising on functionality. In addition, there are rigorous postprocessing methods that can eliminate or neutralize excess solvent. Biomaterials can be made from substrates that are bio-based or nontoxic to satisfy ecotoxicological safety standards. Optimizing processing parameters, limiting solvent usage, and selecting lower-energy drying methods, such as normal pressure drying when feasible, also has the potential to decrease the total energy requirements to manufacture hydrogels and aerogels. Furthermore, as technology continues to progress, in the future, more circular manufacturing models may be possible through waste-to-aerogel valorization pathways or closed-loop solvent recycling systems, such as using industrial byproducts from rice husk or crustacean shells. Much like the previously stated high-performance textile functions, the environmental impacts of hydrogels and aerogels create major challenges. However, through sustainable material selection, green chemistry, innovating new processes, and understanding lifecycles, the environmental impacts can be more effectively reduced and hydrogel-/aerogel-based textile products can become more aligned with sustainable development [4, 5].

11.3 Recycling and Upcycling Strategies

The increasing use of hydrogels and aerogels for textile applications has raised serious concerns regarding their disposal and reuse. Applications for hydrogels and aerogels vary from intelligent textiles, wound healing, insulation, and environmental protection. While hydrogel- and aerogel-based materials have an incredible range of uses, they often comprise complex hybrid architectures that integrate designed gel matrices with textiles, either natural or synthetic [6]. These combinations present serious challenges to traditional recycling practices that generate environmental and economic inefficiencies. As sustainability issues become more globally relevant, new strategies to promote reuse, reprocessing, repurposing, and recycling textile products of hydrogels and aerogels will become vital. This subsection evaluates the existing challenges in recycling hybrid materials, explores a selection of recent upcycling strategies and reuse research, and discusses how upcycling and reuse studies may be incorporated with circular economy frameworks to promote consistent environmental stewardship and material circularity.

11.3.1 Recycling

Hybrid materials are likely to pose some serious recycling complications because they consist of textiles bonded with hydrogels or aerogels. They often consist of layers that are both physically and chemically bonded, such as polyester or cellulose fibers and inorganic or polymeric aerogels. The material layers are difficult to separate without destroying the functional parts of the textiles that they are composed of. The properties of the inorganic or bio-based gels intermixed with organic fibers add complexity to the interfaces, making extraction and recovery difficult. Hence, this requires higher energy or solvent use during separation or dismantling and results in lower amounts of recovered material. Taken together with high unusable plastic content that also contributes to durable properties, hydrogels and aerogels, particularly those containing networks of covalently cross-linked polymers, are often chemically stable and nonmeltable, making them resistant to mechanical recycling, thermal recycling, and overall recycling. Due to the absence of standardized processes for disassembling hybrid materials, downcycling and incineration are common, resulting in negative effects on the ecosystem and the economy. The advanced character of many aerogels, prioritizing mechanical strength and stability ahead of recyclability, makes these challenges even worse and frequently results in dumping them in landfills after a single-use cycle. Because industrial models frequently lack effective procedures for these hybrid materials, conventional mechanical recycling is insufficient, increasing the risk of waste disposal or downcycling [7–10].

A major step toward the adoption of the circular economy for plastics has been taken with the effective upcycling of low-cost polystyrene (PS) trash into high-performance, closed-loop recyclable aerogels, as evidenced by recently reported science research studies. This novel method first aminates conventional PS, which is usually regarded as nonrecyclable, by a straightforward postfunctionalization phase to create a polyfunctional amine. Poly(styrene-imine) aerogels (PSAs) are produced when this amine combines with aldehydes to form a cross-linked imine network that is finally dried. In addition to their exceptional mechanical strength and thermal stability, these PSAs have high porosity (>86%), low bulk density ($<0.19\,g/cm^3$), hydrophobicity (water contact angle ≈135°), and excellent thermal insulation qualities ($\lambda = 18.0$–$20.1\,mW/m{\cdot}K$). Critically, the incorporation of dynamic imine linkages allows the aerogels to undergo quantitative depolymerization under mild acidic aqueous conditions, enabling recovery of the original monomers in high yield and purity. These recovered precursors can be reused to synthesize new aerogels with nearly identical performance, establishing a true closed-loop recycling pathway. This method not only enhances the sustainability of polymeric materials but also adds significant value to waste PS by transforming it into durable, advanced aerogels suitable for thermal insulation and other demanding applications [11].

11.3.2 Upcycling

Upcycling strategies for aerogel- and hydrogel-based materials focus on reprocessing production waste or postconsumer textiles into functional materials with added value. In contrast to recycling, upcycling frequently focusses on transforming materials or repurposing rather than complete material breakdown. For example, ionic liquids can be used to dissolve cellulose-rich textile waste (such as cotton and denim) and generate aerogels with high porosity and absorbency that can be used for filtration or insulation. Similarly, hydrogel powders made from industrial waste can be cross-linked enzymatically or with safe substances such as citric acid to be used again as textile coatings or moisture-absorbing layers [12–14].

Recent developments have shown that without the need for the extraction or chemical manipulation of biopolymers, tissues made from nature and agricultural waste, such as fruit pulp, vegetable residues, and mushrooms, can be directly transformed into nanostructured aerogels by taking advantage of their natural mesoporosity. Aerogels having ultralow densities (0.01–0.12 g/cm^3), substantial specific surface areas (108–446 m^2/g), and exceptional oil sorption characteristics are produced using this environmentally friendly method, which requires little processing beyond water washing, ethanol solvent exchange, and supercritical CO_2 drying. In addition to having the same textural characteristics as traditional cellulose or pectin aerogels, these bio-derived aerogels can be used practically to structure oleogels for either food or cosmetic uses [15].

In textile recycling, cotton and regenerated cellulose fabrics have been successfully dissolved in ionic liquids mixed with DMSO, followed by coagulation and supercritical or freeze-drying to yield lightweight, porous aerogels. The morphology and porosity of these aerogels can be finely tuned by adjusting the solvent system and drying method, demonstrating the impact of phase separation mechanisms on final material structure. Furthermore, the prilling (jet-vibration) technique has enabled the upcycling of viscose textiles into highly spherical cellulose aerogel beads, with excellent surface area (~400 m^2/g), low density (~0.08–0.12 g/cm^3), and verified biocompatibility. These processes highlight the feasibility of converting textile waste into value-added aerogel materials suitable for filtration, insulation, biomedical, or cosmetic applications—while aligning with circular economy goals by diverting waste from landfills and reducing dependency on virgin resources. These techniques demonstrate that production and postconsumer waste can be reintegrated into the value chain with minimal energy input and without compromising performance [16].

Several promising case studies illustrate real-world recycling and reuse of aerogel and hydrogel materials. In industrial insulation, waste PS-based aerogels have been successfully reprocessed into new insulation-grade aerogel panels. These materials maintain their structural and thermal properties after recycling, offering a closed-loop pathway. In environmental remediation, monolithic carbon aerogels doped with metals (e.g. Fe, Au, Pt, and Cu) have demonstrated excellent reusability as catalysts—capable of maintaining more than 70% of their initial activity after five cycles without the need for reprocessing. Similarly, covalent organic frameworks (COF) aerogels and alginate-based X-aerogels have shown high sorption capacities and stable performance after multiple reuse cycles, particularly in removing organic pollutants and heavy metals from aqueous environments. Hydrogel powders have also been reused in textile applications through simple regeneration techniques such as pH adjustment, solvent washing, or humidity modulation, enabling up to 10 reuse cycles for some desiccant-type hydrogel aerogels [5, 17–21].

In addition to wasting precious resources, conventional disposal techniques for hydrogel and aerogel products, such as waste disposal and burning, also run counter to sustainability objectives. The linear "produce–use–discard" paradigm must give way to one focused on reuse, reprocessing, and molecular recovery to include these materials into circular economy models. Recent advancements in "design-for-recycling" aerogels provide dynamic, bidirectional covalent linkages that enable materials to regenerate into monomers or oligomers after depolymerizing in mild environments. Then, by reusing these recovered building blocks, aerogels with comparable qualities can be created. Another important strategy is to create aerogel nanostructures that can be put back together after disassembly using noncovalent interactions (such as hydrogen bonds and electrostatic forces). Such recyclable aerogel technologies, while still in their infancy, have the potential to be included into closed-loop textile production, promoting resource conservation and lowering reliance on petrochemical feedstocks. According to the fundamentals of the circular economy, these techniques promote material innovation and responsibility for the environment by including recyclability into the design phase [22–24].

11.4 Biocompatibility Concerns in the Use of Hydrogels and Aerogels for Textiles

Thorough biocompatibility testing is necessary for integrating hydrogels and aerogels into textile materials, particularly for wearable, biomedical, or skin contact applications. Given the intimate contact with skin, potential dangers include exposure to hazardous chemicals, allergic responses, and cell toxicity. Assessing these factors is essential for determining whether these materials can be used in medical, cosmetic, or advanced textile products.

Hydrogels and aerogels are often used in wound dressings, wearable sensors, drug delivery patches, and thermal regulation textiles. Their porous structure, high water content (in hydrogels), and surface functionality enhance comfort and performance. However, their intimate and prolonged contact with the skin necessitates careful scrutiny. Any leachable component, such as unreacted monomers, residual solvents, or additives, can cause irritation or systemic toxicity. Studies have demonstrated that natural polymer-based hydrogels (e.g. alginate, chitosan, and gelatin) exhibit excellent skin compatibility due to their biodegradability and bioresorbability. Conversely, synthetic hydrogels (such as polyacrylamide or PVA) may require additional purification steps to remove toxic residues. Similarly, silica-based aerogels, though structurally beneficial, can shed fine particles that may pose inhalation hazards during manufacturing or prolonged wear [25–27].

Biocompatibility testing typically involves in vitro cytotoxicity assays (such as MTT or Alamar Blue tests), in vivo dermal irritation assessments on animal models or human volunteers, and allergenicity screening. For instance, in a study by Janvikul et al. [28], chitosan-based hydrogel dressings showed no cytotoxic response on L929 fibroblasts for in vitro biocompatibility. Dermal irritation tests conducted on humans using cellulose aerogel pads showed no irritation or sensitization, confirming the material's dermatological safety [29].

Chemical cross-linkers such as glutaraldehyde, epichlorohydrin, or carbodiimide can significantly affect the biocompatibility of hydrogels. Glutaraldehyde, although effective for cross-linking, is known for its cytotoxicity and must be either avoided or used with thorough posttreatment to remove residues [30]. Enzymatic or physical cross-linking (e.g. using genipin or UV radiation) is gaining popularity for safer hydrogel systems [31]. Nanoparticles incorporated into hydrogels and aerogels (such as ZnO, Ag, or TiO_2) for antimicrobial or functional enhancements also raise concerns. While many nanoparticles exhibit antimicrobial activity, their concentration, release rate, and interaction with skin cells must be optimized. For instance, silver nanoparticles at high concentrations have been shown to cause oxidative stress and genotoxicity in skin cells [32]. Therefore, dosage control and encapsulation techniques are essential to balance efficacy and safety.

REACH (Registration, Evaluation, Authorization, and Restriction of Chemicals), a European Union regulation, ensures substances used in materials do not pose health or environmental risks. Any hydrogel- or aerogel-based textile product must meet REACH compliance, especially regarding restricted substances such as formaldehyde or heavy metals. The certification ensures that the textile and its components are free from harmful levels of substances that could affect human health. This standard is particularly relevant for baby wear and sensitive-skin applications.

11.5 Consumer Acceptance and Education

In addition to technical capabilities, customer perception, education, and engagement are critical to the successful commercialization of textiles based on hydrogel and aerogel. Even with the quick development of sophisticated gel-based materials with established properties (such as moisture control, antibacterial activity, and heat regulation), public awareness and confidence are still low. The main obstacles and solutions for closing the gap between scientific knowledge and consumer acceptability are examined in this section.

The discrepancy between effectiveness, as perceived by consumers, and performance, as assessed by science, is one of the main issues. When it comes to properties such as breathability, water retention, and antimicrobial protection, hydrogels and aerogels frequently perform better than traditional materials. However, due to a lack of tactile familiarity or visual clues, end consumers could find it difficult to recognize these benefits. For instance, research indicates that customers value ease of care, softness, and skin feel more than scientifically proven qualities such as controlled release or heat resistance [33]. Even when smart textiles are shown to properly control body temperature, people might not notice an improvement in comfort unless the effect is quick and obvious [34]. To verify performance impressions, developers should incorporate sensory design aspects (texture, stretchability, and color signals) and provide customer trials or comparison demonstrations.

Because of their connotations with wetness, stickiness, or fragility, gel-based materials in wearing textiles may cause consumer skepticism, particularly in countries where hydrogel or aerogel technologies are not widely used. In a study by Jeong et al. [35], concerns were voiced over the washability and longevity of smart clothing impregnated with hydrogel. The belief might be further undermined by worries about toxicity or skin irritation, particularly in delicate applications such as wound healing or baby wear. Resistance can be lessened by open communication supported by biocompatibility certifications (ISO 10993, OEKO-TEX) and visual evidence (e.g. before-and-after wear tests, washing endurance videos). To lessen opposition to innovation, behavioral science also suggests using social proof techniques such as evaluations or influencer endorsements [36].

Despite the sustainability advantages offered by hydrogels and aerogels (such as reduced weight transportation and biodegradable formulations), customers frequently do not know about those advantages until they are made explicit. For example, consumer involvement in sustainable fashion behavior is restricted by unclear care labels and a lack of sustainability narratives. In straightforward language, brands should emphasize durable attributes, maintenance recommendations, and eco-benefits. This involves the use of care films, infographics, and quick response (QR)-linked instructions for material procurement, end-of-life disposal, and cleaning. Augmented reality (AR) applications, NFC tags, and QR codes are examples of smart labeling technologies that provide a creative means of educating customers at the point of usage. Product care instructions, sustainability reports, and video demonstrations may all be accessed using QR codes [37]. Using multimedia storytelling tools increases transparency and trust by allowing customers to interact with the product during its entire life cycle [38].

11.6 Economic Feasibility and Life Cycle Analysis

Although hydrogel- and aerogel-based textiles provide potential functional advantages, including antimicrobial protection, controlled moisture management, and thermal insulation, their cost viability has a big impact on their uptake. The cost structures, life cycle evaluations, and systemic issues that manufacturers—especially small-to-medium-sized enterprises (SMEs)—face are examined in this section, along with potential sources of financial assistance.

Complex manufacturing procedures, frequently including several steps such as polymerization, cross-linking, aging, solvent exchange, and drying (supercritical or freeze-drying), are required to produce hydrogels and aerogels. Both operating time and energy usage rise during these phases. For instance, the production of silica aerogels is expensive, mainly because of the supercritical drying step, even if they are lightweight and thermally efficient. According to reports, the exorbitant expense of producing silica aerogel renders it inappropriate for use in bulk textile applications without modification [39]. Similar to this, natural polymer hydrogels (such as chitosan, alginate, and xanthan gum) are more economical to get raw materials, but their high water content and drying energy still limit their scalability [8]. To cut costs, scientists are looking into bio-based substitutes (such as cellulose-based aerogels) and atmospheric pressure drying. To effectively scale up hydrogel manufacturing, continuous manufacturing is also being explored as an alternative to batch processing [40].

For hydrogel- and aerogel-integrated textiles to be more affordable than traditional textiles, they must exhibit value-added functionality. Despite the greater material cost, hydrogel-based wound dressings were found to provide quicker healing and fewer problems, resulting in total savings in medical expenditures [41, 42]. Aerogel hybrids perform better than polyester filling in textile insulation in terms of thermal resistance-to-weight ratios, which results in less material used per product and longer-term energy savings. However, the initial material cost remains a bottleneck for mass market adoption. Adoption can be promoted by targeting premium applications (e.g. technical sportswear, medical textiles, and protective clothing), where performance justifies premium pricing.

The life cycle cost (LCC) analysis of hydrogel- and aerogel-integrated textiles evaluates the total economic impact from raw material extraction to end-of-life disposal. Given the high energy demands and solvent-intensive processes involved in aerogel production, LCC must account for raw material costs—precursors such as tetraethyl orthosilicate (TEOS) and sodium silicate significantly influence costs. While sodium silicate is cheaper and more eco-friendly than TEOS, its high-temperature production (~1300–1600 °C) increases energy expenses. Biomass-derived precursors (e.g. cellulose, starch, and algae) offer cost advantages but may require additional processing. In drying methods, supercritical CO_2 drying, though efficient, is energy intensive. Freeze-drying, while less solvent dependent, consumes substantial electricity. Ambient drying is cost-effective but may compromise aerogel properties. Similarly, solvent recovery and recycling, especially ethanol and methanol usage in aging/solvent exchange stages, contributes to costs. Implementing solvent recovery systems (e.g. rotary evaporators) can reduce expenses by 30–35%. Industrial-scale production reduces costs by ~40–90% compared with laboratory-scale production due to optimized energy use and bulk material procurement. Disposal costs vary: silica aerogels can be landfilled safely, while biodegradable aerogels (e.g. lignin-gelatin) decompose naturally, reducing waste management expenses [5].

The commercialization of hydrogel and aerogel technologies in textiles faces significant challenges for small-to-medium enterprises (SMEs) and startups. One of the most prominent barriers is the high initial capital cost associated with advanced processing equipment. For instance, supercritical drying systems and solvent recovery units—essential for aerogel synthesis–require substantial investment, often beyond the reach of small firms. Moreover, the production process, particularly for aerogels, involves complex steps such as gelation, aging, and controlled drying, which demand specialized technical expertise. Startups and early-stage companies often lack dedicated R&D facilities or trained personnel, making it difficult to optimize these processes efficiently.

Numerous governmental and private financing channels have been established to encourage innovation and commercialization to overcome these structural and economic obstacles. Among the most effective forms of assistance are subsidies and financial assistance from the government. The expenses of capital investment, pilot production, and regulatory certification are partially covered by these programs. Early-stage development can also be facilitated by public–private partnerships. International development organizations and green innovation funds are becoming more interested in projects that use closed-loop solvent recycling or convert industrial biowaste into aerogels because they support these objectives.

11.7 Conclusion

The use of hydrogels and aerogels in the textiles could shake the industry by enabling lightweight, high-performance, and multifunctional textiles. However, to achieve that success, several overlapping issues need to be addressed. Their environmental issues remain a serious concern due to energy-consuming processing methods, the use of toxic agents, and poor biodegradability of traditional formulations. There are issues with circularity as recycling and upcycling remains limited, particularly, within complicated hybrid material systems. Consumer acceptance requires improved communication of both performance and sustainability in operational textiles. Biocompatibility must be fully assessed to provide adequate assurance of safety for devices in contact with skin. Economic difficulties stemming from high processing expenses and limited levels of scalability hinder commercial appeal to the majority around the world, especially in developing countries and SMEs.

Solutions to tackle these issues is through using natural and renewable bio-based alternative raw materials, implementing green cross-linking methods, recovering solvents and energy sources, using dynamic covalent chemistry that permits recyclability, and integrating circular economies through waste valorization. This chapter introduced several areas that support these solutions, as well as facilitated the multifaceted shift to sustainable hydrogel and aerogel textiles through improvements in funding sources, consumer educational resources, lifespan assessments, and collaboration between industry and academia. At a systems level, to progress forward, a balance must be established between materials innovation, sustainability, economics, safety, and the end user for textiles based on hydrogel and aerogel. By aligning academic, research, government policy contexts, and industrialized bases, the textiles industry might change what a new generation of ecofunctional materials can look like for a new sustainable future.

References

1 Vrabič-Brodnjak, U. (2023). Hybrid materials of bio-based aerogels for sustainable packaging solutions. *Gels* 10 (1): 27.

2 Aldaghi, S.A., Costamagna, M., Perucca, M. et al. (2024). Environmental impact of polyurethane-based aerogel production: influence of solvents and solids content. *Resources* 13 (10): 138.

3 Agbna, G.H.D. and Zaidi, S.J. (2025). Hydrogel performance in boosting plant resilience to water stress – a review. *Gels* 11 (4): 276.

4 Karatum, O., Bhuiya, M.M.H., Carroll, M.K. et al. (2018). Life cycle assessment of aerogel manufacture on small and large scales: weighing the use of advanced materials in oil spill remediation. *Journal of Industrial Ecology* 22 (6): 1365–1377.

5 Turhan Kara, I., Kiyak, B., Colak Gunes, N., and Yucel, S. (2024). Life cycle assessment of aerogels: a critical review. *Journal of Sol-Gel Science and Technology* 111 (2): 618–649.

6 Nguyen, P.T.T., Do, N.H.N., Goh, X.Y. et al. (2022). Recent progresses in eco-friendly fabrication and applications of sustainable aerogels from various waste materials. *Waste and Biomass Valorization* 13 (4): 1825–1847.

7 Dong, C., Hu, Y., Zhu, Y. et al. (2022). Fabrication of textile waste fibers aerogels with excellent oil/organic solvent adsorption and thermal properties. *Gels* 8 (10): 684.

8 Ahmed, E.M. (2015). Hydrogel: preparation, characterization, and applications: a review. *Journal of Advanced Research* 6 (2): 105–121.

9 Huang, L.-J., Lee, W.-J., and Chen, Y.-C. (2022). Bio-based hydrogel and aerogel composites prepared by combining cellulose solutions and waterborne polyurethane. *Polymers* 14 (1): 204.

10 García-González, C.A., Blanco-Vales, M., Barros, J. et al. (2025). Review and perspectives on the sustainability of organic aerogels. *ACS Sustainable Chemistry & Engineering* 13 (18): 6469–6492.

11 Xiong, Y., Wang, C., Dong, L., and Tomović, Ž. (2025). Closed-loop recyclable high-performance aerogels derived from polystyrene. *Advanced Functional Materials* 35 (19): 2420472.

12 Gan, L., Zhang, D., Yue, X. et al. (2022). A recyclable and regenerated aerogel membrane derived from waste plastic for emulsion separation. *Journal of Environmental Chemical Engineering* 10 (5): 108221.

13 Li, T.-T., Wang, Z., Ren, H.-T. et al. (2021). Recyclable and degradable nonwoven-based double-network composite hydrogel adsorbent for efficient removal of Pb(II) and Ni(II) from aqueous solution. *Science of the Total Environment* 758: 143640.

14 Borzova, M., Gauvin, F., and Schollbach, K. (2025). Upcycling waste mineral wool into ambient pressure-dried silica aerogels. *ACS Sustainable Chemistry & Engineering* 13 (7): 2955–2965.

15 Gibowsky, L., De Berardinis, L., Plazzotta, S. et al. (2025). Conversion of natural tissues and food waste into aerogels and their application in oleogelation. *Green Chemistry* 27 (17): 4713–4731.

16 Négrier, M., El Ahmar, E., Sescousse, R. et al. (2023). Upcycling of textile waste into high added value cellulose porous materials, aerogels and cryogels. *RSC Sustainability* 1 (2): 335–345.

17 Jha, M.K., Das, P.P., Gupta, S. et al. (2023). Complementing the circular economy with a life cycle assessment of sustainable hydrogels and their future prospects. In: *Sustainable Hydrogels* (ed. S. Thomas, B. Sharma, P. Jain, and S. Shekhar), 489–503. Elsevier.

18 Feng, S., Yao, L., Feng, M. et al. (2024). Sustainable regeneration of waste polystyrene foam as cooling coating: building cooling energy saving, CO_2 emission reduction and cost–benefit prospective. *Journal of Cleaner Production* 434: 140361.

19 Merli, F. and Buratti, C. (2024). Properties and energy performance of wood waste sustainable panels resulting from the fabrication of innovative monolithic aerogel glazing systems. *Construction and Building Materials* 438: 137310.

20 Lv, Q., Zhu, X., Zhou, T. et al. (2024). Multifunctional and recyclable aerogel/fiber building insulation composites with sandwich structure. *Construction and Building Materials* 423: 135902.

21 Dharmapriya, T.N., Lee, D.-Y., and Huang, P.-J. (2021). Novel reusable hydrogel adsorbents for precious metal recycle. *Scientific Reports* 11 (1): 19577.

22 Pathan, S. and Bose, S. (2018). Arsenic removal using "green" renewable feedstock-based hydrogels: current and future perspectives. *ACS Omega* 3 (5): 5910–5917.

23 Zhao, T., Zhang, T., Xu, Z., and Zhao, Y. (2022). Emulsion-based, flexible and recyclable aerogel composites for latent heat storage. *Journal of Colloid and Interface Science* 627: 72–80.

24 Wang, C., Eisenreich, F., and Tomović, Ž. (2024). Aerogel-to-sol-to-aerogel (ASA) process for recycling, repairing, reprogramming of high-performance organic aerogels. *Advanced Functional Materials* 34 (28): 2314447.

25 Stergar, J. and Maver, U. (2016). Review of aerogel-based materials in biomedical applications. *Journal of Sol-Gel Science and Technology* 77 (3): 738–752.

26 Kopecek, J. (2009). Hydrogels: from soft contact lenses and implants to self-assembled nanomaterials. *J. Polym. Sci., Part A:Polym. Chem.* 47 (22): 5929–5946.

27 Boateng, J.S., Matthews, K.H., Stevens, H.N.E., and Eccleston, G.M. (2008). Wound healing dressings and drug delivery systems: a review. *Journal of Pharmaceutical Sciences* 97 (8): 2892–2923.

28 Janvikul, W., Uppanan, P., Thavornyutikarn, B. et al. (2007). Fibroblast interaction with carboxymethylchitosan-based hydrogels. *Journal of Materials Science. Materials in Medicine* 18 (5): 943–949.

29 Torres-García, R., Flores-Estrada, J., Cauich-Rodríguez, J.V. et al. (2022). Design of a polyacrylamide and gelatin hydrogel as a synthetic extracellular matrix. *International Journal of Polymeric Materials and Polymeric Biomaterials* 71 (4): 266–277.

30 Ulubayram, K., Aksu, E., Gurhan, S.I.D. et al. (2002). Cytotoxicity evaluation of gelatin sponges prepared with different cross-linking agents. *Journal of Biomaterials Science. Polymer Edition* 13 (11): 1203–1219.

31 Li, Z., Lu, F., and Liu, Y. (2023). A Review of the mechanism, properties, and applications of hydrogels prepared by enzymatic cross-linking. *Journal of Agricultural and Food Chemistry* 71 (27): 10238–10249.

32 Ahamed, M., AlSalhi, M.S., and Siddiqui, M.K.J. (2010). Silver nanoparticle applications and human health. *Clinica Chimica Acta* 411 (23–24): 1841–1848.

33 Laitala, K., Klepp, I., and Henry, B. (2018). Does use matter? Comparison of environmental impacts of clothing based on fiber type. *Sustainability* 10 (7): 2524.

34 Tang, K.H.D. (2025). Advances in thermoregulating textiles: materials, mechanisms, and applications. *Textiles* 5 (2): 22.

35 Jeong, J., Kim, Y., and Roh, T. (2021). Do consumers care about aesthetics and compatibility? The intention to use wearable devices in health care. *SAGE Open* 11 (3): 21582440211.

36 Claudy, M.C., Garcia, R., and O'Driscoll, A. (2015). Consumer resistance to innovation – a behavioral reasoning perspective. *Journal of the Academy of Marketing Science* 43 (4): 528–544.

37 Niinimäki, K., Peters, G., Dahlbo, H. et al. (2020). The environmental price of fast fashion. *Nature Reviews Earth and Environment* 1 (4): 189–200.

38 Bassano, C., Barile, S., Piciocchi, P. et al. (2019). Storytelling about places: tourism marketing in the digital age. *Cities* 87: 10–20.

39 Garrido, R., Silvestre, J.D., Flores-Colen, I. et al. (2019). Economic assessment of the production of subcritically dried silica-based aerogels. *Journal of Non-Crystalline Solids* 516: 26–34.

40 Koebel, M.M., Huber, L., Zhao, S., and Malfait, W.J. (2016). Breakthroughs in cost-effective, scalable production of superinsulating, ambient-dried silica aerogel and silica-biopolymer hybrid aerogels: from laboratory to pilot scale. *Journal of Sol-Gel Science and Technology* 79 (2): 308–318.

41 Kaspar, D., Linder, J., Zöllner, P. et al. (2015). Economic benefit of a polyacrylate-based hydrogel compared to an amorphous hydrogel in wound bed preparation of venous leg ulcers. *Chronic Wound Care Management and Research* 2: 63–70.

42 Sood, A., Granick, M.S., and Tomaselli, N.L. (2014). Wound dressings and comparative effectiveness data. *Advances in Wound Care* 3 (8): 511–529.

12

Future Horizon

Bushra Mushtaq and Sheraz Ahmad

National Textile University, School of Engineering & Technology, Department of Textile Technology, Sheikhupura Road 37610, Faisalabad, Pakistan

12.1 Sustainable Hydrogel and Aerogel Textile Innovations

Hydrogels and aerogels are the emerging frontiers in functional textile innovation, especially in the context of sustainability, biomedical, and intelligent material systems. Their novel physicochemical characteristics, high porosity, low density, water sensitivity, and biocompatibility positioned them as a key enabler of future-oriented next-generation textiles capable of addressing urgent environmental and technological needs [1, 2].

Future technologies are becoming more multifunctional and adaptive. Textiles using hydrogel technology are being developed to dynamically modulate humidity and temperature, adapting to alterations in body activity or environmental conditions [3]. At the same time, textiles incorporating aerogels provide possibilities for high-performance thermal insulation, fire resistance, and mechanical robustness, capabilities of specific interest in aerospace, defense, and high-performance athletic apparel [4, 5].

Another notable direction of current research is the synthesis of smart hydrogel and aerogel composites that can be combined with biosensors, drug delivery units, and energy-harvesting devices [6, 7]. Such materials are the center of wearable electronics and health monitoring systems, where the textiles are not merely passive coverings but active interfaces that engage with biological, mechanical, or environmental signals.

Technological advances in 3D and 4D printing have also enabled the development of tailored hydrogel/aerogel architectures with the ability to exert accurate control over material function, structure, and responsiveness [8]. These are fast-tracking the transition toward textile systems for human health and sustainable development.

12.2 Collaborations and Interdisciplinary Research

To unlock the full potential of hydrogel and aerogel fabrics, a strong interdisciplinary platform is critical. Future development relies on intense synergy among:

- Polymer chemists who can engineer molecular architecture optimized for responsiveness or degradation patterns.
- Textile engineers and materials scientists, who can enable fiber-level integration and production scaling.
- Medical and biological scientists, who confirm biocompatibility and therapeutic efficacy in medical environments.
- Environmental scientists who evaluate life cycle sustainability, biodegradability, and end-of-life effects.
- Business developers and entrepreneurs who convert laboratory innovation into marketable products.

These collaborations can be encouraged through research institutes, cross-industry innovation centers, and public–private collaborations. International R&D programs such as the European Union's Horizon Europe, the National Science Foundation (NSF) Smart and Connected Health program, or the UK Research and Innovation (UKRI) Sustainable Textiles program offer funding and structural support for such interdisciplinarity.

Besides, open-source materials databases, artificial intelligence–enabled materials discovery platforms, and computational modeling software provide researchers and producers with a common platform for exploring new formulations, forecasting performance parameters, and fine-tuning material combinations across a wide range of textile applications [9].

12.3 Potential Market Trends

The global textile industry is poised for significant transformation, and hydrogel and aerogel technologies are expected to drive innovation across several market sectors:

- **Sustainable Apparel and Eco-Fashion**
  Increasing consumer consciousness and legislative pressures have driven demand for biodegradable and circular materials. Hydrogel-treated moisture management fabrics and aerogel-insulated lightweight insulating fabric are poised for increased adoption in this market [10].
- **Medical and Health-Care Textiles**
  Drug-delivery, vital sign monitoring, and wound healing accelerating hydrogel textiles are gaining considerable interest in therapeutic and diagnostic applications. The smart textile market for health care is estimated to reach more than USD 3 billion by 2026 [11].

- **Extreme Environment and Protective Gear**
  With rising climate extremes, defense and industrial industries demand materials that provide better thermal regulation, flame resistance, and impact absorption. Aerogel textiles fulfill those demands while being ultralight and permeable [12]. Hydrogel and aerogel composites allow fabrics to move, temperature or perspiration, responding to needs for intelligent, ergonomic sportswear. Embedded wearable sensors in hydrogel matrices also aid in real-time performance analytics [13].
- **Circular and Regenerative Product Systems**
  As the market moves toward circular economy patterns, hydrogel and aerogel fabrics are being designed to be recyclable, compostable, nontoxic, and degradable according to emerging eco-labeling and Extended Producer Responsibility (EPR) policies [14].

12.4 Regulations and Safety Standards

To enable responsible innovation and consumer confidence, regulatory standards need to adapt to meet the special challenges and advantages of hydrogel and aerogel textiles. Key areas are as follows:

- Safety testing for skin and inhalation contact, especially for nanomaterial-enhanced aerogels.
- Testing protocols and standards for biodegradability and degradation in the environment, including marine and landfill conditions.
- Regulations on medical-grade hydrogel applications, especially in wound dressings and wearable diagnostics.
- Eco-certifications like GOTS, OEKO-TEX®, or Cradle to Cradle, which confirm the nontoxic and sustainable character of these high-tech materials.
- Life cycle analysis systems that combine upstream production, end use by the consumer, and downstream recycling or end-of-life disposal to provide a complete impact assessment.

In addition, industrial groupings and global organizations need to agree on terminology, testing, and reporting specifications to facilitate product development, conformity, and international trade of functional smart textiles.

12.5 Conclusion

The coming decade will see the coming together of biological function, intelligent responsiveness, and green engineering in textile technology, facilitated largely through hydrogels and aerogels. These provide a platform for engaging with global issues such as climate resilience, health care, waste in materials, and consumer trends for smarter and greener products. To realize their full potential, stakeholders should prioritize collaborative research, innovation of the supply chain, and ethical commercialization. The aerogel textile and hydrogel revolution is not only technologically viable but also ecologically required and economically sound. Through the development of a collaborative environment that connects academia, industry, and government, hydrogel and aerogel textiles will further develop as essential elements of a sustainable, adaptive, and intelligent textile future.

References

1 Caló, E. and Khutoryanskiy, V.V. (2015). Biomedical applications of hydrogels: a review of patents and commercial products. *European Polymer Journal* 65: 252–267.

2 Maleki, H., Durães, L., García-González, C.A. et al. (2016). Synthesis and biomedical applications of aerogels: possibilities and challenges. *Advances in Colloid and Interface Science* 236: 1–27.

3 Fang, Y., Chen, G., Bick, M., and Chen, J.J.C.S.R. (2021). Smart textiles for personalized thermoregulation. *Chemical Society Reviews* 50 (17): 9357–9374.

4 Hsieh, J.-C., Li, Y., Wang, H. et al. (2022). Design of hydrogel-based wearable EEG electrodes for medical applications. *Journal of Materials Chemistry B* 10 (37): 7260–7280.

5 Meador, M.A.B. and Nguyen, B. (2023). Improving elastic properties of polymer-reinforced aerogels. In: *Springer Handbook of Aerogels* (ed. M.A. Aegerter, N. Leventis, M. Koebel, and S.A. Steiner III,), 791–808. Cham: Springer.

6 Thirumalai, D., Santhamoorthy, M., Kim, S.-C., and Lim, H.-R.J.G. (2024). Conductive polymer-based hydrogels for wearable electrochemical biosensors. *Gels* 10 (7): 459.

7 Chenani, H., Saeidi, M., Rastkhiz, M.A. et al. (2024). Challenges and advances of hydrogel-based wearable electrochemical biosensors for real-time monitoring of biofluids: from lab to market. A review. *Analytical Chemistry* 96 (20): 8160–8183.

8 Roh, Y., Lee, Y., Lim, D. et al. (2024). Nature's blueprint in bioinspired materials for robotics. *Advanced Functional Materials* 34 (35): 2306079.

9 Raccuglia, P., Elbert, K.C., Adler, P.D. et al. (2016). Machine-learning-assisted materials discovery using failed experiments. *Nature* 533 (7601): 73–76.

10 Sajovic, I., Kert, M., and Boh Podgornik, B. (2023). Smart textiles: a review and bibliometric mapping. *Applied Sciences* 13 (18): 10489.

11 Ruckdashel, R.R., Khadse, N., and Park, J.H.J.S. (2022). Smart E-textiles: overview of components and outlook. *Sensors MDPI (JEAR)* 22 (16).

12 Wu, B., Qi, Q., Liu, L. et al. (2024). Wearable aerogels for personal thermal management and smart devices. *ACS Nano* 18 (14): 9798–9822.

13 Dutta, T., Chaturvedi, P., Llamas-Garro, I. et al. (2024). Smart materials for flexible electronics and devices: hydrogel. *RSC Advances* 14 (19): 12984–13004.

14 Silander, D. (2019). The European Commission and Europe 2020: smart, sustainable and inclusive growth. In: *Smart, Sustainable and Inclusive Growth* (ed. C. Karlsson, D. Silander, and D. Henning), 2–35. Edward Elgar Publishing.

13

Conclusion

Yasir Nawab[1,2] *and Sheraz Ahmad*[1]

[1] *National Textile University, School of Engineering & Technology, Department of Textile Technology, Sheikhupura Road 37610, Faisalabad, Pakistan*
[2] *University of Kamalia, QAED Academy, Rajana Rd Kamalia, Pakistan*

The exploration in the above chapters has been a detailed account of how hydrogels and aerogels have revolutionized the world of textiles. These porous materials have found their way not only as a new kind of additive but in fact as the basis of a new paradigm in the field of textile science and technology, from their basic synthesis to their use in the latest generation of smart and protective clothing. This overall conclusion is a synthesis of the main themes, reconciliation of the challenges, and a look into the future that these materials are about to bring. The distinct and complementary properties of hydrogels and aerogels respond to the most urgent needs of the contemporary world, namely, the necessity of sustainability, the desire to achieve high performance, and the need for smart functionality. The usage of hydrogels and aerogels in textiles is a paradigm shift in material science as the textile industry is leaving conventional fabrics and heading toward multifunctional systems that react intelligently to environmental and physiological requirements. The fact that they are incorporated into the textiles, be it a natural, synthetic, or blended fiber system, offers a flexible platform that is not bound to the restrictions of traditional weaving, knitting, or nonwoven processes. Aerogel fabrics are the lightest fabrics with the best insulating value of any material in the world and are defining protective clothing and extreme environment apparel. The hydrogel textiles, with their active moisture balance and responsive behavior to stimuli, open a new age of active comfort and innovative biomedical applications, including smart wound dressings and wearable health monitors.

One of the strongest and repeated motifs is the complementarity of these two kinds of materials. The future is not in using either material or the other, but the art of skillfully blending them. The potential of hybrid systems offers the possibility of truly multifunctional and adaptive textiles: the moisture responsiveness of a hydrogel is combined with the thermal insulation and flame-retardant qualities of an aerogel seamlessly. Such fabrics were capable of smartly controlling the microclimate of a wearer as the physiological and environmental parameters changed. Although the opportunities of hydrogel and aerogel fabrics are considerable, their way to the mass market is not easy. The major barriers are scalability, brittleness, the cost of production, recyclability, and environmental impact. Their application is also complicated by technical problems, including the brittle nature of aerogels and fluid leakage in hydrogels. To overcome these impediments, it is important to transition to green chemistry and the concept of a circular economy.

The future of this is seen in bio-based plant and animal-derived polymers, energy-efficient production, and recycling through dynamic covalent chemistry. To achieve sustainable and functional textile innovation, these challenges will require interdisciplinary cooperation between scientists, engineers, designers, policymakers, and industry stakeholders. In the future, the combination of hydrogels and aerogels with the new technologies, including the use of artificial intelligence, nanotechnology and bio-inspired design, will potentially transform the textile industry. The materials of the future will be lighter, stronger, and more adaptive, with functions such as real-time sensing, energy harvesting, end-of-life sustainability, and so forth. Of more importance, the fact that innovation should be aligned to both ecological and economic sustainability is that these innovations are viable and necessary. As stressed throughout the book, to achieve this vision, profound interdisciplinary collaboration is needed. The integration should be optimized, and material scientists, textile engineers, biologists, data scientists, designers, economists, and policymakers should collaborate to make the products safe and biocompatible, consumer-ready, and supported by favorable regulatory environments. To sum up, hydrogels and aerogels are not only new textile materials but a vision of the future of intelligent, sustainable, and human-centered design. With a combination of performance and responsibility, they place textiles at the forefront of solutions to global challenges, including health care and climate resilience. The next few decades will probably see them evolve into mainstream elements of daily living and bring a new age of eco-functional, adaptive, and high-performance textiles.

Index

a

acrylic-based hydrogels and aerogels 41–42
acrylic fiber-based hydrogels 71–73
adaptive clothing systems 176–177
advanced cushioning system 103–104
aerogel-based functional accessories
  belts and straps 100
  eyewear 100
  jewelry 100
  smart and responsive biodegradable accessories 101
  3D printed aerogel accessories 100
aerogel-elastomer composites 100
aerogel fabrics 227
aerogel-infused nonwovens 196
aerogel materials 160–162
aerogel polyester chip 40, 41
aerogels 3
  preparation, sol-gel process 16–17
aerogels, defined 35
aerogel synthesis 161, 201
aerogel textile composites 22, 202–203
aerogel textile innovations 223
aerogel textile integration techniques 166–167, 201–202
aerogel-textile wound dressings
  alginate aerogels 118
  coating techniques 119
  fiber-based aerogels and integration 119
  in-situ sol-gel synthesis-textile impregnation 119
  natural biopolymers 117
  organic (polymeric) 117
  particle integration/embedding 119
AeroSF 39
alcogels 17, 201
alginate aerogels 118
  drying methods 20
alginate hydrogels 13, 99
alumina aerogels 189
aluminum hydroxide aerogel 22
ambient pressure drying 18, 19, 142, 201
animal-derived polymers 228
animal fiber-derived biopolymers
  advantages 136
  historical development 136–137
antibacterial properties 86–87
antibiotics properties 86
aramid nanofibers (ANF) 199
Aspen Aerogels 23
AT-PTR (aerogel-functionalized textile passive thermal regulator) 83
AZM-SS/PVA hydrogel 44

b

bacterial cellulose/chitosan hybrids 200
ballistic-resistant textiles 193–194
belts and straps 100
bio-based plant 228
bio-based textiles 90
biocompatibility
  and cytotoxicity 144–145
  sustainability and 177–179
  testing 214–215

biodegradability 89–90
biodegradable fashion accessories
  aerogel-based functional accessories 100
  challenges in 98
  hydrogel-based leather alternatives 98–99
bio-intelligent hydrogels 105
biomaterials' biodegradability 144
biomedical applications 146–149
biopolymer aerogels 17
biosensors, and diagnostic devices 148
blended-fiber hydrogels and aerogels 73–74
blended hydrogels and aerogels 42–43
  composite textile substrates 45–46
  natural-synthetic blends 43–44
boron nitride (BN) aerogels 190

c

carbide-based aerogels 190
carbon aerogels 127, 161
carbon nanotube (CNT) 172
carbothermal reduction 190
carboxymethyl cellulose (CMC) 38, 60
cellulose aerogel pads 215
cellulose aerogels 118, 191
cellulose-based aerogels 62
cellulose-based hydrogel composites 86
cellulose-based hydrogels 58
  with Ag/AgCl nanoparticles 21
cellulose-based superabsorbents 38
cellulose nanocrystals (CNCs) 44
cellulose nanofiber aerogels 20
cellulose nanofibril (CNF) 172
cellulose possesses 57
cellulose-PVA hydrogels and aerogels 43–44
chemical cross-linkers 215
chemical cross-linking 135, 140
chemical grafting 165
chemically crosslinked hydrogels 15–16
chemical protection 195–197
  aerogels of 192–193
chemical stimuli 175–176
chitosan aerogel fibers 23
chitosan (CS)-based aerogels 191
chitosan-based hydrogels (CS-Gels) 178
chitosan-gelatin composite films 145
circular economy 228
closed-loop recycling 103
collagen extraction 139–140
composite aerogels 16
composite textile substrates
  functional coatings 45–46
  laminated/coated fabrics 45
conductive aerogels 169
conductive aerogels and hydrogels
  electrical and electrochemical properties 170–171
  e-textiles, applications for 171–173
  material design and composites 167–170
consumer acceptance 215–216
consumer education 215–216
conventional textiles, characterization of 2
cotton-based hydrogels
  aqueous solution hydrogels 58
  CMC solutions 61
  microcrystalline cellulose 58–59
  succinic acid 60
cotton-based hydrogels and aerogels 37–38
cotton cellulose 37
cotton fiber reinforced poly(vinyl) alcohol hydrogel composite 21
cotton waste 38
covalent crosslinking 15
cross-linked hydrogels of self-healing poly(acrylamide) clay 42
crosslinking agents 191
crosslinking, and network structure 145
cross-linking method 3
cryogels 19
crystallization 14
cytotoxicity, biocompatibility and 144–145

d

dextran-based hydrogels 15
dialdehyde sodium alginate 99
dip-coating methods 166
dip coating process 116
direct ink writing (DIW) 100
drug delivery systems 147
  wound dressings and applications
    aerogels for 121–122
    hydrogels for 121
drying technology comparison 141
dry-wet spinning technique 172

e

eco-friendly clothing 97–98
economic feasibility 216–218
electromagnetic interference (EMI) shielding 190
electron-conductive hydrogels 168
electrospinning 20, 116–117, 202
electrospraying 202
electrospun nanofiber layer 167
electrospun polyacrylonitrile (PAN) fibers 167
environmental impact assessment 210–211
environmentally friendly method 213
enzymatic degradation 144
epichlorohydrin (ECH) 61
e-textiles
  aerogel in 172–173
  hydrogel in 171–172
eyewear 100

f

fabrication techniques 55
fiber-based aerogels and integration 119
flame-resistant garments 198
flame-retardancy of textiles 87–89
fluorine-free silane monomer 164
free radical polymerization 15
freeze-drying 19, 61, 135, 141, 201
FTIR spectroscopy 145
functional textiles
  antibiotics and antibacterial properties 86
  biodegradability 89–90
  flame-retardancy 87–89
  moisture management 82–84
  thermal insulation 84–86

g

gelatin methacryloyl 200
gelation 17
gold nanoparticles (AuNPs) 176
graphene aerogels 192
graphene-impregnated aerogel 102
green chemistry 228

h

health-care textiles 55
hemostatic and antibacterial applications 148
holocellulose nanofibrils/cellulose aerogel fibers (HCAFs) 90
hybrid alginate cryogels 19
hybrid silica-titania aerogel 22
hydrogel and aerogel
  characteristics of 136
  circular and regenerative product systems 225
  definition of 135, 136
  extreme environment and protective gear 225
  medical and healthcare textiles 224
  regulations and safety standards 225
  research 224
  sustainable apparel and eco-fashion 224
hydrogel and aerogel-infused footwear
  advanced cushioning system 103–104
  application of 104
  environmental impact, conventional footwear 103
  moisture management and hygiene 104
hydrogel-based composites 21
hydrogel-based leather alternatives 98–99

hydrogel coating 164
hydrogel/fabric composite flame-retardant material 22
hydrogel integration methods 164
hydrogel materials 158–160
hydrogels 3, 35
  preparation 13
hydrogels and aerogels
  collagen extraction 139–140
  current market landscape and clinical impact 137
  extraction and purification 138–140
  keratin extraction 138–139
  multidisciplinary convergence 137
  preparation and fabrication methods 138–142
  silk fibroin extraction 138
hydrogel-textile composite wound dressings (HTCWD) 114, 115
hydrogel-textile integration techniques 163–166
hydrogel wound dressings
  dip coating process 116
  electrospinning 116–117
  in situ polymerization 115–116
  sol-gel process 116
  solvent casting 116
hydrophobicity 196

i
impact resistance, aerogels of 192–197
industrial by-products 211
injectable hydrogels 148
inorganic aerogels 189–190
in situ-forming gels 148
in situ incorporation 202
in situ polymerization 115
in situ sol-gel synthesis-textile impregnation 119
in vivo testing 144
ion-conductive hydrogels 168
ionizing radiation 15

j
jewelry 100

k
kapok/polyester composite aerogel 70–71
keratin 39–40
keratin-derived aerogels 145
keratin extraction 138–139
  from wool 66–67
Kevlar safeguarding suit 197
Knudsen effect 189

l
laboratory test methods 194–195
laminated/coated fabrics 45
lamination process 202
lifecycle analysis 216–218
lightweight protective gear 197–198
lightweight thermal-insulating shear stiffening gel (LT-SSG) 197
low density monolithic silica gels 19
lyophilization 141

m
magnesia aerogels 189
mechanical stability 143
medical and healthcare textiles 224
mesoporous microparticles 175
metal aerogels 190
microcrystalline cellulose 59
moisture management 35, 36, 82–84
  and hygiene 104
molybdenum carbide (MoC) aerogels 190
multifunctional Kevlar safeguarding suit (MKSS) 197
multiresponsive aerogels 200
multi-step solvent exchange process 18

n
natural-fiber hydrogels and aerogels 37–40
  cotton 57–62
  cotton-based systems 37–38
  silk 62–66
  silk-based systems 39
  wool 66–68

wool-based materials 39–40
natural hydrogels 12, 159
natural material-based hydrogels 115
natural polymer hydrogels 12, 37
natural-synthetic blends
cellulose-PVA 43–44
silk-PLA 44
wool-polyurethane 44
N-carboxyethyl chitosan 176
needle-punched nonwoven layer 167
nitride based aerogels 190
non-covalent interactions 214
non-solvent phase separation 192
North Face's Aerogel Trail GTX® boots 104
nylon-based hydrogels and aerogels 41

o

one-step solvent exchange process 18
ophthalmic and dental applications 148
organic aerogels 191–192
organosoluble polyimide salt 174
oxide-based aerogels 189

p

padding 166
padding technique 201–202
passive daytime radiative cooling (PDRC) 192
pectin aerogels 17
personal thermal management (PTM) 192
phase-change hydrogels 101–102
physical cross-linking 135, 140
physically crosslinked hydrogels 13–15
physical stimuli 173–175
PLA-dextrin hydrogel 69
plant-based hydrogels 96
plant fiber-based hydrogels and aerogels
future perspective 129
properties and application areas of 112–113
smart bandages 124
synthesis methods, properties, applications 112
plasma grafting 165
poly(acrylamide) hydrogels 42
poly(acrylic acid) hydrogels 42
polyester-based hydrogels 68
polyester-based hydrogels and aerogels 40–41
polyester (polyoligoglycerol maleate) based-hydrogel synthesis 68, 69
polypyrrole (PPy) 168
polyurea aerogels 191
polyurethane (PU) aerogels 192
polyvinyl alcohol (PVA) hydrogels 14
polyvinyl alcohol tannic acid hydrogels (PVA-TA) 14
porosity and pore size 143
porous aerogel, heat transfer 188
porous materials 227
protective gear 225
protective textiles 187, 198–200
protic ionic liquid (PIL) 67

r

recycling 212
reduced graphene oxide (RGO) 172
repellency ability of aerogels 23
resorcinol-formaldehyde (RF) aerogels 17, 191
resorcinol-to-catalyst (R/C) ratio 191
resulting glyhydrogel textile dressing 21

s

self-assembly 14
sensor applications 127–128
shape memory alloys (SMAs) 177
shape-memory polymer aerogels (SMPAs) 200
shear thickening fluid (STF) 194
silica aerogels 16, 55, 189
silica-polymer composites 161
silicon carbide (SiC) aerogels 190
silk-based hydrogels 63, 64
silk fibroin 100
silk fibroin extraction 138
silk fibroin hydrogels 39
silk-PLA hydrogels and aerogels 44

silk sutures 136
silver (Ag) aerogels 190
silver nanoparticles 164
SiO_2 aerogel-coated fabrics 45
skin regeneration 146
smart and protective clothing 227
smart and responsive biodegradable accessories 101
smart bandages
  applications 124–125
  future perspective 125
  materials used in 123–124
  structure levels 123
smart textiles 55
sodium alginate/carboxymethyl chitosan hydrogel beads 14
sodium carbonate 191
sol-gel method 166
sol-gel process 116
sol-gel processing of aerogels 16–20
sol-gel technique 58
sol-gel transition control 140–141
solvent exchange process 18
sound absorption characteristics, of silica aerogel polyester blankets 22
sportswear
  aerogel in athletic apparel 102
  biomechanical support 102
  closed-loop recycling 103
  environmental impact of 101
  moisture management systems 101
  temperature control 101–102
  3D knitting 103
stimuli-responsive hydrogels 13, 16
stimuli-sensitive hydrogels and aerogels 173–176
supercritical carbon dioxide ($scCO_2$) 20
supercritical drying 61, 135, 141, 201
supercritical fluid drying 18
supercritical fluids 19
superhydrophobic aerogels 199
superhydrophobicity 166
supramolecular hydrogels 127
surface morphology 144
sustainability, and biocompatibility 177–179
sustainable fashion
  biodegradable fashion accessories
    aerogel-based functional accessories 100
    challenges in 98
    hydrogel-based leather alternatives 98–99
  eco-friendly clothing 97–98
  future perspectives and challenges 105–106
  hydrogel and aerogel-infused footwear
    advanced cushioning system 103–104
    application of 104
    environmental impact, conventional footwear 103
    moisture management and hygiene 104
  hydrogels and aerogels role 96
  manufacturing breakthrough 105
  need for 95–96
  scope and objectives 96
  sportswear
    aerogel in athletic apparel 102
    biomechanical support 102
    closed-loop recycling 103
    environmental impact of 101
    moisture management systems 101
    temperature control 101–102
    3D knitting 103
    waterless dyeing 102
sustainable hydrogel 223
sustainable waste treatment methods 90
swelling behavior 142–143
synthesis techniques, hydrogels and aerogels
  blended fibers 73–74
  natural fiber
    cotton 57–62
    silk 62–66
    wool 66–68
  synthetic fibers
    acrylic fiber-based hydrogels 71–73
    polyesters 68–71

synthetic-fiber hydrogels and aerogels 40
  acrylic-based systems 41–42
  nylon-based systems 41
  polyester-based systems 40–41
synthetic polymers 12

t
technical textiles 157
Tessmannia burttii Harms (Caesalpiniaceae) stem 99
textile applications, complementarity in 162–163
textile evolution, historical timeline of 2
textile industry xv
textile innovations 20–23
textile integration 200–203
textile recycling 213
thermal bonding 167, 202
thermal insulation 36, 40, 84–86
  mechanism of 188–189
  properties of SiO_2 aerogel 45
thermal stability 143
thermoplastic nanocomposites (TPCs) 177
thermoplastic polyurethane (TPU) 192
three-dimensional structure of hydrogels 55
tissue engineering scaffolds 147
titania aerogels 189
traditional bandages 123
triisocyanates 72

u
upcycling strategies 212–214
UV irradiation 165
UV shielding application 22

v
vinyl hybrid silica nanoparticles (VSNPs) 42
volume phase transition temperature (VPTT) 173

w
waterless dyeing 102
water vapor transmission rate (WVTR) 145
wearable electronics and health monitors 128–129
  applications 126–127
  materials used in 126
wet gels drying methods 19
wet silica gels 17
wool-polyurethane hydrogel and aerogel 44
wound dressings
  aerogels for 117–120
  applications 21
  hydrogel-textile composites 114–117
  properties 113–114
wound healing 146

x
X-ray diffraction (XRD) 145

z
zirconia aerogels 189